# LEAN CONSTRUCTION

## FOR SUPERIOR PROJECT PERFORMANCE

(Enhancing Delivery, Quality, Profitability, Safety & Sustainability)

OrangeBooks Publication

1st Floor, Rajhans Arcade, Mall Road, Kohka, Bhilai, Chhattisgarh 490020

Website: www.orangebooks.in

---



*Most images have been taken by the authors or sourced with permission from organizations such as LivSpace Pvt. Ltd., Odisha Works Department, and Konkan Railways. For enhanced clarity, some visuals and excerpts have been adapted from public sources and AI tools (e.g., Napkin AI, MS Designer, ChatGPT). These have been credited in the Acknowledgements and linked appropriately.*

**First Edition, 2025**

**ISBN:** 978-93-7426-394-5

**Price:** Rs. 1450.00



Printed in India

# LEAN CONSTRUCTION

## FOR

## SUPERIOR PROJECT PERFORMANCE

(Enhancing Delivery, Quality, Profitability, Safety & Sustainability)

**Dr. Subhash Rastogi**

**Er. Manoranjan Misra**

**Er. Tony Jacob**

OrangeBooks Publication

www.orangebooks.in

*This book is dedicated to all the students/ professionals*
*who strive to build with fairness, efficiency, and responsibility —*
*may Lean be their ally.*

*And also, to our respective families —*
*Sadhana, Siddharth, Sandeep and Shubhangi;*
*Bandana, Abhipsa and Anubhav;*
*Misha and Tiya —*
*who inspired us to keep sharing our knowledge*
*with as many people as possible, and who supported us selflessly.*

*Dr. Subhash Rastogi*
*Er. Manoranjan Misra*
*Er. Tony Jacob*

## What the Industry Leaders Say

I have had the privilege of experimenting with Lean Construction methodologies across multiple projects in the pharmaceutical sector, and my experiences have been a mix of excellent successes and valuable lessons from failures. This book offers a practical guide that aligns quite well with my learnings, especially in standardizing Lean implementation processes globally. I look forward to seeing how this work contributes to further refining Lean practices and improving future project outcomes.

**François HOSOTTE**

**VP Global Engineering, Technology and Safety, of a Swiss Multinational Pharmaceutical Company, France**

From the early days of Lean implementation in 2013, we've seen first-hand the remarkable results it can deliver, both in terms of efficiency, product quality and client satisfaction. This book not only captures the essence of Lean Construction but also provides actionable insights that are essential for anyone looking to integrate these principles into their projects. It's a must-read for professionals globally who wish to leverage Lean Construction for competitive advantage, something we have been doing and benefitting since long.

**Jaiprakash Aildasani**

**Chief Operating Officer-Global, Space Matrix, Singapore**

Implementing Lean Construction principles in the middle of a critical, national mega-project like the 21KM sea bridge proved to be a game changer for us. We started seeing projects as a series of repetitive smaller projects, like Piling, pile cap and Pier etc. processes, as mini project of one span, repeated several times. Go for very deep analysis for mini-project for improvements in an out-of-the way and get the benefits multiple times. This book offers a comprehensive approach, to accelerate project timelines and improve efficiency. The real-world examples and tools they provide, are a must-read for professionals aiming to implement Lean in large-scale projects. It has covered all aspects of Project management

**Sanjay Patil**

**Head-Bridges BU, Larsen & Toubro Ltd, Mumbai, India**

Since 2017, we have strategically applied Lean Construction methodologies across government infrastructure projects, consistently delivering measurable improvements—most notably in accelerating project timelines and enhancing public satisfaction. The dedicated chapter on Lean in the government sector particularly stood out, offering a nuanced perspective on navigating challenges such as governance mechanisms and accountability frameworks. It clearly illustrates how Lean principles can be adapted to drive efficiency and impact within the public domain. This book serves as a valuable guide for leaders aiming to embed operational excellence into government project delivery.

**Sameer Hota**

**Engineer in Chief and Managing Director**
**Odisha Bridge & Construction Corporation, India**

Lean Construction Practices along with Total Quality Management" offers a compelling and practical roadmap for organizations committed to excellence. The book effectively underscores the critical role of Total Quality Management and Lean principles in achieving high-quality outcomes. Particularly noteworthy are the chapters on de-bottlenecking using Value Stream Mapping and the predictive strategies for defect prevention—both of which provide actionable insights for enhancing productivity.

The author's emphasis on integrating safety and sustainability within quality frameworks reflects a holistic understanding of modern construction challenges. This book is not only informative but also transformational—an essential read for professionals and organizations aiming to drive continuous improvement and operational efficiency without compromising quality.

**Varughese Mathew**

**Chief Operating Officer – Sobha Group, Dubai,**
**Vice Chair: International Cost Management Standard, UAE**

## What the Academic Leaders say

This book on Lean Construction offers a comprehensive, practical framework for enhancing project performance across progress, quality, safety, and profitability. Its insights will be invaluable for PGP students, construction professionals, and government executives, especially given the government's role as the largest infrastructure investor. The dedicated chapter on the government sector and its relevance to the real estate industry makes this book a timely resource for driving efficiency, innovation, and stakeholder value in construction.

**Dr. Chandan Chowdhury**

**Executive Director,**
**Punj Lloyd Institute of Infrastructure Mgt.,**
**Practice Professor, Indian School of Business, India**

I went through several chapters of the Book. I can see a massive exercise here from Dr Rastogi and his co-authors, covering a very wide canvass for the various topics. There must have been a very deep and wide research exercise behind this interesting compendium! My sincere complements to team on the wide and interesting coverage of the various topics. Truly the focus seems to be on the project execution agencies, who can learn a lot from this compilation! I hope this will bring a sea-change in the construction industry with this wide-ranging book. It provides a window to the young engineers into what is possible as well as a master reference book for the involved professionals for action plans. Best of Luck in your laudable endeavour.

**Prof. N. Raghavan**

**Professor of Practice, BTCM division,**
**Indian Institute of Technology Madras, Chennai, India**

I have gone through the book on Lean Construction, and I believe it offers an insightful approach to improving construction project management. The integration of Lean methodologies with real-world practices will benefit MBA students and executives focusing on operations and project management. The 16+ case studies, more than 150 figures, a special chapter covering how to adopt Lean Construction in Government Sector, are of immense value. I would recommend this book as an essential read for those looking to enhance their knowledge of efficiency and waste reduction in construction projects.

**Prof. Ramkrushna Padhi**

**Associate Professor,**
**Indian Institute of Management, Sambalpur, India**

I had the privilege of reviewing this book on Lean Construction and found it to be a comprehensive and insightful resource for both academic inquiry and practical implementation. The authors present Lean practices in construction with clarity and depth, effectively bridging theory and application. The inclusion of a global perspective enhances the relevance of the content across diverse construction contexts worldwide. Given the increasing emphasis on infrastructure development in the Middle East, this book is particularly timely and valuable. I strongly recommend it to scholars, practitioners and policymakers seeking to advance construction project management through Lean principles in the middle east region and beyond.

**Dr. Mohammed Zein Elabidin Babiker Elshafie**

**Professor, Qatar University, Qatar.**

Having taught Lean Construction for many years and carrying out Lean implementation consultancies, I find this book to be a solid resource for both students, faculties and practitioners. It effectively bridges the gap between theory and practice, making Lean methodologies more accessible. I would highly recommend this book to those in construction management programs and professionals seeking to understand and implement Lean principles in their projects.

**Dr. Venkatesan Renganaidu**

**Professor, Dean of IQAC at NUCS- Hyderabad**

# FOREWORD 1 (Academic Leader)

**Prof. K. N. Satyanarayana**

**Director, I I T Tirupati**

Member, Empowered Committee for Industrial and International Collaborations, Govt of India.

Expert Member, Board of Management of Building Materials & Promotion Council

Founding member,

In recent years, the construction sector has witnessed a growing interest in Lean methodologies as a means to improve efficiency, quality, and value delivery. As an academic and practitioner with a deep interest in Lean Construction, I am pleased to see a book that not only integrates theory and practice but also provides meaningful direction for future progress.

The book, ***"Lean Construction for Superior Project Performance (Enhancing Delivery, Quality, Profitability, Safety & Sustainability)"*** is a thoughtful and comprehensive text that captures the essence of Lean philosophy while addressing the unique challenges of construction in emerging economies like India. The authors successfully demonstrate how Lean Construction 4.0—integrating traditional Lean principles with digital innovations—can serve as a transformative force across both Real Estate and Linear Infrastructure projects.

The book's emphasis on collaborative planning systems, waste elimination, proactive quality control, and stakeholder alignment reflects a mature understanding of systemic challenges in project delivery. Chapters such as "Creating the Big Room" and "Value Stream Mapping" translate abstract concepts into practical tools for industry application. The inclusion of more than 15 real-life case studies enhances the credibility and applicability of the ideas presented.

What particularly stands out is the chapter on public sector implementation. By combining academic rigor with field interviews and ground realities, the authors offer a rare blend of insight that will be valuable to scholars, students, policymakers, and professionals alike.

I strongly recommend this book as a core reference in engineering and management curricula, as well as a practical guide for infrastructure practitioners aiming to embrace a more productive, sustainable, and collaborative future.

***– Prof. K.N. Satyanarayana***

# FOREWORD 2 (Senior Government Leader)

**Sanjay Kumar Shukla, IAS**

**Additional Chief Secretary, Govt. of M.P.**

Urban Development and Housing Department,
Planning, Economics and Statistics Department,
Aviation Department

India is undergoing a transformational journey in urbanisation and urban development. The rapid growth of Tier 1 and Tier 2 cities, combined with the demand for quality infrastructure, has placed the construction sector at the centre of the nation's growth strategy. Our ability to build faster, better, and safer will define our trajectory toward becoming a global economic power. However, challenges like delays, cost overruns, safety lapses, and poor quality remain widespread, especially in public infrastructure projects.

The book ***"Lean Construction for Superior Project Performance (Enhancing Delivery, Quality, Profitability, Safety & Sustainability)"*** is a significant and timely contribution toward addressing these issues. It offers not just a diagnosis of the prevailing problems in traditional construction project management systems, but also practical solutions rooted in Lean Construction 4.0 principles. The authors present a unified and holistic approach that combines innovative planning tools, stakeholder collaboration frameworks, digital technologies, and waste elimination methods to drive superior performance.

What makes this book especially relevant to Government and public sector entities is its strong emphasis on actionable recommendations. The chapter dedicated to public sector implementation—based on in-depth interviews with senior government officials, contractors, and department heads—offers innovative grounded strategies for change. It inspires confidence that transformation is not just desirable but achievable.

This book is a must-read for Government officers, urban development bodies, infrastructure agencies, and stakeholders of the construction sector, with a blueprint for achieving faster execution, better utilization of public funds, and higher stakeholder satisfaction. I commend the authors for this visionary work and encourage policymakers and practitioners alike to draw from its insights for building a stronger, smarter, and more efficient India.

***– Er. Sanjay Kumar Shukla, IAS***

## FOREWORD 3 (from Global PM Leader)

**Ben Breen, PMI-CP™**

**Vice President, Gigaprojects**

Project Management Institute, USA

**Ben Breen, PMI-CP™**

The construction industry is undergoing a profound transformation worldwide, driven by the increasing demand for efficiency, sustainability, and digital integration. In today's fast-evolving landscape, **Lean capabilities are no longer just a competitive advantage—they are a necessity for organizations striving for excellence** and to guarantee a more sustainable future for all.

This book, ***"Lean Construction for Superior Project Performance (Enhancing Delivery, Quality, Profitability, Safety & Sustainability)"*** as a **comprehensive and practical guide** for industry leaders, project managers, and construction professionals. It provides a structured approach to enhancing value for stakeholders by minimizing inefficiencies, fostering collaboration, and leveraging digital technologies for superior project outcomes. The case studies featured in this book illustrate real-world examples of successful Lean adoption across various countries, making it highly relevant to the global construction community.

As an Engineer by practice and Director of Construction at the Project Management Institute (PMI), I have witnessed firsthand the transformative power of Lean methodologies in construction projects. Organizations that embrace these principles achieve **higher productivity, optimized resource management, and significantly reduced waste**—ultimately delivering greater project value to key stakeholders.

My familiarity with the author, Tony Jacob, extends beyond professional recognition—I have witnessed firsthand Tony's dedication to advancing the construction project management profession. As one of the first individuals to earn both the PMI Certified Practitioner (PMI-CP)™ and the PMI Senior Construction Professional (PMI-SCP)™ credentials, Tony has demonstrated a deep commitment to the industry. His extensive experience as an active PMI volunteer and Construction Ambassador reflects his passion for knowledge sharing and continuous improvement. Through his work, he has contributed significantly to shaping the future of Lean construction methodologies, making his perspective in this book particularly valuable to professionals in the built environment.

The insights in this book will be particularly valuable for construction professionals, project managers, and policymakers looking to enhance project efficiency, improve collaboration, and drive sustainable outcomes. By applying Lean principles, organizations can streamline processes, reduce waste, and optimize resource utilization, leading to **higher productivity and better project outcomes**. The real-world case studies provide practical strategies that can be adapted to diverse construction environments, equipping industry leaders with the knowledge to navigate challenges, implement digital transformation, and strengthen risk management practices. It is an essential read for those dedicated to building a more efficient, safe, and sustainable future.

***– Ben Breen***

# PREFACE

If you have ever managed or worked on a construction project, you know the struggle is real, at the work-site, at the head-quarter and at all the Contractors', Consultants' office. Deadlines slip, budgets balloon, quality issues pop up unexpectedly, and safety and sustainability? Well, that sometimes takes a backseat to urgency. On top of that everyone seems to be putting lots of (unpaid) overtime and still is stressed. Sound familiar? We thought so too.

This book is born out of more than 100 years of combined hands-on experience, consulting, and countless conversations with project managers, engineers, architects, and site supervisors. We've seen first-hand how traditional construction project management struggles to deliver efficiency. Though the industry is evolving, but many projects still rely on outdated methods that lead to unnecessary waste—of time, money, and resources. This also reduces the Value to the many key stakeholders of the project. That's where Lean Construction 4.0 comes in.

**Our motive behind writing this book is simple:** to bridge the gap between theory and real-world application. We wanted to create something that speaks to those readers who want to make their projects smoother, faster, and more profitable. The book will also be useful for postgraduate students who are the future Managers, and who are looking to build a strong foundation in modern construction management.

What you'll find here isn't just textbook theory. This book is packed with practical insights, real-world case studies, and actionable strategies to help you implement Lean Construction principles effectively. There are more than 50 small examples and 15 detailed Case Example included to make the concepts clear. We're talking about faster project completion, higher quality, reduced costs, and improved safety—all without the usual chaos. The recommendations also include, what can be adopted immediately, what can be helpful in medium and long-terms.

### What's Inside the Book?

Each chapter tackles a key aspect of Lean Construction and its application in real projects. There are many case-studies available, which are the real experience and experimentation of the authors. Here's a quick snapshot of all the 12 Chapters, what you'll find:

**1. Role of Construction Sector for Global Economic and Social Growth:** Why construction is more than just bricks and mortar—it's a global economic powerhouse and a driver of economical, societal and environmental development. It also focusses on the various key stakeholders of the project, and their Value expectations from the project.

**2. Problems with Current Construction Project Management System (PMS) and Need for Lean Construction 4.0:** The challenges of traditional construction management and why a Lean approach can provide solutions to those and therefore is the future. One detailed Case Study is provided for better understanding

**3. Collaborative Planning System (CPS):** A single and simple, smarter way to collaborate to plan, execute, monitor, and control projects for minimizing surprises and risks, better coordination and communication among the relevant stakeholders, resulting in better quality, faster project progress, avoiding ant budgets overflow, fewer delays, and lesser stresses for the people involved. One detailed case Example is included in this

**4. Enhancing 'Value & Profits' for All Stakeholders by Eliminating Eight Construction Resource Wastes:** This identifies and defines the 'Value (including Profits)' for various stakeholders. Then it guides how to eliminate or reduce. The efforts and budget spent on the activities, which are not adding values to the key stakeholders. 'Hunting' and eliminating the biggest sources of waste in construction improve efficiency tremendously. Two detailed Case Studies are discussed in this.

**5. Creating 'The Big Room' for Better Collaboration & Communication:** Here we discuss how bringing all relevant stakeholders into one space—physically or virtually, and focusing on various causes for inefficiencies and finding joint solutions for those can resolve coordination and communication issues to a great extent. One case example is provided in detail.

**6. Value Stream Mapping (VSM) for De-bottlenecking Processes and Speeding Up Project Progress:** In this we explain practical tool to visualize and eliminate roadblocks in construction workflows, which helps in reducing the process or Activity duration even more than 50%. Two detailed Case Examples are included here

**7. Proactive Defects Prediction & Prevention for Better Quality Management: In this we ask, w**hy wait for defects to appear, instead predict those and prevent those to occur. This saves a lot, on rework delays, and provides the good quality first time, every time. One Case Example is explained in details here.

**8. Make-Ready the Work for a Safer Project Environment:** This explains with examples how structured preparation can make sites safer and more productive. One example is discussed in details in this.

**9. Leveraging Digital & Other Technologies for Superior Performance:** This chapter tells about various digital technologies like, Virtual & Augmented Reality, BIM, IoT, AI, Drone, Robotics etc can bring drastic change and benefits in reshaping construction. One detiled Case Example is included here.

**10. Lean Construction in Government Sector:** This chapter lists special risks and unique challenges Government and Public Sector Construction faces. This may include Political Risks, Social Risks, Government Risks, Consultants Risks etc and also suggests new innovative ways of resolving those. Two detailed Case Examples are explained here.

**11. Five S: An Approach to Eliminate Human Stress and Wastage of Searching Time: This chapter explains a** simple but powerful Japanese technique to organize construction sites and workflows. This leads to higher construction outputs and reduces stresses of the workmen. One Example is discussed in details here.

**12. Recommendations for Effective & Speedy Implementation: This chapter explains a** step-by-step guide to bringing Lean principles into action in your projects. This also recommends how you can modify your Contracts with Contractors and Consultants to ensure better results of adopting Lean Construction.

## Final Thoughts

Lean Construction isn't a buzzword, it is a game-changer. Whether you're leading a multi-million-dollar project or just starting in the field, this book will help you **work smarter, not harder**.

We hope this book gives you **practical tools, fresh perspectives, and the motivation** to take your construction projects to the next level. Let's build better, faster, and with fewer headaches!

Dr. Subhash Rastogi

Er Manoranjan Misra

Er. Tony Jacob

# Table of Contents

*Lean Construction is not a buzzword, it is a game-changer. Whether you're leading a multi-million-dollar project or just starting in the field, this book will help you work smarter, not harder.*

# CHAPTER 1

# CONSTRUCTION DRIVING GLOBAL SOCIO-ECONOMIC GROWTH

# 1. Role of Construction Sector

The construction sector has always been a cornerstone of human progress. It not only provides the physical infrastructure that supports modern life but also drives economic growth, employment, and social development. From housing and healthcare facilities to transport networks and industrial plants, construction shapes the way societies function and evolve. Its role extends beyond building structures — it creates opportunities, connects communities, and sustains development across generations.

## 1.1 Role of Construction Sector for Global Economic and Social Growth

The history of construction is almost as old as humanity itself. From mud roads and huts to mega highways and skyscrapers, its evolution reflects mankind's growing mastery of materials and technology. Today, the sector spans a wide range of projects: commercial facilities such as hotels and hospitals, residential developments from villas to apartments, and infrastructure for road, rail, water, and air transport, as well as industrial and energy systems. Importantly, it covers not only new facilities but also the expansion and maintenance of existing ones — structures that touch our lives every day, year-round.

Global investment in construction has been tremendous and continues to rise. According to the global research report ID: 5519701 [1], *Construction Market – Global Opportunity Analysis and Industry Forecast 2022–2030*, the market was valued at USD 7.28 trillion in 2021 and is projected to reach USD 14.41 trillion by 2030, with a CAGR of 7.3%. Revenues come from entities such as organizations, partnerships, and sole traders engaged in building and engineering projects, including highways and utility systems.

Yet, while demand has grown rapidly, construction management and technologies have not kept pace. Inefficiencies remain widespread and have not been addressed in a focused manner. Given the scale of investment, it is critical to adopt newer approaches and strategies that can significantly improve project performance in terms of progress, quality, safety, sustainability, and profitability. Stakeholders — builders, clients, governments, investors, contractors, consultants, and end-users — all require greater value from these investments.

Emerging challenges add urgency: evolving technologies, diverse clientele, and rising expectations in the new millennium demand fresh solutions. This chapter therefore highlights both the economic and social importance of construction and the need to re-examine existing practices. By addressing inefficiencies and embracing innovation, the sector can deliver superior outcomes for all stakeholders and ensure that future growth is sustainable, inclusive, and resilient.

## 1.2 Huge Contribution of Construction Sector for Economic and Social Growth

The construction sector perhaps is and will remain, one of the most important sectors in the global economy in the coming couple of decades. The importance is in terms of following:

- Economic growth (the growing total global investment envisaged in this sector)
- Social growth (for the needs of growing Population, growing Urbanization and growing Employment)

According to a recent research report from Research & Markets of 2022 - "Construction Market-Global Opportunity Analysis and Industry Forecast 2022-2030" global research report ID: 5519701 "[1], drivers for Growth Opportunities are primarily the increasing urbanization, and growing population across the globe. This is creating additional needs for new residential and commercial buildings and other urban infrastructure. This also includes some rural and social infrastructure, especially in the third-world and developing countries.

India is no exception to this growth pattern. In fact, India needs the focus on construction sector, proportionately even more than the global average. Fortunately, the Indian Government is concentrating on defining and consolidating the policies that will ensure time-bound creation of world-class infrastructure in the country [2].

This is also reflected through the National Infrastructure Pipeline (NIP) Initiative. NIP is the investment plan unveiled by the Indian Government for enhancing infrastructure in identified sectors for a period of five years from 2020 to 2025. This is estimated to be approximately INR 102 lakh crore (USD 1.4 trillion). This plan is a major contributor to help India reach its aspirational plan of a USD 5 trillion economy by 2025.

For this, the funding is planned in a Public Private Partnership (PPP) mode. The recommended contribution by the central government, state governments and private sectors is expected to be around percentage ratio of 39%, 39% and 22% respectively. Besides this, the private sector is also expected to invest separately in residential and commercial buildings, given the growing needs of urbanization in India. India has also an ambitious growth plans of approximately INR 304 lakh crore (about USD 3 trillion) of investment in the Indian infrastructure sector till 2040 to sustain the country's continuous development.

**Phase 1** of many mega programs have already started. These mega programs include Bharatmala (INR 5.35 lakh crore for Phase 1) and Sagarmala (total outlay of INR 8 lakh crores). Bharatmala involves construction of major expressways, highways and roads connecting all major metros and A-class cities in India. Sagarmala is planned to connect all major ports in India. Other mega programs include programs such as smart cities mission. 'Housing-for-All' needing about 6 million houses in next 5 years, Swachh

Bharat Abhiyan, etc. All these programs aim to build world-class infrastructure in the country.

This major investment will also generate a huge amount of employment in construction sector. It is estimated that it will need about need 70 lakh skilled project managers over the next 10 years for projects in key industries like roads, railways, IT, and manufacturing. Moreover, the workmen with different skills requirements will be in the range of several crores.

With this information, you must have realised that the construction sector is really one of the most important sectors for the growth of economic and social needs of the global economy. For India also, this is probably the sector with the biggest growth planned. We are sure, you must be feeling lucky to be in this sector for your career.

## 1.3 Two Main Pillars of Construction: Real Estate Projects and Infrastructure Projects

You will see that the construction sector is primarily divided into two major Industries - real estate and infrastructure. We would like to classify these two industries further into five groups, as depicted in the Figure 1.1.

One reason for this classification is that, a large number of construction projects' products are inhabited by people. These products are commonly referred to as 'buildings'. People stay in these buildings for a short duration (like in hotels or resorts), or long duration (like in villas and apartments). Therefore, besides functional requirements, the aesthetic finish and quality is a particularly important aspect for them. This is more so in the present era when people's individual choices are of great importance. People's choice is shifting from mass-production to personalization, also called mass-customization, especially for residential buildings. You will recall the famous quote, "Beauty lies in the eye of the beholder". What is great-looking house to one, may not be so for others. Therefore, we must stay focused on the design and finishes of the products based on the choices of the customers. You may appreciate this better if you relate it to the beauty and the look-and- feel of construction work-items like tiles, electrical fixtures (lamps, switches), false ceiling, wallpaper, etc. Here, besides the safety of the people (relating to the strengths of various work items), the feel-good factor, like ambience, is also equally important. Many newer 'design and build' solutions are being developed for such cases to satisfy personalization needs.

A good way to differentiate the quality issues of construction for habitat and non-habitat components, is to talk about the construction components in terms of Buildings and non-Buildings. In the Buildings, where people occupy the space for longer-duration, the quality of finishes and overall look-and-feel is equally important as that of structural strength and people's safety. This sense of aesthetics also impacts their quality of life and the general experience of the people living there. The other classification can be named as the non-buildings, where there are almost no people for long time, they just pass-through it, and therefore quality of aesthetic finishes is not so important. However, the functional strength quality is extremely important.

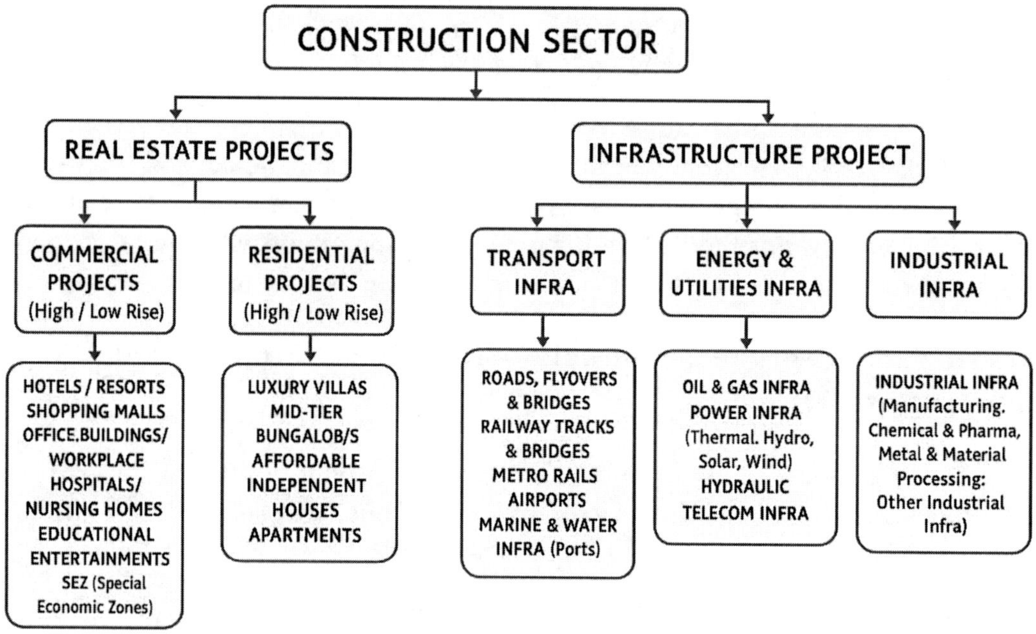


**Figure 1.1: Two PILLARS of Construction Sector**

For buildings, we can consider different kinds of real estate buildings, like buildings for residential purposes, buildings for commercial offices, shopping malls, etc. This list will also include the 'building' parts of the large infrastructure projects. For instance, Railway station building, metro station buildings, airport commercial buildings, and sea ports commercial buildings, etc.

In the case of non-buildings, we can consider the 'transport' sector of roads, railways, metro flyovers, bridges, airstrips, etc. We also consider utility services of water, power, drainage etc. These types of projects can also be named as the linear infrastructure projects, because of the nature of the longitudinal shape of these projects.

**Figure 1.2: Buildings where look-and-feel and 'finishes' are also important, besides structural strength & safety**

However, in all types of Construction, the performance parameters of great importance are: total project cost (budget & profits), project completion duration, project work safety, project quality (for both real estate and infrastructure projects), sustainability and flexibility of individual project needs- that is, any personalization required. Achieving and improving this performance measure is one of the toughest problems with the construction sector. Although the contents of this book cover the entire construction sector, there is a slightly higher focus towards real estate construction projects. The next section explains briefly how construction evolved to present version of Construction 4.0, on the lines of Industry 4.0, leveraging Lean Philosophy and omni-present several digital technologies.

## 1.4 Evolution of Strategies for Construction Sector, moving from 'Construction 1.0' to 'Construction 4.0'

The global economy has evolved significantly over the last century. The first economic growth was driven by Agriculture, and then, from early 20th century to mid-seventies, the economy was driven by manufacturing all kind of products to improve quality of life. Later, the growth was driven by service economy, like logistics, IT services, hospitals, etc. Currently, it is driven globally by the disruptive digital technologies (such as social media, mobiles, analytics, cloud, IoT, 3D printing and many other proprietary emerging technologies).

Terms like Digital India, Digital USA, Digital Germany are common, reflecting the digital economy's rise as new growth driver. Just as this shift impacted the manufacturing industry, these changes in economies focus have also influenced the type of products, services and business processes – including those in the construction industry.

More and more of adoption of Lean philosophy and various digital technologies is underway to achieve superior project performances and resolve the performance issues mentioned earlier. Figure 1.3 below visually depicts the changes in strategies for managing construction projects over the last century.

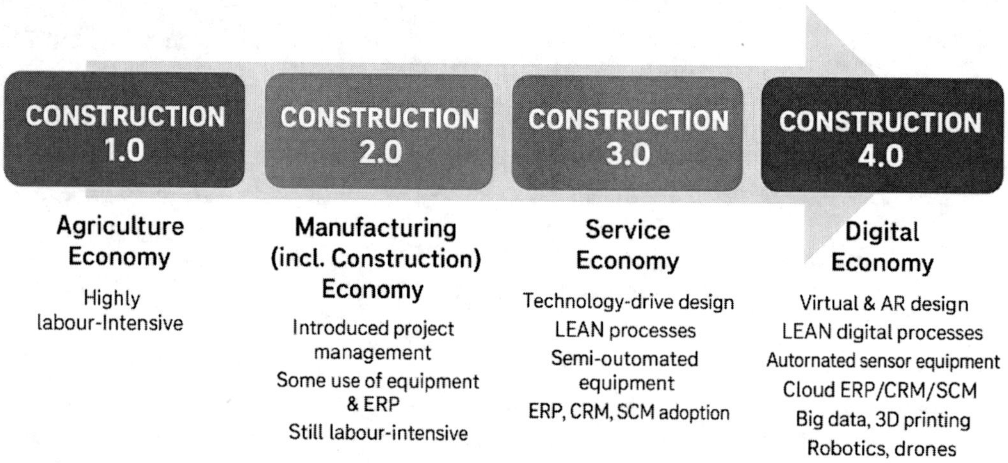


**Figure.1.3: Evolution of Construction Sector during last 100 Years**

The evolution of changing strategies from one generation to another, in terms of processes, practices, technologies, capabilities and value (e.g. Rastogi 2017) [1], can broadly be summarised as follows:

- **Construction 1.0:** The construction and business processes involved were predominantly labour oriented. Buildings were very similar in designs at a high level. There were not many specialized disciplines such as architecture, project management consultants, or EPC main contractors, which are very common presently.
- **Construction 2.0:** The mid-20th century saw the emergence of many specialised disciplines (e.g. architecture & design, project management). Some construction automation also evolved for the taller buildings and faster progress using mechanical & automated construction equipment. However, the site work remained predominantly labour-oriented. Some automation in business process was achieved through softwares like ERP (Enterprise Resources Planning, e.g., Oracle, SAP), project management software (MS-Project, Primavera). But performances and productivity were still quite low, resulting in significant cost and time overruns.

- **Construction 3.0:** The late 20th Century saw technology-driven architecture & design (e.g., AutoCAD, BIM- Building Information Modelling), the adoption of Lean philosophy in construction systems, Lean Integrated Delivery System (IDS), and more automation and specialized equipment (e.g. slab concreting using pumps). Newer software like CRM and Enterprise-wide Project Management (EPM) tools also emerged. These developments led to reasonable improvements in project performance parameters such as construction quality, project completion duration, and safer working conditions. However, many stakeholders, including builders, contractors and end-users were still not fully satisfied.
- **Construction 4.0:** In the 21st century, major transformations have taken place – and continue to grow exponentially – to bring about vastly improved project performance:
    - Technology became the driving force, for both construction processes and business processes. Popular tools include WhatsApp, mobile devices/iPad for real-time video communications and documents sharing including drawings. Big data analytics now includes structured and unstructured data (text, voice, image and video). Other widely adopted technologies include artificial intelligence, virtual and augmented reality, 3D printing, and Internet of Things (IoT)/ sensors- all now affordable and widely available.
    - Evolving user needs and expectations: The behaviour and expectations of end-users, especially millennials, changed drastically. Tech-savvy, impatient, globally-connected and demanding, they expect world-class quality at frugal costs and highly personalised, memorable home-buying experiences.
    - On one hand, real estate construction focus is shifting to 'mass customization' at agile speed, especially in the urban areas. The building designs are getting complex and uniqueness is in great demand. On the other hand, in the rural areas, 'mass production' at high speed is needed for the enormous number of affordable houses required.
    - Increased focus on sustainability: Green construction has gained renewed importance due to global climatic challenges – unlike in the past, when it was low priority.

All this is becoming possible due to the increasing application of Lean philosophy and digital technologies, which is the core of Construction 4.0.

As per our observation, most countries - including India- are currently operating at best at the 'Construction 3.0' level. However, now there is an urgent need to jump to Construction 4.0, because of new challenges. The need is also due to not being able to attain the performance parameters expected currently. This new generation of Construction 4.0 also poses its own new challenges. However, at the same time, these also offer innovative new solutions to resolve them.

Some of these new challenges for India, under Construction 4.0 are shown below:

**Figure 1.4: Key Challenges in Current Construction 4.0 Scenario**

Besides these, there are a few more, such as:

- Real estate regulation and development bills introduced, with several restrictions and severe penaltiesfor delays and poor quality.
- Increasing focus on sustainability from the government.
- Presently, highly impractical planning, and inefficient project construction and business processes.
- Presently, a very high amount of rework and wasteful construction and business activities. These are not yet being taken seriously.

We have provided details of 'construction 4.0', which includes a strong focus on Lean philosophy, and various digital and other technologies are explained in forthcoming chapters. This will bring superior performance in terms of faster project progress, better quality, avoiding over costs/ budgets and greater safety.

Because of all these, if Indian construction practices do not change significantly, it would not be possible to achieve the goals set for the next few years. In fact, the research article Navigating the digital future: the disruption of capital projects)[1], from McKinsey capital projects & infrastructure (2017) talks about such impact. This global research report mentions that if the construction productivity could catch-up with the productivity of the total economy, the industry's value added could rise by whopping USD 1.6 trillions a year, globally. This alone can meet world's annual infrastructure needs or boost global GDP by 2%. A reassessment of current practices is required in India, as well as in the rest of the world, to best leverage Lean philosophy and digital and other technologies for achieving superior project performances.

## 1.5 Better Value for All Stakeholders, involved in Construction in Current & Future Times

Like other businesses, in construction the customer is the primary focus, as they pay for the product or service created. Customer 'value' may be expressed in terms of price, quality, availability, or after-sales service expected from a construction product, such as a house. The customer may be an individual seeking ownership of an apartment, a small organisation building a shopping mall to lease to shops, or a large entity such as the government creating a highway for public use.

In recent years, attention has also shifted to the 'end-users' who benefit from the constructed product but may not directly pay for it. Most government projects, especially in social or rural infrastructure, follow this pattern. In such cases, it is important to consider the value perceived by end-users, ensuring the facility is easy to use for the purpose it was created and funded by the government.

Earlier, businesses recognised mainly two key stakeholders. One was investors, shareholders, or promoters who provide capital to create and run the business. For them, 'value' was measured in terms of returns, dividends, and revenue growth.

Over the years, this created many new problems in the world from the construction business. The problems were primarily related to environmental hazards, especially contributed by the construction and chemical industries. The contractors often left large amounts of debris or construction waste at the site itself or disposed/dumped it into some nearby pond, lake, or valley in order to increase their ROI. In fact, many lakes and rivers were damaged due to this debris disposal, and subsequently dried up. Also, many trees were cut, and forests were destroyed to construct new buildings and infrastructure. Hence conservation of natural resources (like trees, forests, oil & gas, rivers, and lakes etc.) and environmental preservation became a particularly important aspect in construction projects. Due to this, the government and society became important stakeholders in the construction sector.

Another important and unusual aspect of construction is that it depends to an extremely high extent on the services of the consultants, contractors and sub-contractors, right from feasibility study to design, construction and maintenance of the facilities. Therefore, taking care of the business interests of all those business partners is also important for the sustainability and growth of the construction business.

For this purpose, we recommend five groups (Rastogi, Laghate, 2021) [3] of stakeholders for the sustainability and growth of the construction industry. These five groups are also shown in Figure. 1.5 below. The figure also lists what is the meaning of "Value" for each of these five groups. For the sustenance and growth of the construction business, some of the most important 'value' expectations from different stakeholders are listed below.

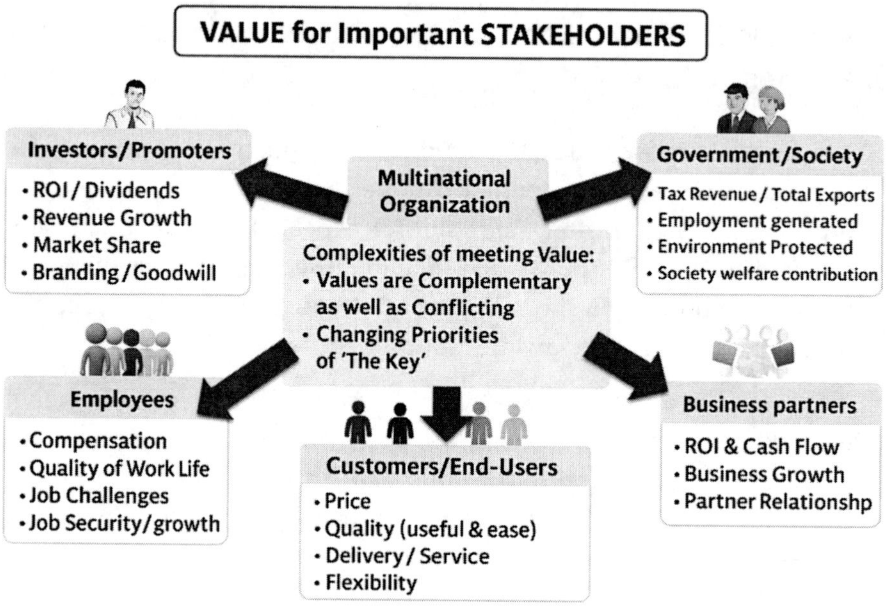


**Figure 1.5: Important Stakeholders for Construction Business & meaning of 'Value' for themselves**

### 1.5.1 Customer/ End-User

Value for the customers is always in terms of high quality of products & services at low prices, and its availability when it is required. However, the concept of value has changed over the years. Now the customer wants 'unique' designs of buildings or interiors, a unique and pleasant experience of buying or owning an asset, and ease of using the buildings or the infrastructure. Unique buildings also mean they need some personalisation for the change of design and build based on their own expectations, as the project progresses based on their more updated knowledge and (global) exposure. As the project progresses, they not only expect that the changes suggested by them will be accepted, but also that they will be implemented extremely fast and possibly at no extra cost, or at very low additional cost. This is especially true for the millennial new-age customers, who are the largest segment of end-users now. You may realise the problems of the construction industry because of this change in customer expectations.

### 1.5.2 Government/ Society

The construction industry is one of the biggest contributors to pollution and environmental damage, due to which many statutory regulations have evolved and many more are still evolving. The government also wants this sector to generate several new employment opportunities, besides supporting societal development. In fact, in India, it has been mandated by the government to spend a certain percentage of the annual revenue for the development of the society around the project's geography. Similarly, for infrastructure projects like road construction, it is required to plant new trees which are ten times more in number than the trees removed because of the project. This is to ensure ecological balance. Adhering to these requirements might increase the cost of construction; hence, in the interest of the customers, the industry must find and embrace innovative new solutions. These solutions may be both in terms of management science, like Lean Construction 4.0, and by adopting various newer technologies, including digital and construction. We will focus more on these in the rest of the book.

### 1.5.3 Investors/ Promoters/ Shareholders

The meaning of 'value' for this category of stakeholders is centred around profits, which is linked with the dividends and the ROI. Even banks and other financial institutions, who provide investments for construction projects, are conscious of the ROI of the projects so that they can recover their investments. This also helps in reducing their NPAs (Non-Performing Assets). Revenue growth and market share growth are other important components of value for them. As you have seen in earlier sections of this chapter, the construction industry is a fast-growing sector, hence satisfying these needs will not be exceedingly difficult. However, taking care of the newer needs of customer delight, in terms of customers' and end-users' quality expectations, a better adoption of new progressive management practices and technologies will enable them to attain higher goals.

### 1.5.4 Business Partners

Most of the work across the lifecycle of the construction projects is outsourced to expert consultants, a large number of consultants, contractors and specialist sub-contractors and suppliers. Even sales and marketing functions are also often outsourced. In such situations, there are significant value expectations from business partners in terms of faster responses, better consultants' and contractors' knowledge, honouring their time commitments, their trustworthiness (i.e., not engaging in any fraudulent activity) etc. However, business partners perceive their 'value' in terms of timely payments to them, their own profits, long-term relationships for assured future business etc. Both these needs can be simultaneously attained only by better management practices and adopting appropriate technologies.

### 1.5.5 Employees

Employees are the ones who operationalise the strategies of the companies into reality, and therefore are especially important stakeholders. Their concern for value to them is also equally important as for the other stakeholders. For them, 'quality of work life' is especially important, besides the compensation they get for their contribution to the organisations. This simply means they want a good job challenge, a good boss, good colleagues, and a good relationship with subordinates. Job security and career growth are other important factors for them. Embracing the newly emerging management practices and use of emerging technologies provides the opportunity to make their quality of work life much better, thus providing better value to the other stakeholders of the organisation. They also want continuous learning of new knowledge to keep their market value high.

In this discussion, you will also observe that there is a clash of values among some of the stakeholders. For instance, if promoters increase the compensation of the employees, their profits and hence the dividends are adversely affected. If the organisation must achieve the value for both these important stakeholders simultaneously, it must conduct business differently. That is, it must adopt and embrace newer management science concepts and newer technologies in business and construction processes. That is the reason Lean Construction 4.0 is critical to adopt for this sector early.

Therefore, the focus of all the later chapters in this book is on the Lean Construction 4.0 strategies to be followed for the best innovative project management practices in construction, and the technologies to optimally satisfy the 'value' needs of the key stakeholders.

## 1.6 Changes of Expectations for Various Stakeholders after Covid-19 Pandemic

New global virus threats like COVID-19 have had a significant negative impact on the construction sector. Two main reasons were responsible: workers were not allowed to gather in groups, and economic activity slowed as people avoided interaction. This reduced money flow and investments in construction. Some of these impacts and possible responses are outlined below.

For investors and promoters, such viruses reduce demand for commercial buildings, as seen during 2019–2021. With more people working from home, office space requirements decline, and demand for airports and hotels also falls as businesses shift to virtual meetings and communications. Travel and hotel stays reduce, and project postponements or cancellations become common. Yet this decline can be partly offset by new opportunities. Construction companies should collaborate with digital firms to create smart solutions — for example, using virtual reality mock-ups to help customers finalise finishes. This requires both construction expertise and digital skills.

Employment in construction is also hit by reduced demand and postponed projects, along with safety measures limiting crew sizes. Jobs have suffered worldwide. However, reskilling can open new opportunities. As remote work grows, many homes lack proper layouts for effective work. This creates demand for affordable redesigns and interior solutions to address space and noise constraints. Architects and builders must innovate to meet this need.

Consultants and contractors must design safer processes to minimise virus spread. Measures include limiting workers on site, enforcing distancing, and splitting crews into shifts. They must still meet deadlines with fewer workers, which calls for strategies such as Lean philosophy.

Government and society also shape the industry. New regulations on environment, health, and safety present challenges but also opportunities. By adopting Lean Construction Project Management and leveraging digital technologies, the sector can better address future threats. This book focuses on these strategies.

In summary, because of continuous changes in the environment, opportunities, crises, technology and expectations from various stakeholders, there is a strong need to sincerely look at new ways of managing construction projects. The rest of the book discusses the relevant Lean Construction 4.0 project management approach, which needs to be adopted quickly across the globe to address new issues, accelerate projects and provide better quality and safety in construction activities at lower cost. This is even more important for India and other fast-developing countries.

## 1.7 References

1. *Fuchs, S., Nowicke, J., & Strube, G. (2017). Navigating the digital future: The disruption of capital projects. McKinsey Capital Projects & Infrastructure.*
2. *Research & Markets. (2022). Construction market—Global opportunity analysis and industry forecast 2022–2030 (Report ID: 5519701).*
3. *Rastogi, S., & Tewari, G. (2018, September 27–30). Lean digital thinking framework for addressing the problems of Construction 4.0. Proceedings of International Conference ICIEIND 2018 on Industry 4.0, SOA University, Bhubaneswar.*
4. *Rastogi, S., & Laghate, A. (2021). Quality management of construction in the 21st century: Leveraging LEAN philosophy and digital technologies. Innovative LEAN Mentoring Academy, Chapter 1.*
5. *Niti Aayog, Government of India. (2019). Report of the Task Force on Project/Programme Management.*
6. *Mohanty, R. P., & Rastogi, S. C. (1986). An action research model for productivity management. International Journal of Operations & Production Management, 6(2).*
7. *Rastogi, S. (2015, August). Construction 4.0: New generation construction in a hyper-connected world. The Master Builder.*

# CHAPTER 2

# Limitations of PMS & Need for Lean Construction 4.0

# 2. Traditional Project Management Process

This chapter examines the limitations of traditional Project Management Systems (PMS) in construction projects. While PMS frameworks such as IPECC provide necessary structure, they often fail to deliver superior results in terms of time, cost, quality, and stakeholder satisfaction. To understand why, we first review the traditional PMS process before exploring the need for Lean Construction 4.0 as a complementary approach.

## 2.1 Traditional Project Management Process for a Typical Construction Project

The prime focus of a project management system (PMS) is to deliver the project scope on time, within budget, and with required quality, while meeting client requirements such as environmental, safety, and health standards. A typical PMS follows the IPECC processes—Initiate, Plan, Execute, Monitor & Control, and Close—as outlined in PMBOK. The Plan, Execute, and Monitor & Control phases align closely with the PDCA cycle, and in some organizations the Monitoring & Controlling group is referred to as 'Delivery.'

Small, short-duration projects may be managed as a single phase, but larger and more complex projects must be divided into clear stages for effective planning, execution, and control. For major undertakings such as manufacturing facilities, infrastructure like roads and bridges, or large residential and commercial developments, phased execution ensures better outcomes.

The typical lifecycle of a construction project is shown in Figure 2.1, with six phases: Concept, Design, Procurement, Construction & Handing Over, and Facility Maintenance & Decommissioning. Since maintenance is primarily an operational activity rather than a project, this book will not focus on that phase.

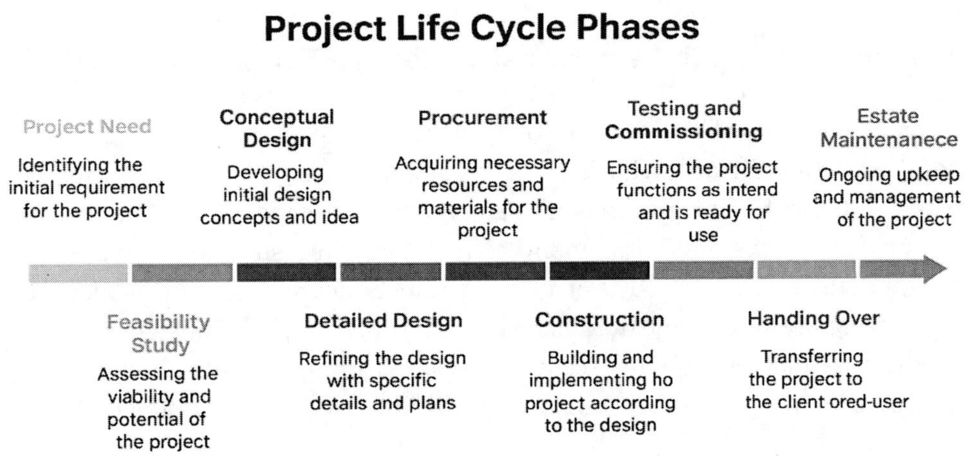


**Figure 2.1: Lifecycle of a Typical Construction Project Divided into Phases**

Based on the complexity of the project, the feasibility studies may be considered as a separate phase or part of conceptualisation. The Design & Engineering phases can be further detailed into Schematic Design, Construction Design Documentation etc. The testing, commissioning & handing over may be considered as separate phases in complex engineering projects, whereas they are treated as one phase in real estate and linear infrastructure projects. However, each phase needs to be managed using project execution steps of Initiation, Planning, Executing, Monitoring & Controlling, and finally, Closing (IPECC). This is shown as an illustration in Figure. 2.2.

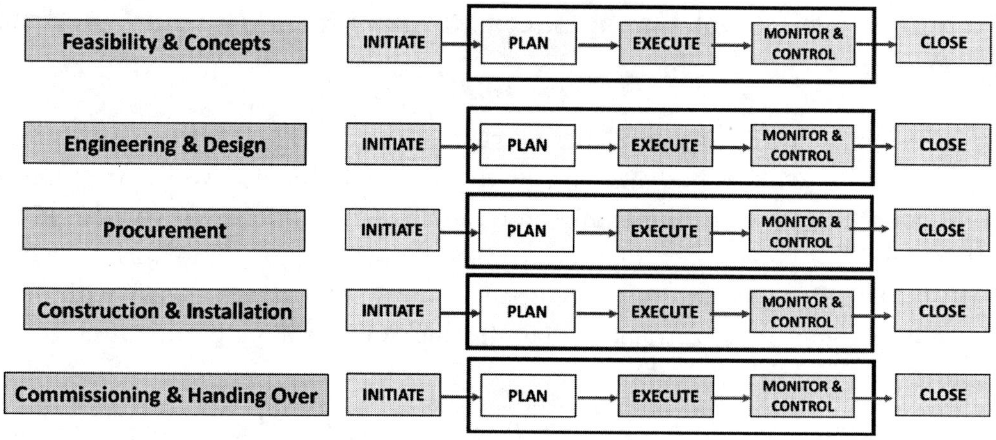


**Figure 2.2: IPECC in each phase (Example: Construction)**

The PMS manual of respective organisations and the project management plan of the project will have a detailed elaboration of the project management processes followed. However, it has been observed globally, and also corroborated by past projects that the authors have been involved in, that these IPECC processes in each phase are 'necessary conditions but not sufficient' to provide all the performance parameters expected by all key stakeholders in the current era of Lean Construction 4.0, as explained in Chapter 1.

## 2.2 Need to Augment to Lean Construction 4.0 Tools for Superior Results

As mentioned above, the IPECC process group is only a 'necessary but not sufficient' condition for achieving a project's superior performance in terms of project progress, its quality (including finishes), safety of people, especially workmen, and above all, the profits to the builders, main contractors and suppliers who invest their tremendous efforts. Almost 70% of global projects are not able to achieve expected performance. Many studies by McKinsey, KPMG, AGC, Autodesk Construction and others have found that delivery, quality, cost, safety, environmental sustainability and profit targets are not achieved, despite following the IPECC framework of PMI, USA.

Besides, in the current world, for the sustainability of business itself, it also needs to embrace many new goals like corporate social responsibility (CSR), better business partnership relations (with both suppliers and sellers), better work-life balance for employees, and also better environmental protection. This is in addition to the current project performance requirements of being on time, on budget, of required quality, and with target profits. Even these common performance goals are not attained in more than 70% of construction projects. This makes the job of PMS even more challenging. Historically, some of the most commonly reported issues in projects are listed in two categories below. These have been endorsed by various global researches.

### 2.2.1 Day-to-day Issues Faced with Traditional PMS, Impacting Most of the Project Performance Results Adversely

1. **Cash Flow Shotage:** This is one of the biggest problems mentioned in most projects. There is always a shortage of money available compared to the money required for payments to main EPC contractors and other sub-contractors, suppliers and workers in time. This shortage leads to further inefficiencies in project progress due to delayed payments.

2. **Delayed Projects:** Projects are mostly found to be behind the originally base-lined schedule, prepared at bidding time. All kinds of reasons (excuses?) are mentioned to justify the inefficiencies of the project management (PM) process for those delays. The delivery targets are revised several times and pushed farther for most projects. The McKinsey Global Institute Report [2] has found that 77% of projects around the globe are 40% or more behind schedule.

3. **Budget Overruns:** Projects mostly go over budget from the base-line approved budget. Many reasons including delays and poor cash flow are given to justify the inefficiencies of the PM processes. This often results in getting revised budgets approved. The KPMG Report [3] quantifies that 69% of projects have more than 10% (even up to 80%) budget overruns globally.

4. **Under-utilized Workforce:** Workers often remain idle for want of a 'required' work front, as the previous task was not completed by another contractor or specialist contractor, despite commitments. Often, the 'required material' was not supplied in time or was of poor quality, despite promises by the procurement team. This is depicted in an interesting animation in Figure. 2.3 below.

5. **Skilled Manpower Shortage:** There is a skilled manpower shortage of workmen across the world, and more so in India. Besides that, often the number of workmen who turn up for work is less than planned. Unplanned absenteeism, not accounting for predictable absenteeism during known festivals, unfavourable weather conditions like extreme heat or rain, and high attrition rates due to demand–supply gaps also contribute to this manpower shortage at site. AGC [4], the global construction association in the USA, reports that 44% of firms indicate labour shortage caused them tremendous delays in ongoing projects. It also states the attrition rate of construction workers is 21.4%, the highest among all sectors. This is also shown in Figure. 2.3 below.

6. **Equipment Shortage:** There are often shared equipment like tower and boom cranes, steel cutting and bending machines etc. required at different locations on the same project. These may not be made available to another group when required. The promised resources might not have been mobilised at the right time or in the right numbers. They might also be unavailable for use due to frequent breakdowns.

7. **Drawings/Information Shortage:** The required drawings are not ready or are not approved at the promised time. There are often errors in the drawings due to haste or lack of required expertise. The changes or RFI queries are not incorporated into the drawings in time, and there may be different versions of drawings available with different stakeholders like the construction site in-charge, quality in-charge and material suppliers etc.

**Figure 2.3 Top Six (out of 12) day-to-day Issues Faced during Construction, Impacting Performance**

8. **Incorrect Time Estimates:** The initial estimate of the project or activity duration is very often inaccurate and unrealistic given the actual available kinds of resources.

9. **Poor Quality & Rework:** There are many quality issues which the team is not able to resolve, or which take much longer than the desired time. There is a lot of rework activity as well. Autodesk Construction Cloud [5] quantifies that 35% of construction professionals' time is spent in non-productive activities such as looking for the right project information, conflict resolution about who will bear the cost of poor quality and rework, and actually dealing with mistakes and rework.

10. **Frequent Change Management:** As the project progresses, there are frequent changes in activity priorities or new changes requested by the client. This is unavoidable, as the complete project may not be very clear in the beginning. However, the number of changes can be reduced with proper discussions and knowledge available among the team. Project Group Analysis Group [6] has quantified that globally, on average, 35% of all construction projects will have major changes. This, by any standard, is very non-productive and impacts profits adversely, besides introducing more delays.

11. **Work Interruptions:** Customers encounter various (unpleasant) surprises and project interruptions (availability of the right resources in time, right information and approvals in time, unknown statutory constraints etc.) at different stages of the project, resulting in project delays.

12. **Productivity Fluctuations:** There is high variation in productivity and the outcome of each work package and trade on a daily basis. This results in a high degree of unpredictability and waiting times at several stages of the project during construction progress. Sourceable [6] global survey has found up to 50% variation in productivity between two groups of workers doing identical jobs on the same site at the same time, and up to a shocking 500% for two groups working on identical work but at different sites.

Figure. 2.4 below shows the summary of the above discussions in a visual form. Based on past experiences of the authors and various other global expert research studies, the following reasons are listed as the root causes of the problems mentioned above. Unless these issues are resolved in some way, the current issues will continue to exist, and current project performance cannot be made superior. Not only that, the new requirements around environment, sustainability, and other needs of various millennial stakeholders also cannot be satisfied.

**Figure 2.4: Balance Six day-to-day Issues Faced During Construction, Impacting Performance**

### 2.2.2 Root-Causes of Poor Performance

- **Parallel Portfolios:** Most builders start a greater number of projects than the realistic cash flow can manage properly. Therefore, due to cash flow problems, each project faces a paucity of funds and hence remains only partly completed, not ready for handover at the originally planned dates. The challenge for management is how to leverage the limited available cash flow and still complete all the projects by the originally targeted dates.
- **Unrealistic Corporate Plans:** The corporate plan for the project is generally quite unrealistic. Planning does not take into account the real constraints of the actually on-boarded EPC contractors, sub-contractors and other consultants involved in the project. The plan is primarily prepared based on the past experiences of the planner in similar projects. Often, the planner's experiences are incomplete for some of the work packages involved. The assumptions made by the planner when preparing the plan (like activity duration or required resources etc.) may also be very different from the reality of the current project. Hence, there is generally a huge mismatch between corporate and construction-site planning. In general, the people who actually execute the work are not involved in preparing the plan, even at a later stage, which creates an unrealistic plan and schedule.
- **Ineffective Governance:** In most construction projects, payments are linked to the material quantities consumed/completed during the billing period (like SFT of plastering in the week, cubic metre of concreting done in the fortnight etc.). Therefore, worker productivity and project output computed periodically cannot be uniform. For example, the volume of concrete work completed in a week can vary significantly depending on whether the work involves constructing columns or slabs. The project plan is also often made in terms of quantities completed for various work packages, for review and control purposes. This is not realistic for the reasons stated above. These completed quantity measurements are useful for the CFO and procurement team for payments, but they do not provide any clear indication of actual project progress or actions required to catch up when not on schedule. One cannot visually relate these figures to site reality, as one does during physical visits. Moreover, governance is normally conducted only to question the project team, not to enable resolution of their real constraints or roadblocks.
- **Errors in Drawings:** These are the main basis for material procurement and construction activities on site, but they suffer from several problems. These include inaccurate BoQs, frequent drawing revisions and missing details. There are also issues with understanding the correct specifications or requirements of the client. In addition, there is very poor collaboration between the site plan and the drawings plan for required drawing releases. Hence, very often, the required

drawings are not available on time. Figure 2.5 below shows some humorous errors in design and/or execution—this is just for some fun.

**Figure 2.5 Some Funny errors in Design and/or Execution**

- **Lack of Collaboration (3C):** There is a high amount of interdependence among several trades involved in the project, but the extent of regular *coordination, collaboration and communication* required among them is generally missing. This 3C is also missing among highly dependent engineering, procurement and construction teams. The KPMG report found that 82% of construction firms feel they need far better collaboration with contractors and consultants.
- **Reactive Work Culture:** The culture of top management is usually reactive, 'control and command' driven for the project, rather than one of proactive 'enablement and support' towards various contractors. The project management approach overall is also primarily reactive—doing post-mortems (progress status reviews) and dealing with crises. The proactive nature of identifying risks or alerts in advance, and finding solutions—including making the required resources and information available in the near future—can enable and support teams to minimise crises or interruptions. This is almost completely missing in most projects. KPMG also reports that 78% of construction companies believe that project risks are increasing, leading to significant delays.
- **Lack of Shared Vision:** There is a lack of shared vision of the project among various stakeholders due to the way contracts are currently prepared. Execution is done in siloed stages by all contractors, sub-contractors and specialist contractors involved. All of them take care of only their part of the contract, rather than keeping the overall project goals and objectives in view. Roles, responsibilities and KPIs also require far better clarity than what is followed presently.

- **Delayed On-boarding:** Quite often, the on-boarding of contractors or consultants is delayed, and they do not get enough time and information to understand the project in sufficient depth to plan and organise their resources properly and in sync with others. This is also true for specialist contractors and suppliers.
- **Easy Approval of Extensiobns:** There are frequent requests for extending project completion target dates, which are often accepted by the client. Generally, clients are not able to assess the genuineness of the reasons provided. No serious attempt is made to improve current processes to reduce project duration and make up for the lost time.
- **Limited Technology Use:** For execution of business processes (project management, quality management, procurement management etc.) and construction processes, very limited technologies are used—either due to unfamiliarity or lack of risk-taking capability for adopting innovative and better solutions. Even digital technologies, which are cost-effective and low risk, are not being tried. The Autodesk Report found that 75% of respondents stated an increasing need for rapid decision-making in the field, but due to lack of technology adoption, they were not able to meet this need.
- **Poor Selection of Contractors/Suppliers:** In order to apparently save money, the vendor with the lowest quotation is selected, who may ultimately cost much more than planned. There are no efforts to develop them or build long-term relationships. KPMG Report [3] says that almost 70% of construction business owners cite poor contractor performance as the single biggest reason for project underperformance. Autodesk & Dodge Data & Analytics [5] say that 66% of general contractors are carrying added costs from overtime or second shifts on at least three-quarters of their projects due to schedule slippage, with 50% needing to extend the project end date. Yet, this very important area is not taken seriously.
- **Unkept Promises:** Promises made by contractors regarding quality and delivery dates are often not honoured. This is mainly because commitments are based on hunches, not micro-planning. It is also due to poor collaboration among various contractors and consultants, and their low accountability. This leads to one of the biggest problems—progress interruptions due to non-availability of the 'promised work front' to other contractors. Autodesk Report [5] found that high-trust construction firms are twice as confident in meeting project deadlines, saving them millions of dollars annually.
- **Blame Game:** For any delays, quality issues or cost overruns, there is mostly fire-fighting and blame shifting among various contractors and consultants, each trying to safeguard themselves. There is very little focus on jointly resolving problems for the overall benefit of the project.

- **Busy Syndrome:** Everyone seems to be in a state of "busy syndrome." People are 'too busy' with routine tasks and unplanned meetings to have any time for improvement—even for critical bottleneck business or construction processes. Figure. 2.6 depicts this in a humorous way.

**Figure 2.6: "Busy Syndrome" commonly found in Construction Projects**

- **Sub-optimal Resource Utilization:** Generally, there is sub-optimal utilisation of different resources like workforce and equipment. These resources produce outputs far lower than their capabilities.
- **Too Many Sub-Contractors:** There are a large number of specialist contractors involved, leading to many quality problems and delays due to poor communication, cooperation and collaboration.
- **Poor Workforce Quality:** The quality of all the people working in the projects, from workmen to managers, is generally poor. Besides, there is no focus on continuous learning or improving their quality. The US Chamber of Commerce [6] found that 62% of contractors report high levels of difficulty in finding skilled workers.

All these issues lead to poor customer satisfaction and also poor contractor and employee satisfaction, in terms of delays, overspending, quality not meeting expectations, stressful working conditions, poor work environment etc. The following section suggests how these issues can be addressed by leveraging Lean philosophy and resolving most of the problems mentioned in 2.2.1.

### 2.2.3 Lean Construction 4.0 'Principles' to Address Root-Causes

A lot has been written about Lean concepts. You can refer to some of the references mentioned earlier. Here, we will cover the most fundamental aspects of Lean philosophy in simple terms. There are five guiding principles of Lean which form its foundation.

**These are:**

- **Challenge the Status Quo:** Always look for a better way of doing things.
- **Go and See:** Visit the place of work to observe and discuss both the good and the bad parts of IPECC—initial designs, planning, execution, monitoring and control. Don't stay in the office to discuss problems and solutions.
- **Continuous Improvement:** Always have a 'quest' for learning and trying superior and innovative new ways of doing things.
- **Respect the People:** Each person has valuable thoughts and must be leveraged for the overall improvement of the projects.
- **Teamwork:** People should take pride in working as part of a team.

Besides these five fundamental principles, there are also five very important keywords or key focus areas in Lean concepts, which should become part of every Lean expert's approach, regardless of the task. These keywords are given below and explained briefly. You will learn more about these in later sections with suitable case studies and illustrations.

- **Value:** We must first define clearly what 'value' the project adds to the organisation, for the ultimate customers or end-users. What features, functions, designs, and products or services are actually required, and what are not? When we focus on value, we start critically observing the practices, features and efforts that do not contribute to the identified value—and therefore should be considered as 'wastes', in terms of time, effort and cost. Section 2.3 discusses this in greater detail with examples.
- **Value Streams:** In order to deliver value, a construction project is conceived, requiring several activities and processes to be completed to obtain the final assets—be it real estate buildings or infrastructure projects. We need to see how the 'work' flows at the place of work—both at the construction site and in other supporting departments like design, procurement etc. These sets of processes, both construction and business, are called value streams, as they add value to the overall project. These value streams need to be executed effectively and efficiently.

- **Flow:** We must ensure that all work in the value streams flows efficiently at the site and within the broader construction process. We need to critically examine the value streams to identify constraints or roadblocks that cause hold-ups, interruptions, or slowdowns. Also, identify what causes rework, which leads to waste of human effort and materials. We need to remove these causes and make the work flow smoother and faster across all value streams, to realise value more quickly and cost-effectively. We may need to consider redesigning traditional processes and ensure that this smooth workflow repeats consistently. This is explained in greater detail in Chapter 6.
- **Pull:** Once all value-adding activities and steps are flowing smoothly, we should work on reducing the time required and accelerate the process. This enables us to start work as late as possible, incorporate value-adding changes suggested by the client, and still complete the task quickly to meet market or project deadlines. It ensures we only perform work when it is actually required by the internal or external customer at the next step.
- **Perfection:** We should continuously strive for improvement. Even if we already have a good construction or business process, tapping into the ideas and knowledge of the people working on the project allows us to eliminate further waste and add more value. This supports daily process improvement.

Based on the above discussion, a workable and understandable definition of Lean Construction 4.0 can be worded as:

"It is a systematic approach to define 'value', identify and eliminate 'wastes', to bring 'speed, predictability and certainty in workflow' through 'pull' planning, and achieve 'perfection' in project work through continuous improvement—leveraging technologies wherever required."

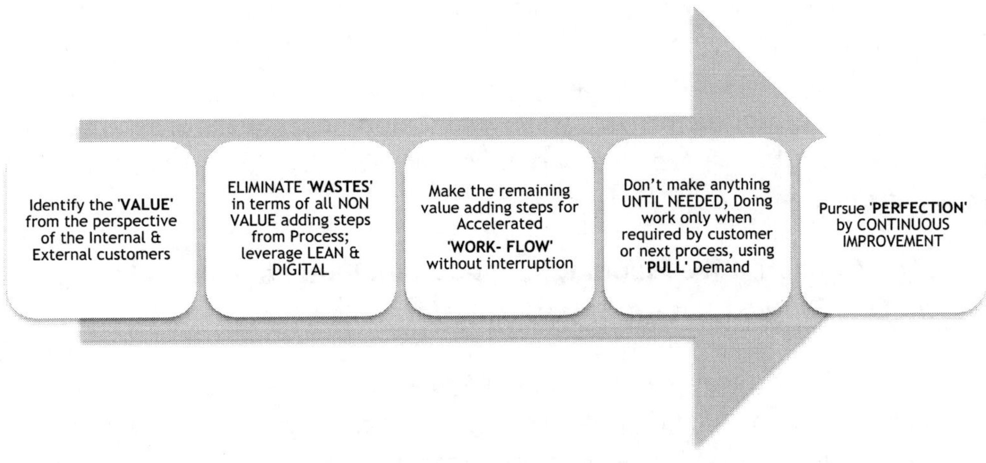


**Figure 2.7: Simple definition of Lean Construction (5 Key words of LEAN are written in BOLD)**

**Because of this focus, Lean Construction 4.0 provides the following superior performance outcomes for projects in terms of:**

- Delivering greater business value from the project
- Accelerating the project and reducing its duration whenever necessary
- Managing project costs better and avoiding revised increased budgets, thereby improving profitability
- Delivering projects more safely
- Minimising unnecessary rework and preventing extra delay and cost due to it
- Creating a more predictable production program for 'steering' the project proactively towards its objectives and goals
- Reducing stress on project management staff
- Helping to improve the efficiency and effectiveness of the overall production process for the project
- Welcoming changes suggested by the customer to enhance value, while still aiming to make timely deliveries through process improvements
- Encouraging innovation among the team and fostering better relationships
- Minimising all kinds of wastage and helping improve sustainability
- Compensating for the weaknesses of CPM/PERT-based unrealistic planning and monitoring

The most important, new and 'core' concept in Lean Construction 4.0 is the Collaborative Planning System (CPS). CPS refers to project planning, scheduling, executing, monitoring and controlling in a better way compared to the traditional PMS. The basic concept of CPS directly addresses one of the biggest weaknesses of the current PMS—that is, the lack of coordination, cooperation and communication among key stakeholders. CPS is the core of Lean Construction. There are also other tools such as Value Stream Mapping, Big Rooms, and identifying and eliminating the Eight Wastes, which help in achieving project goals even more effectively. The philosophy and fundamentals of CPS are explained briefly below. The detailed processes, however, are covered in Chapter 4.

### 2.2.4 CPS: Collaborative Planning, Scheduling, Executing, Monitoring & Control System: One Single Plan for Superior Performance

One of the biggest reasons for failing to adhere to the 'unrealistic master plan' is the present way of managing the project, which focuses more on project planning and control only. That is, first it prepares the unrealistic project plan and schedule without involving the consultants and contractors/suppliers or understanding their real constraints at the time of execution. The main contractor and other sub-contractors do sign off on this plan as agreed. However, that is more for securing the business of the project without realising their own constraints and limitations at different points of time during project progress.

During project execution, the client keeps reviewing those agreed plans periodically to monitor the progress made till then. That means the reviewer asks what was planned, how much was achieved, and then forces the execution team to 'somehow' meet the plan. This review focuses more on the post-mortem of project progress—that is, the recent past—rather than on risks, resolutions, and support needed for the near future. The reviewer does not ask what risks or problems are expected in the near future, or how those will be resolved. Nor is it asked whether any help is required by the execution team from the reviewer to resolve those issues. This approach really does not help much, and therefore most projects face the problems listed earlier in 2.2.1.

This can be compared to driving a car using only the rear-view mirror. The rear-view mirror shows only what has already happened and passed by, but not what is going to come—which is visible only through the front windscreen. In reality, looking through the front windscreen is more important to complete the rest of the journey successfully. Looking in the rear-view mirror is required only occasionally in between.

While driving forward to reach the target destination, the driver must also ensure that the time available and fuel available goals of the journey are not missed, in order to stay within the agreed time and budget. The time and fuel remaining for the forward journey—and what lies ahead on the road, especially if it is a single lane or passes through a forest or hilly terrain—are like the remaining schedule and cost goals for completing the rest of the project successfully.

Figure. 2.8 shows that while the rear-view mirror (post-mortem) is very clear, the front windscreen of the car is hazy. Therefore, driving using only the rear-view mirror may result in missing planned targets or even cause an accident - a failed project. Unfortunately, this is how current projects are monitored and controlled.

**Figure 2.8: Driving Using the Rear-View Mirror of the Car**

Between 'Project Planning' and 'Project Control', the real missing link for superior performance is the Production Planning (or Construction Planning, in the context of construction projects) and Production Control part. This part primarily focuses on how to perform the work in order to steer the project proactively towards the original target, using the front windscreen along with the rear-view mirror.

This means that in order to translate the project plan into reality, you should ideally do what was planned earlier and is required to achieve the schedule and cost targets. But if it is observed that project progress is not on track—i.e. it has slipped on time, cost, scope, and quality targets—you have to do things differently. If the schedule and cost are overrun, then the team must find an alternative way of doing things to achieve the originally committed goals. The team must also innovatively find ways to make up for the missed targets so far. They should identify how to proactively prevent those issues from recurring.

This missing link of production planning and control is filled by the Collaborative Planning System (CPS) (discussed later). This is shown as a symbolic diagram below (Figure 2.9).

**Figure 2.9: The Symbolic Representation of Collaborative Planning System**

Actually, the Collaborative Planning and Scheduling System (CPS) is a collaborative production or construction planning and control system for attaining the committed, baselined project goals of schedule, budget, scope and quality proactively and realistically. This involves the collaboration of all the relevant people, such as design and engineering consultants, contractors, sub-contractors, specialist contractors, and key suppliers. This team collaboratively arrives at an agreeable and feasible plan which does not compromise the commitments made regarding the major milestones agreed upon by the client in the master project plan at the time of sign-off.

CPS complements the traditional PMS system prepared using PERT/CPM (normally leveraging project management software such as Primavera or MS Project). These traditional PMS systems often fail to address the issues listed in 2.2.1 in a structured way. Lean Construction provides a formal, structured framework for removing the root causes listed in 2.2.2 and resolving almost all the problems. Figure. 2.10 shows the framework for the additional steps to be introduced on top of the baselined plan in order to manage the project realistically and to ensure achievement of the committed goals to the senior leadership team.

CPS primarily adds five layers or levels of collaborative planning and scheduling: Must-Do, Should-Do, Can-Do, Will-Do (weekly), and Will-Do (daily). It also adds another five levels of monitoring and control, and one level of continuous improvement. These are described briefly below and will be explained in more detail in Chapter 3.

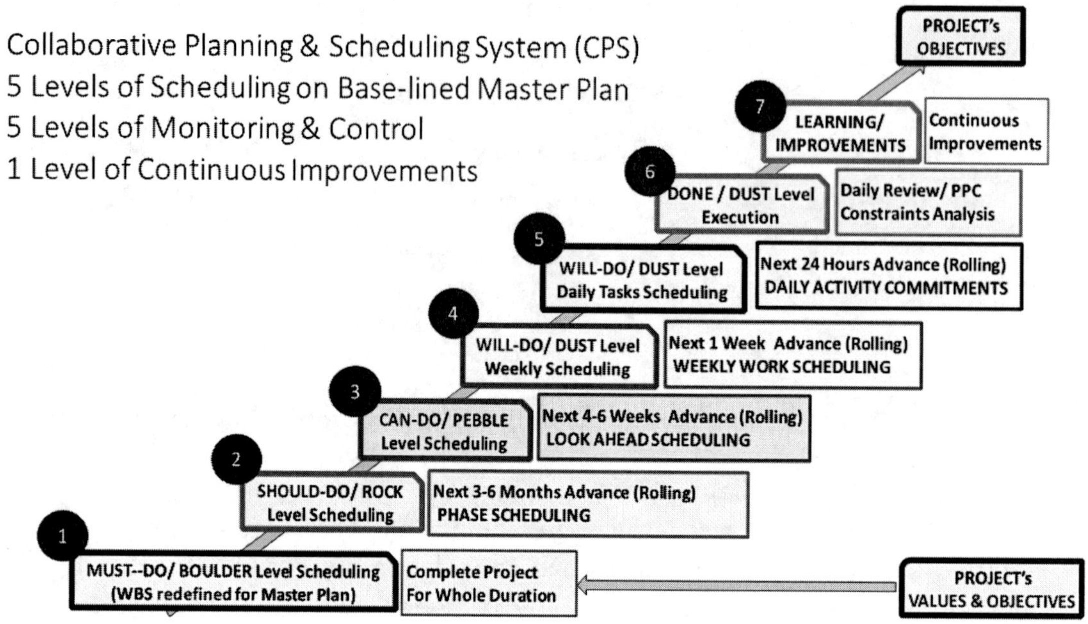


**Figure 2.10: Brief description of CPS**

## Step 1: "Must-Do" / 'Boulder Milestones' Work Structure Planning

In this step, the baselined original master plan (which was signed during bidding) is revisited and revised as per the WBS (Work Breakdown Structure) recommendations of Lean Construction requirements. Instead of considering the entire detailed activity plan, only the major milestones are retained for further planning. The total duration for the revised Lean project plan is also reduced by 10–15%, to retain this 10–15% of the duration as a buffer, but only towards the end of the project.

## Step 2: "Should-Do" / 'Rock' Level Collaborative Scheduling

The project activities for the boulder phase milestones are redefined, if necessary, but are considered only for the next 4–6 months of the project phase, and not the complete project. Considering activities beyond that timeframe is often meaningless due to several uncertainties associated with the distant future. There is an iterative collaborative planning and scheduling process involving all the relevant people, including those from design and procurement. They prepare an agreeable new Lean plan for those rock-level processes and activities. However, they need to ensure that the committed baselined boulder milestones are achieved. The problems or constraints identified in attaining this are listed for proactive resolution. These problems may be related to the availability of manpower, machinery and equipment, detailed drawings, materials, cash flow etc. Based on these agreed plans, the handover dates for various activities (work fronts) are agreed among the next dependent contractors.

### Step 3: "Can-Do" / 'Pebble' Collaborative Scheduling

This is collaboratively derived from the Should-Do schedule covering the next 6–7 weeks, by further detailing the 'pebble' activities required for those 'rock' processes. An updated list of constraints—related to all the resources and information required during this project phase—is identified, reviewed, and resolved. The team may request management support for issues that cannot be resolved at their level.

### Step 4: "Will-Do Weekly" / 'Dust' Collaborative Schedule

From the Can-Do pebble-level schedule, the activities that have no constraints—related to drawings, materials, people, machines etc.—are further broken into dust-level smaller activities or tasks. Those with no constraints are then selected for scheduling in the immediate next week, and a schedule is prepared for the next 5–6 days. Since contractors and sub-contractors were involved in this process, and they have discussed and understood the reality of near-future constraints, they can now commit to completing these activities and tasks with greater confidence.

### Step 5: "Will-Do 24 Hours" / 'Task' Collaborative Schedule

The Will-Do activities schedule, containing the commitments made by various contractors for the immediate next week, is passed on to the foremen of the relevant trades or work packages. The foremen prepare their 'to-do' list of even smaller dust-level tasks to be completed in order to achieve the next 24-hour committed schedule. Any important information about work-front handovers from the previous contractor is verified by the team and confirmed during a short daily joint meeting in the Big Room, where all the relevant project information—daily, weekly, monthly and quarterly plans and progress—is visually displayed. This is discussed in greater detail in Chapter 3 on CPS and Chapter 6 on Big Room.

After all these five levels of schedules are prepared, their execution, monitoring and control are carried out on a day-to-day and weekly basis, as described below.

### Step 6: "Will-Do 24 Hours / Done" Collaborative Monitoring and Control

Daily, foremen and contractors hold a 20-minute meeting to update the status of completion of the previous day's 'to-do' tasks, and to note the reasons for non-completion, if any. They also plan in detail, at the dust level, the tasks for the next day.

### Step 7: Learning / Improvements

The reasons for frequent delays or overspending, identified during Step 6, are reviewed. For repetitive causes or key issues, new ways of doing things are discussed and implemented for future improvements. The team also plans how to make up for any time losses incurred during the previous week or day, in terms of both time and cost.

**In short, CPS (also called The Last Planner / LPS ®) primarily focuses on:**

- While planning and scheduling processes, involve those people who distribute and carry out the actual work. This will ensure a realistic schedule, with greater commitment and a higher possibility of achievement in reality. Planning should not be limited to only the corporate planner.
- Complementing the CPM/PERT-based high-level corporate plan with five additional levels of collaborative planning and scheduling. This addition is required for carrying out the production planning and control discussed earlier. It ensures realistic and doable planning and scheduling, with key risks identified and addressed in advance. This also ensures better commitment from the contractors and consultants involved in engineering, procurement and construction. It further improves the performance of project control.
- These five additional collaborative layers steer the project proactively, rather than simply carrying out a post-mortem of project tracking. This is like driving while focusing on the front windscreen, and occasionally checking the rear-view mirror.
- Making work ready proactively, rather than reacting through regular firefighting to resolve crises. This means preparing the eight key requirements needed to complete activities and tasks: Drawings / Information / Approvals, Materials, People, Equipment and Tools, Previous work completion, Space and Safety requirements, and External Environment (such as weather, religious concerns etc.). This minimises project risks on a regular basis.
- Developing a shared understanding of project value and eliminating possible constraints or interruptions to ensure a smoother flow of work.
- Making educated promises with confidence, rather than receiving unrealistic completion orders from the corporate planner.
- Maintaining a workable backlog of ready-to-do work, to buffer tasks against capacity limitations and time loss.
- Keeping the buffer time as a comprehensive block at the end of the project, rather than distributing buffers across each activity.
- Empowering site personnel to take innovative steps to make up for any delivery date slippage by improving processes through innovation.
- Fostering continuous learning and improvement.

CPS also takes care of many presumptions and realities observed by experts through their research work over the last several years. These are:

- The production (construction) system is not only technical in nature, but also social.

- All plans are forecasts. And all forecasts are wrong. The forecast errors vary with forecast length (time duration) and level of detail.
- Planning is dynamic and does not end until the project is completed.
- Involving those who directly supervise or perform the work in the planning process results in better and more realistic plans, and a greater ability to adapt when required.
- Operational performance (safety, quality, time, cost, profitability, environmental sustainability etc.) varies with the degree of planning detail and preparation.
- Willingness to invest time in planning and preparation is directly proportional to the reliability of workflow from internal suppliers of work for the next stage. If previous contractors do not keep their promises for handover of work fronts, the next contractor will not prepare a detailed plan, as it would be unrealistic.
- Making commitments publicly promotes better commitment and increases the effort to deliver on that commitment.
- The probability that commitments should and do increase when both parties (internal customer and supplier) practise reliable promising.
- An essential prerequisite for reliable promising is that the supplier can say 'No... because...' and 'Yes... provided...' in response to a request.
- Understanding the project objectives, and the current and future state of the project, helps people make better decisions.
- Perfect planning may not be achievable, but it is possible to avoid making the same mistake twice.
- Variations in the production system can be reduced but never eliminated; therefore, buffers are required to absorb them.
- Workflow reliability improves if only sound, sequenced, standard and sized tasks are committed.

Lean Construction 4.0 concepts, with CPS as its core and leveraging relevant technologies, address all these and other root causes listed earlier. The addition of further concepts and methodologies from Lean Construction will help customers attain superior value from project execution. This framework for incorporating Lean philosophy is called Lean Construction 4.0 PMS, as opposed to traditional PMS.

## 2.3 Meaning of 'Values' for Different Project Stakeholders

The key stakeholders in construction projects include clients, who may often be the government or public sector for public utility infrastructure and real estate projects, and corporations for commercial or housing real estate projects. Other stakeholders include builders, investors/bankers, main contractors, consultants, sub-contractors or specialist contractors, material suppliers etc. The value in Lean Construction 4.0 for many of these important stakeholders is provided below in brief bullet points as an illustration. The Lean Construction 4.0 PMS team must define value clearly for their real projects, as value may differ across different types of projects. This may require serious discussion among stakeholders.

### 2.3.1 Meaning of Value for the Senior Management Team of Client, Builders & Investors

The value for senior management team may come from some of the aspects shown in the following diagram (figure 2.11). Table 2.1 shows the meaning of value for residential apartments/ villas for end-users & investors.

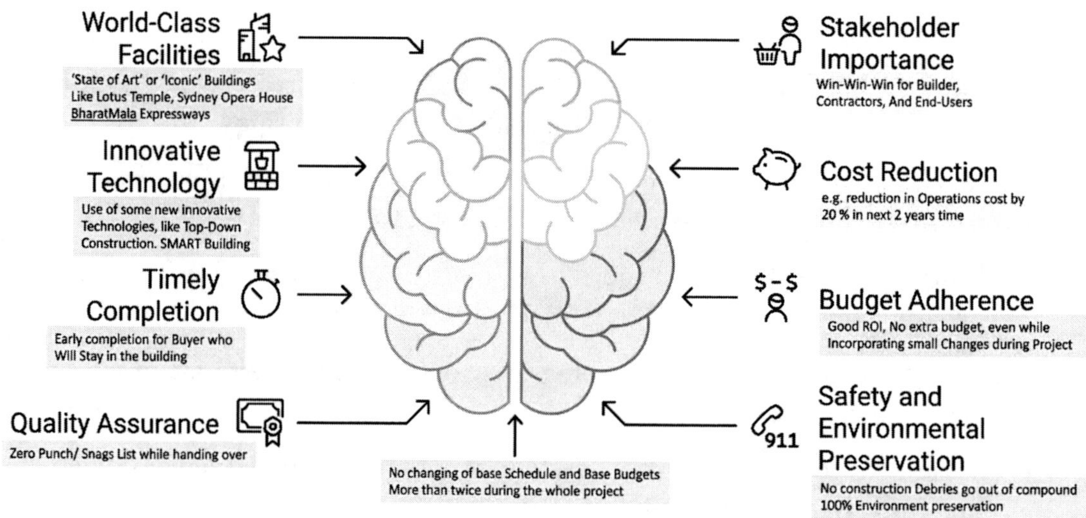


**Figure 2.11: Value for the Client's Leadership Team, Builders and Investors**

## Table 2.1 Meaning of VALUE for Residential Apartments/ Villas for End-Users & Investors

| REAL ESTATE VALUE CALCULTOR | | | |
|---|---|---|---|
| **VALUE** | | **PARAMETERS** | **UNIT OF MEASURE** |
| USE VALUE | CUSTOMIZATION | Customization provided | # of items allowed to select from 5 listed items |
| | | (Kitchen Platform height, Tiles, Interior paint, Light Fixtures, etc.) | |
| | Common Facilites | | INR/Sq Ft invetsed for these facilities |
| | | Swimming Pool | Square feet / number of Apartments or Villa |
| | | Sauna / Spa | Square feet / number of Apartments or Villa |
| | | Landscaped Garden | Square feet / number of Apartments or Villa |
| | | Power Backup | KVA / 1000 Sq Ft |
| | | Gym | Square feet / number of Apartments or Villa |
| | | Joggers Track | Length of Track / number of villa or apartment |
| | | Cycling Track | Length of Track / number of villa or apartment |
| | | Kids Play Area | Square feet / number of villa or apartment |
| | | Badminton Court | Number of Courts / number of villa or apartment |
| | | Tennis Court | Number of Courts / number of villa or apartment |
| | | Snooker | Number of Tables / number of villa or apartment |
| | | Billiards Table | Number of Tables / number of villa or apartment |
| | | Dance / Yoga Studio or Center | Square feet / number of villa or apartment |
| | | Visitors Parking | Number of parkings / number of villa or apartment |
| | ECO BALANCE | Solar Energy Use | KVA / 1000 Sq Ft |
| | | Rain Water Harvesting | % water being recycled for reuse in gardens etc |
| | | Kitchen / Bathroom Water Recycling | % water being recycled for reuse in gardens etc |
| | | Sewage Water Recycling | % water being recycled for reuse in gardens etc |
| | | Number of Water Sensors used for Water Consumption Control | # of sensors used / number of villa or apartment |
| | | Number of Light Sensors used for Power Consumption Control | # of sensors used / number of villa or apartment |
| | Private Facillities in Apartment/Villa | | INR/Sq Ft invetsed for these facilities |
| | | Private Garden | Square Feet / Apartment or Villa |
| | | Terrace | Square Feet / Apartment or Villa |
| | | Spa / Jacuzzi | Number of Jacuzzi / Apartment or Villa |
| | | Theatre Lounge | Square Feet / Apartment or Villa |

| | | | |
|---|---|---|---|
| | | Elevator | Number / Villa or Apartment |
| | | Servant Room | Square Feet / Apartment or Villa |
| | | Digital Security System (Video/Audio) | Premium / Indian brand |
| | | Quality of Life (Noise level) | dB level in that area |
| | | Quality of Life (Pollution) | PPM level in that area |
| ESTEEM/ BRAND Value | General Looks and Aesthetics Appeal – External Building | Brand of ARCHITECT | Global / Indian Premium / Reputed |
| | General Looks and Aesthetics Appeal – Interiors on Model Apartment | Brand of ARCHITECT & INTERIOR DESIGNER | Global / Indian Premium / Reputed |
| | Brand of Key material | Brand of GENERAL CONTRACTOR | Global / Indian Premium / Reputed |
| | | Bathroom Fittings | Global / Indian Premium / Reputed |
| | | Number of bath tubs / Integrated Showers Systems | Global / Indian Premium / Reputed |
| | | Kitchen Fittings | Global / Indian Premium / Reputed |
| | | Light Fixtures (LED / CFL) | Global / Indian Premium / Reputed |
| | | Electrical Switches | Global / Indian Premium / Reputed |
| | | External Wall material | Façade / Premium Tiles / Paint |
| | | Floor material | Italian Marble / Marble / Vitrified Tiles |
| | OTHER BUYERS | Social Status of Other Buyers / Investors in the project | Businessmen / MNC Professional |
| | NOVELTY | Newness / Novelty factor / Signature iconic Design | Unique / Very Good / Good |
| COST Value | | EMI Convenience (15 years base) | EMI / Current Rent |
| | | Total Budget / Cost | Rate INR / Sq Ft |
| | | Discounts offered | % |
| | | Cash Flow / Payment Terms (specially due to historical delays) | Historical % Project Delay |
| | | Level of Statutory approval (Land, Project, Architecture etc) | % (number approved / number required to be approved) |
| | | Payment by Cheque | % by Cheque of the total amount |
| | | Cost Reduction due to special Technology / Management | % |
| Time Value | | Time promised to Deliver | Months |
| | | (Historical performance) of Delay in possession of projects | % or Months |
| PLACE/LOCATION Value | Proximity to key Facilities (tour need priority in the bracket) | | Average distance to be traveled for these 11 Facilities |

| | | | |
|---|---|---|---|
| | (Critical/Very High/High) | International School | Distance in Kilometers |
| | (Critical/Very High/High) | Premium Hospital | Distance in Kilometers |
| | (Critical/Very High/High) | Big Mall / Hypermarket | Distance in Kilometers |
| | (Critical/Very High/High) | Theatre / Multiplexes | Distance in Kilometers |
| | (Critical/Very High/High) | Office | Distance in Kilometers |
| | (Critical/ Very High/High) | Main Road (60 feet and above) | Distance in Kilometers |
| | (Critical/Very High/High) | Airport | Distance in Kilometers |
| | (Critical/Very High/High) | Railway Station | Distance in Kilometers |
| | (Critical/Very High/High) | Wellness Center | Distance in Kilometers |
| Exchange Value | | Amenities Realization, expected | % (based on CREDAI / Govt plans) |
| | | Speed of Resale | # of days |
| | | Socio-Economic Development of area | % of population > x annual income or Professionals |

## 2.3.2 Meaning of 'Value' for EPC/ Main Contractors

The value for EPC main contractors may come from some of the following items, as shown in the diagram (Figure 2.12) below:

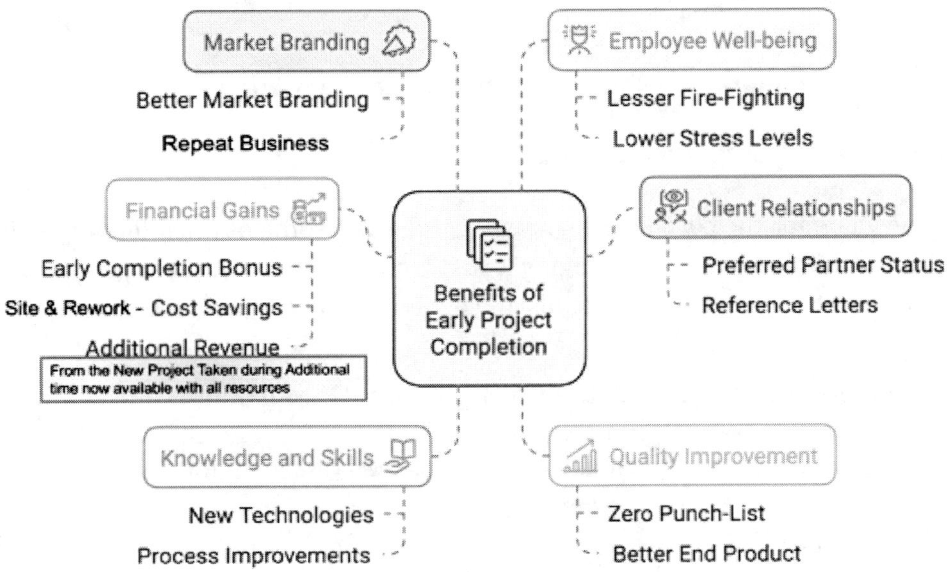


**Figure 2.12 Value for EPC/ Main Contractors**

### 2.3.3 Meaning of 'Value' for all Sub-Contractors & Suppliers

The value for the sub-contractors will be similar to the EPC main contractor, shown in the diagram below (figure 2.13).

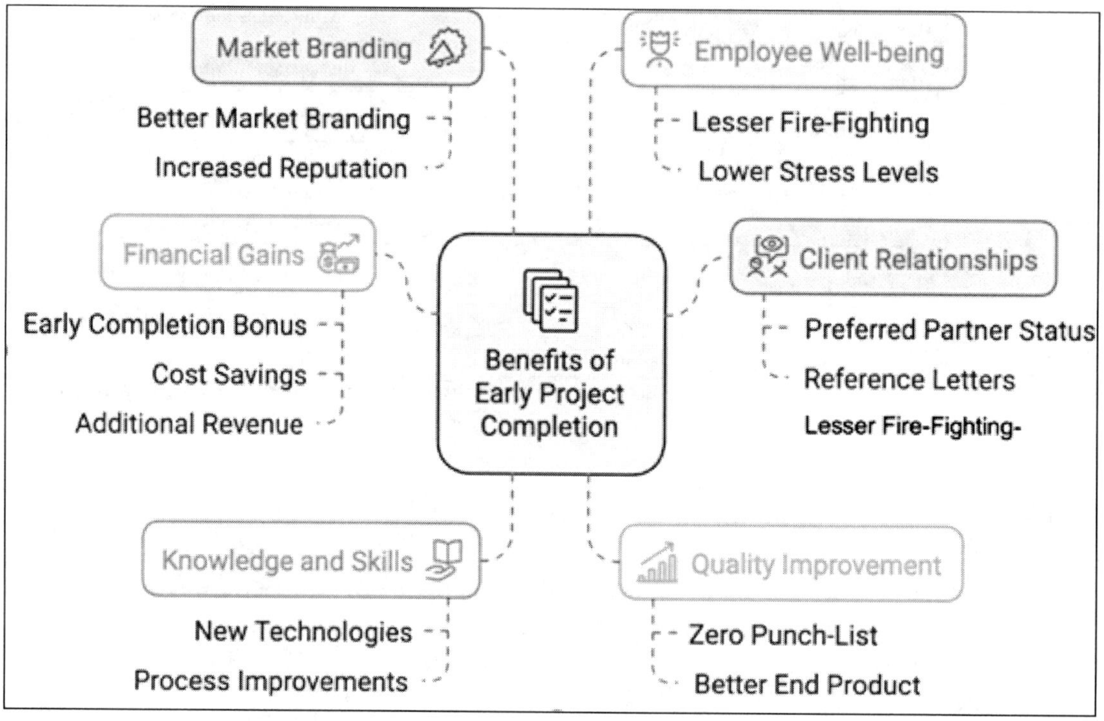


**Figure 2.13 Value for Sub-Contractors & Suppliers**

### 2.3.4 Meaning of 'Value' for Core PMS Team

The value for the core PMS team will come from some of the following points. In the current scenario of a fast-changing construction industry, most people are also looking forward to enhancing their skills and acquiring new knowledge etc.

### 2.3.5 Realisation of Superior Value through Lean Construction 4.0

As can be seen from previous discussions, Lean Construction 4.0 - and especially the CPS framework - when combined with the right technologies - functions in a way that enables most of the value for the client and other stakeholders to be achieved. It has been estimated through global research that Lean Construction has saved up to 10–20% of project costs and has accelerated project completion, delivering better quality end

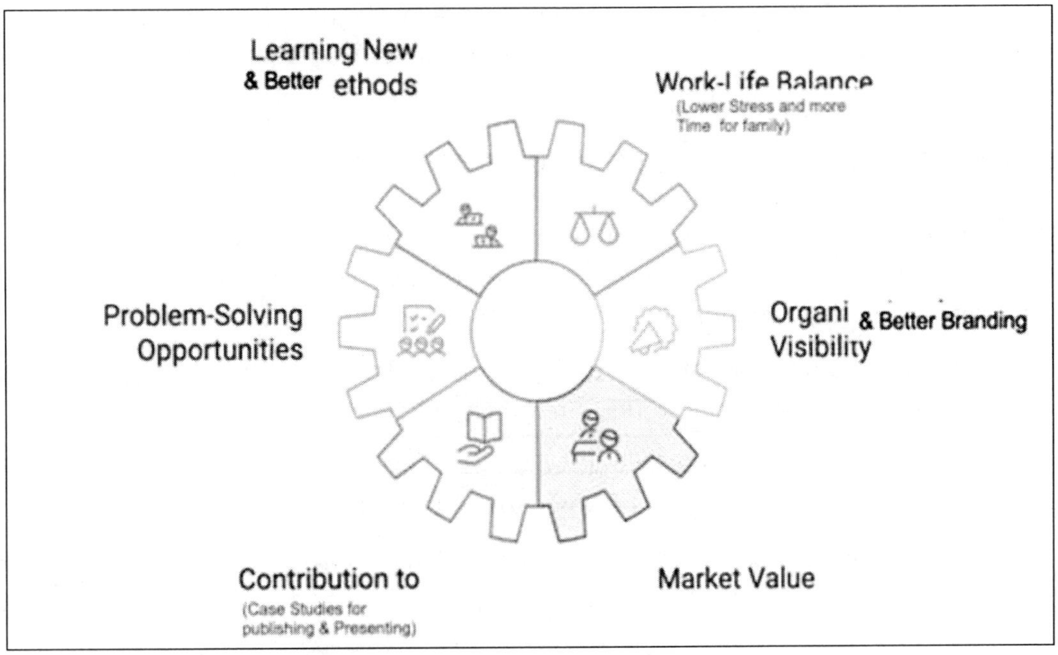


**Figure 2.14: Value for Core PMS Team**

results. It has also contributed significantly to environmental sustainability by consserving all resources, including natural ones. Therefore, this concept is gaining acceptance and popularity across the world.

## 2.4 Other Useful Tools for Lean Construction 4.0

In Lean Construction, besides CPS, some additional dimensions and tools are also embedded into the traditional PMS to make it more valuable to customers, end-users and other stakeholders, and to minimise various kinds of waste traditionally associated with PMS.

The following are some of the key Lean tools that are particularly useful in Lean Construction. Detailed explanations of these tools are provided in the later chapters of the book:

- Waste Identification and Elimination
- Value Stream Mapping (VSM)
- The Big Room
- Defect Management (Prediction and Prevention)
- Learning from Schedule/Budget Overruns and Continuous Improvement
- 5S

Figure. 2.15 below shows, through a simple diagram, how these tools can be integrated with the traditional PMS. Most of these tools are explained in later chapters of the book.

## 2.5 Lean Construction 4.0: A Different Way of Working and Work Culture

Lean Construction 4.0's way of working and work culture has several significant differences from traditional project management practices. Some of these were already discussed in previous sections. The Lean Construction 4.0 work culture emphasises:

1. Providing superior value to customers and other key stakeholders, rather than simply maintaining the status quo or just meeting predefined objectives.
2. Promoting continuous learning and improvement, with a focus on identifying and eliminating the eight types of waste that do not add value to the customer.
3. Working collaboratively with a cross-functional approach, aimed at achieving the common goal of the project. This approach reduces dependency on rigid hierarchies and breaks down departmental silos.
4. Valuing employee empowerment and involvement in decision-making processes, giving team members opportunities to contribute ideas and suggestions for improvement.
5. Encouraging learning from mistakes and treating them as opportunities for progress—fostering a culture of experimentation and innovation. In contrast, traditional systems often discourage risk-taking and punish failure.
6. Fostering open communication and transparency throughout project planning and execution, cultivating a culture of trust and shared knowledge.
7. Promoting accountability and ownership, instead of allowing a blame culture or finger-pointing to take root.
8. Investing in employee development, with a focus on upskilling through training, experimentation, and innovation.
9. Innovatively leveraging technology, particularly cost-effective digital technologies, to improve project performance metrics such as value, cost, speed, quality, sustainability, and safety.

**Table 2.2** briefly outlines the typical performance issues and deficiencies of traditional PMS, and how Lean Construction 4.0 PMS introduces a fundamentally different way of working to address them.

The next section highlights some important considerations for ensuring the successful adoption of Lean Construction 4.0.

## TABLE 2.2 Comparison of Traditional PMS Deficiencies vs. Lean Construction 4.0 PMS Approaches

| Lean Construction 4.0: | | | | |
|---|---|---|---|---|
| A differant 'way of working' and 'work culture' from traditional one. | | | | |
| | Project PERFORMANCE Measures | Current C3:C18PMS Practices Current PMS Practies | Deficiency of Current Practice | LEAN Construction 4.0 PMS Way of Working |
| 1 | Project DELIVERY Delays | Project Mgt Tool driven (MS-project/ Primavera); | No concern for plan's feasibility; it fails in almost 80% of Projects | CPS: Master Plan is revisited and replanned involving those who execute the plan, without changing original duration. |
| | | Contract-based 'push' planning. | No management Enabling to Contractors | CPS: Enables removing major Constraints, Pro-Actively. |
| | | | Fails to micro-plan activities, at various stages of project, as too many interdependencies & uncertainties | CPS: 5 levels of additional collaborative micro-planning, taking care of the most current (resources & other) Constraints and resolving those fast and often innovatively. |
| | | | 'Control' is the success strategy for PM | CPS: Shift from Reactive to Pro-Active Production Planning. |
| | | | | VSM: Critical Processes/ Bottlenecks/ Delayed Activities duration is reduced, by improving process, to complete in time. |
| | | | | Technologies (incl Digital): leveraged for faster delivery. |
| 2 | Project FINANCIAL CLOSER/ Disputes | Punching Defect detection done at last moment, delaying HandOver | All above four Points | Defect Management: Focus on prediction of defects in advance and taking measures to Prevent critical ones. |
| | | Hand-over docs delayed; | Lack of collaborative partnership; | CPS: Continuous collaboration, coordination & Communication for faster and defects-free project completion. |
| | | Several Disputes for change control and finishes | Lack of focus on prevention of Defects, | VSM: Critical Processes/ Bottlenecks/ Delayed Activities duration is reduced, by improving process, to complete in time. |
| | | | Lack of Documents planning as a part of regular project planning | CPS: Continuous and timely collaboration, coordination & Communication to keep the relevant Documents ready in advance. |

| | | | | |
|---|---|---|---|---|
| 3 | Poor QUALITY of REAL ESTATE Projects | Focus on Quality Control, rather than Quality Assurance; | Poor understanding of aesthetic/ non-functional requirements & Big-Picture; | Defect Management: Focus on prediction of defects in advance and taking measures to Prevent critical ones. |
| | | Cumbersome Quality Processes, more for compliance than benefits | Poor workmanship; | Defect management: Pro-active just-in-time, just enough daily process training or the workmen, for better workmanship. |
| | | Poor/ confusing communications among teams | Old checklists, not updated frequently, and difficult to use in reality; Lots of paperwork | Defects Management: Most up-to-date Dynamic Checklists, and making Waste Observation Walks pro-actively, using App. |
| | | Rework-cost built in quotations (?) | Frequent Design changes/ errors ; | CPS: Minimizes errors, and also fixes fast. |
| | | | No ownership by anyone; | Big Room: Bring people together for better Team work, Commitment and Accountability. |
| | | | Low Focus on prevention of Defects, only correction | Defect management: Focus on prevention of defects, by on-the-spot training, leveraging Tools / Jigs & Fixtures/ and various other 14 tools. |
| 4 | Project BUDGETS & COST Over-runs impacting PROFITABILITY | Large Project Delivery DELAYS | All the above six points mentioned in Sr # 3 above | VSM & Value Engineering: Timely & innovative process improvements are done to save money somewhere at early stage and consume this saving for later Change Management. |
| | | Frequent Design Changes (often avoidable); Drawing errors | | CPS: Continuous collaboration, coordination & Communication minimizes; Developing Innovative new Process using VE and VSM to reduce Cost. |
| | | Frequent (often avoidable) and delayed change controls; Rework | | Defect management: Focus on prevention of defects, by on-the-spot training, leveraging Tools / Jigs & Fixtures/ and various other 14 tools. |
| | | | | Waste Hunting & Eliminations: Those processes and efforts, which do not add value to customers are identified and removed. |
| | | | | Technologies (incl Digital): leveraged for faster more cost-effective processes based delivery. |

| | | | | |
|---|---|---|---|---|
| 5 | Clients' proposed CHANGES RESISTED | Client often need to change requirements for deriving greater VALUE as project progresses | Resistance to change by all, including Contractors (too much initial-scope driven); Extra Time and Cost demanded | VSM/ VE/ Technologies: Changes suggested are welcome and incorporated, because VALUE is the main Focus. VSM & Value Engineering (VE) is used to develop new Innovative processes to absorb the additional costs into original approved Budgets. |
| 6 | EMPLOYEES & BUSINESS PARTNERS Relations | Client, Architect, Designer, Procurement Team, Project Management Consultant, General Consultants, all speciality Consultants, employees all work in Silos | All of them are highly stressed, and keep fire-fighting and go on blaming one-another for failures/ delays/ snags etc. | Big Rooms & CPS: These Continuous collaboration, coordination & Communication minimize the stresses and over-work for employees and timely payments to business partners because of better Cash-Flow management. |
| 7 | Other Focus Areas | NA | NA | Fouses on Suatainability ;<br>Fouses on Value for all Stakeholders;<br>Believes in SMART WORKING rather than Hard Working<br>Achieves BETTER TEAM WORK<br>Achieves better Safety of workmen |

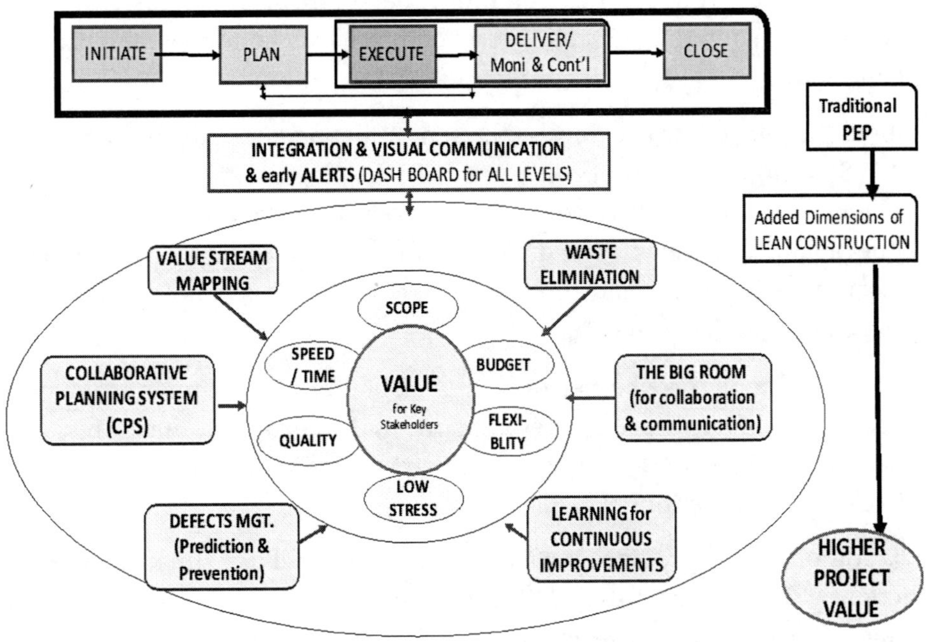


**Figure 2.15: Integration of Lean Construction Tools with Traditional PMS System to Provide Higher Project Value**

### 2.4.1 Lean Implementation Governance & Critical Success Factors

It is emphasised here that Lean implementation is a transformation initiative for the organisation—one that fundamentally changes the 'way of working' on projects. It is not a simple implementation of yet another quality management process for conformance. Therefore, it is important to recognise that the following points must be emphasised when integrating Lean Construction with the traditional PMS. Only then can the initiative succeed and deliver its full range of benefits.

- The senior leadership team should define an overall long-term Lean vision, strategy, and implementation blueprint for the company that embraces the Lean Construction 4.0 PMS.
- For any new project, conduct a dedicated workshop on the importance of Lean concepts for the company. This can be a 90–100-minute session involving the core Lean team identified for the project, representing the company, EPC main contractor, Lean expert(s), and, if possible, 2–3 major sub-contractors such as Civil, HVAC, MEP etc. The senior management team must emphasise the Lean vision, the role of core team members, and the additional responsibilities and expectations from them.
- Special clauses must be introduced (as suggested later in Section 10.2) in the formal legal contract of the EPC contractor.
- The senior leadership team should also arrange regular monthly open-house meetings, newsletters, or video messages from senior management team members to reinforce the common goal of project success. This should be coordinated by the HR managers of both the company and the EPC main contractor.
- A monthly 45-minute review by the senior management team (Lean Construction 4.0 Steering Committee) with the core Lean project team must assess Lean implementation effectiveness and any issues, providing direction and support. Additionally, the team should spend 15 minutes monthly with Lean expert(s) to discuss performance and the support required. This can be aligned with other steering committee meetings.
- HR should monitor whether the culture of 'internal customer' is developing, and provide support where needed. Ensure that appropriate incentive, reward and penalty mechanisms are introduced. Some suggestions for this are provided later in the book in Section 10.4.

Traditional PMS provides structure but fails to deliver superior outcomes. Lean Construction 4.0 complements PMS by embedding value, flow, and continuous improvement into project governance, ensuring resilience and excellence in modern construction. Chapter 3 details the process, tools and templates involved in CPS.

## 2.6 ANNEXURE I: CASE-STUDY Achieving Execution Excellence Through Lean Construction

**Project Scenario:**

A design-build workplace fit-out project of 70,000 SFT was to be completed from a shell building, for an American commercial building. This involved the main trades of MEP, civil finishing and interior design, covering all the work packages required for completion. Due to certain legal requirements, the client had to shift to this new office within a maximum of 90 days. Therefore, the asset needed to be handed over within that time frame.

Following handover, there would be a six-week de-snagging period under warranty, during which work would take place only on weekends to minimise disruption for the client's employees and avoid inconvenience in the workplace.

This appeared to be a mammoth task, as similar projects executed by the company in the past had taken an average of 115 days to close. Given the strict 90-day completion requirement for a ready-to-use workspace, the challenge seemed extremely tough.

This was the main reason why the EPC main (design-build) contractor chose to adopt Lean Construction principles—to accelerate project completion, reduce the number of snags, ensure early handover, and successfully meet the warranty obligations. Figure. 2.16 shows some of the finished workplace designs.

**Figure 2.16: The Finished Workplace Design of the MNC**

## Key Problems and Challenges identified:

Past project experiences and initial discussions were reviewed with participation from the key management team of the design-build main contractor, the American client, selected key sub-contractors, and the PMO consultant. The following problems and issues were identified during a half-day workshop held just before the project began. These needed to be addressed from the early stages, given the tight project schedule:

- Low labour productivity due to limited skills and unavailability.
- Poor planning and frequent work interruptions during project execution.
- A high amount of non-utilised time for both workers and key equipment.
- Low trust and integrity among sub-contractors, often blaming each other for delays and poor workmanship.
- Management and project teams were constantly busy and under pressure, with no time to focus on performance improvement through better processes, relevant technologies, or continuous improvement.
- Poor communication, coordination and collaboration among sub-contractors.
- Too many contractors, consultants and sub-contractors involved—around 120 in similar past projects—leading to excessive specialisation, division of labour, and coordination challenges.
- Contract-based execution encouraged siloed functioning, rather than promoting alignment toward common project goals.
- Measurable and controllable KPIs were not defined for various roles, undermining project success tracking.
- Claims and disputes arose frequently due to several deviations, changes, and rework during execution.
- Recurrence of the same snags—historical snag analysis of past projects revealed repeated issues in various rooms and workplaces, with no concrete plan for prevention.
- Ineffective sequencing of work, with little attention to correct sequencing between trades, resulting in numerous snags and rework.
- Mediocre supervision—contractors' supervisors often had limited ability to interpret drawings.

Many of these challenges have already been discussed earlier in this chapter.

## Lean Construction Solutions tried:

Several best practices and key Lean Construction tools were implemented, in addition to conducting a couple of joint workshops for vendors and the client. Despite the challenges, the project successfully achieved the following results—translating into a return on investment (ROI) of more than 600% in tangible benefits. In addition, there were many intangible benefits.

- The ready-to-use premises were handed over in 80 days, well ahead of the 90-day deadline.
- Project delivery time was reduced by 33%, compared to the historical average of 115 days for similar projects.
- Internally, 95% of purchase orders were released within 2 weeks, versus the usual lead time of 4 weeks.
- Number of snags was reduced by more than 50% compared to previous projects.
- Project closure time was halved, resulting in a very satisfied customer who provided a formal letter of appreciation.
- Several best practices and templates were developed for use in future projects.

**Lean Construction Solutions Implemented:**

As the time available was very limited, the following solutions were quickly discussed, agreed upon, and implemented—effectively addressing the challenges listed earlier. A Collaborative Planning System using the Big Room approach was adopted to align all stakeholders and facilitate faster decision-making. Efforts were directed toward the elimination of wastes to streamline execution and reduce non-value-adding activities. Defect prediction and prevention strategies were introduced, particularly to address issues observed in similar past projects. Continuous involvement from the client ensured timely approvals and quick feedback loops, contributing to smoother progress. Additionally, the use of simple communication technologies helped reduce miscommunication and enhance coordination among contractors and teams.

**1. Collaborative Planning System with Big Room approach:**

- **Design Execution Team Workshop:** Held to understand the critical areas and complex portions of the project's key requirements, ensuring that design could be translated into reality.
- **Procurement Tracker:** Developed to enable collaboration among the design, procurement, finance, site teams, and suppliers. It helped in releasing the right purchase orders quickly, initiating advance payments, and eliminating errors, omissions, and rework.
- **CPS Five-Level Planning:** Used to convert the milestone-based macro project plan (MS Project) into a micro-plan for a two-week lookahead period. It included "make-ready" planning for the next day and week, helping to proactively resolve constraints (e.g. approvals, manpower, materials) for smoother progress.
- **Weekly 'Make-Ready' Meetings with Sub-Contractors:** Conducted to review the lookahead schedule and ensure that resource-related constraints (e.g. manpower, materials, tools) were resolved in advance, minimising delays.

- **'Ask from Management' Escalation Mechanism:** Used to escalate and resolve issues beyond the site team's authority. Success stories included the de-scoping of one electrical contractor and consistent lift availability.
- **Daily Stand-Up Meetings:** Organised to drive execution speed on site by improving communication and cooperation across interdependent teams.
- **Daily Task Diary for All Foremen:** Each foreman carried a daily task sheet in their pocket to focus on planned day's work, aiding timely completion and faster progress.
- **Performance Review Dashboard:** Displayed project progress visually to all stakeholders, increasing transparency and applying constructive pressure to stay on track.
- **Daily Partnership Conference Calls with Client and PM:** Introduced midway through the project to resolve emerging issues, prepare for next-day work, and ensure smooth coordination through improved communication and collaboration.
- **Lean Seminar for Client and Sub-Contractors:** Conducted to build shared understanding among all stakeholders and align them toward a common project goal.

**2. Defects Prevention/Reduction:**

- **UABC** (User Accepted-Testing Based Construction): A new concept was introduced to simplify communication for frequently repeated items (e.g. doors). A single document was created for each item, containing all essential information needed by the workers—eliminating the need for referencing multiple drawings, which workers often do not consult. This helped sub-contractors avoid snags related to both aesthetics and functionality.
- **Root Cause Analysis for Switch Sockets:** Drawing on data from over 200 switch-socket snags encountered in a previous project with the same sub-contractor, the team redesigned the switch-socket installation approach. In this project, switches were installed on wooden panels instead of gypsum partition sheets, effectively preventing the recurring issue.
- **First Piece Inspection:** For repetitive installation items (e.g. junction boxes and modular partitions), a 'first piece inspection' was conducted. This allowed for early feedback on workmanship, reducing the likelihood of rework across subsequent units.
- **Enhanced Client Experience Through Mock-Ups:** In addition to the standard workstation mock-up, a mock-up of modular partitions was presented to the client for informed decision-making. The supplier also installed two different carpet sets in separate rooms to allow employees to physically experience and select their preferred option—an initiative that was highly appreciated by the client.

- **Fortnightly Quality Visits by the Project Director:** This initiative was aimed at enhancing client satisfaction. The Project Director made two formal site visits to critically inspect workmanship and progress against the client's expectations. Corrective actions were taken promptly, based on feedback from these visits.
- **Mid-Course Internal Snagging:** Conducted proactively during the project to reduce the number of defects presented to the client at handover. This approach significantly reduced the final snag list. At handover, the snag list contained 270 items, including 150 raised by the client, compared to the usual 1,000+ seen in similar projects. This allowed for faster project closure.

**3. Faster Handover:**

- **All Quality Enhancement Initiatives (e.g. UABC, internal snagging, first piece inspection):** These interventions significantly reduced the number of snags and enabled faster resolution during the six-week warranty period.
- **Planning Project Closure Activities (Documents, Drawings, etc.) in Regular Lean Planner Pre-Snagging:** Rather than waiting until snagging to start closure activities, incorporating these into the Lean Planner kept vendors alert and ensured timely submission of all required documents. As a result, almost all vendors submitted their documentation on time.
- **Negotiating with the Client to Forego Low Business-Critical, High Rework Snags:** The project team provided strong support to the client and their project manager through regular interaction and proactive engagement. Their trust-building approach led to a more reasonable snag list. By identifying and resolving many issues upfront, they created significant value for the client.
- **Empowering the Site Team and Using Contingency Fund for Timely Corrections/Improvements:** Although not formally announced, the project team took initiative. For example, after securing approval from the designated SPOC, they switched to using wooden frames for floor junction boxes—greatly improving quality and reducing rework due to floor undulations.
- **Fostering a Better Partnership with the Client:** This was particularly important in such a fast-track project. The Project Director made special efforts to build a strong working relationship with the client and their project manager. A daily 15-minute conference call during the second half of the project helped accelerate issue resolution.

This case provides a well-rounded demonstration of Lean Construction 4.0 PMS implementation. Although there was limited opportunity to introduce new technologies, the Lean philosophy alone made a significant impact. It became the first project delivered by the design-build main contractor ahead of schedule and with significantly improved quality. The American client recognised this achievement with a formal letter of appreciation.

# CHAPTER 3

# COLLABORATIVE PLANNING SYSTEM (CPS)

# 3. Collaborative Planning System

Collaborative Planning System (CPS) is a modern approach to project management that integrates planning, scheduling, execution, monitoring, and control. It ensures all stakeholders work together seamlessly to achieve superior progress, cost, safety, and quality goals.

## 3.1 Need for CPS Complementing PERT/CPM based Planning System

While traditional planning methods like PERT and CPM provide structure, they often fail to capture the realities of execution. CPS complements these methods by adding collaboration, transparency, and adaptability to improve project outcomes.

### 3.1.1 Collaborative Planning System (CPS):

CPS is about collaborative planning, scheduling, execution, monitoring, and control. It integrates micro-level scheduling, proactive monitoring, and risk management to minimise variances and achieve project goals. A system becomes effective when applied across all stages—from pre-construction to occupancy—with stakeholder involvement. CPS eliminates silos, defines milestones, and creates early interface points. Teams jointly identify roadblocks, develop solutions, and maintain a feedback loop for improvement. CPS bridges between macro and micro levels, enabling unified teamwork and reliable delivery.

### 3.1.2 The Last Planner System® (LPS®):

Developed by Glen Ballard and Greg Howell, LPS® is a collaborative planning system widely used in construction due to its ability to create reliable and predictive schedules. It emphasises the role of the "last planner", such as foremen and supervisors, who create ground-level plans based on real-world conditions and prerequisites. Plans developed collaboratively are more reliable than those prepared in isolation.

### 3.1.3 Challenges of Traditional Project Management:

Traditional CPM-based project management in construction focuses on a contract-driven, push-based culture, resulting in siloed working, inefficiencies, and poor communication among stakeholders. Each party prioritises its own contractual obligations over the project's overall objectives, leading to wasted effort, low customer satisfaction, and unpredictable project progress. Variability in daily outputs—such as concreting or plastering—further illustrates this inefficiency.

### 3.1.4 Lean Construction Project Management:

Lean Construction emphasises collaborative planning and execution among engineering, procurement, and construction teams, along with the client. It aims to enhance workflow reliability, accelerate project progress, and maximise value for the customer while minimising waste. Lean planning best practices include:

- Developing micro-level plans closer to the time of execution
- Involving execution teams in planning
- Resolving resource constraints collaboratively to "make the work ready"
- Making reliable promises based on detailed planning, not assumptions
- Learning from failures and taking preventive actions

This approach complements traditional planning methods such as CPM/PERT by addressing their limitations and making schedules more realistic. Lean Construction proactively reduces risks, ensuring adherence to schedules and budgets. Further details on CPS are provided in the next sections, with a case study included in Annexure.

## 3.2 Construction Project Phases and Integration of CPS

The following sections describe in greater detail the CPS, especially in the context of real estate or linear infrastructure projects.

### 3.2.1 Different Phases of Construction Projects

The typical life cycle (also known as 'phases') of a construction project is shown in figure 3.1 below. It further expands the planning (which also includes scheduling, execution, monitoring and control) and closing process groups, for the different phases of a typical construction project. These phase durations, however, can be different for different sizes and levels of complexity of projects.

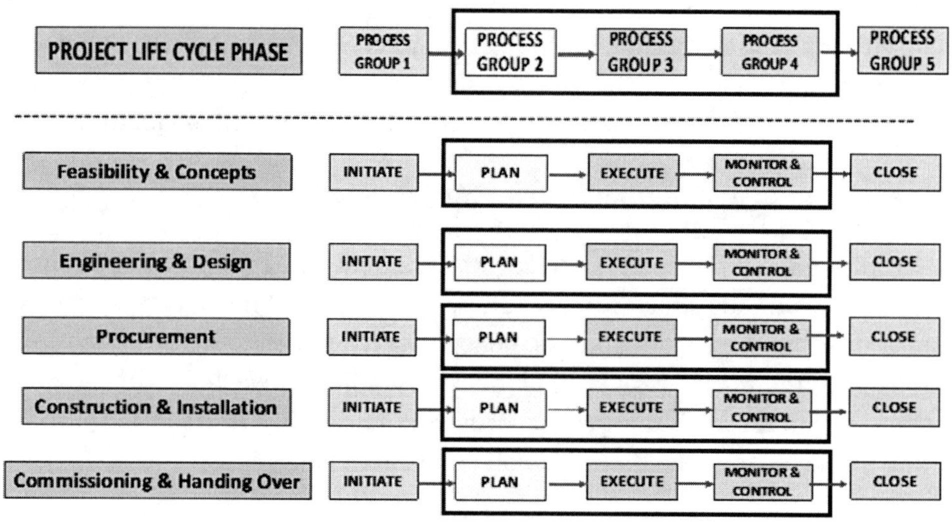


**Figure 3.1: Typical Life-Cycle Phases of a Construction Project & Major Project Management Processes in each Phase.**

In order to explain the complete processes and steps involved in CPS, one project phase of building construction is chosen here as an illustration. This also describes the deliverables and people involved, and the templates to be used in each step. The other phases need to follow the same procedures.

### 3.2.2 Building Construction Phase & Integration of CPS & Other Tools

Figure 3.2 below shows how CPS is integrated with each of the project phases. What actually happens in CPS is explained in section 3.4 below and other later sections.

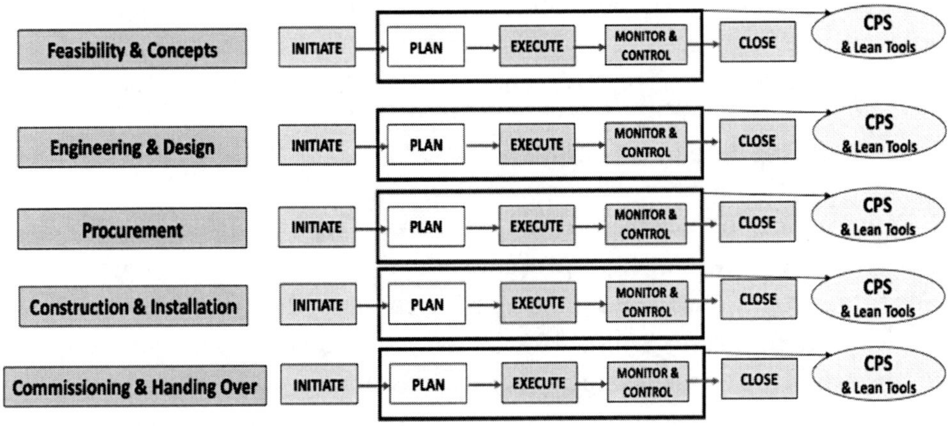


**Figure 3.2: All Construction Life-Cycle Phases integrated with CPS & Lean Tools**

## 3.3 'Five Levels' of CPS Production:

One of the key reasons for introducing five levels is to incorporate the production/construction planning and control system in the project planning and control system. This was briefly explained in section 2.2.4 earlier, and section 3.1 above. Figures 3.3, 3.4 and 3.5 explain this in greater detail.

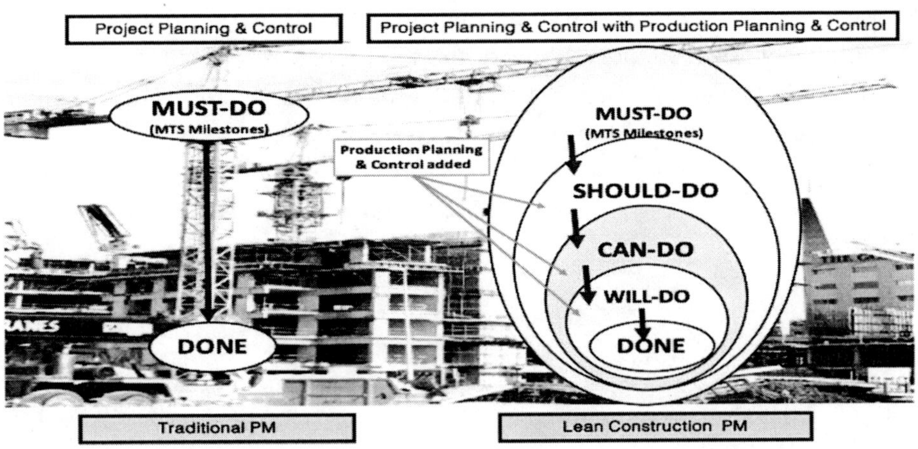


**Figure 3.3: 5 Steps of CPS, embedded into Traditional PM to make it Lean PM**

### 3.3.1 Collaborative Production Planning in Lean Project Management:

Figure 3.3 illustrates how Lean project management introduces three additional collaborative levels—should-do, can-do, and will-do—beyond the traditional post-mortem approach of project planning and control. These levels involve blue-collar workers in defining production methods, which were previously overlooked in micro-planning, leading to unrealistic project schedules. By incorporating these levels, Lean planning proactively identifies and removes hindrances, ensuring smoother execution.

Eight prerequisites, shown in figure 3.4, are essential for starting and completing any construction activity. These include elements like materials, tools and approvals. Missing even one of these prerequisites hinders progress. Ensuring their continuous availability is critical for steering projects towards achieving their goals.

### 3.3.2 Impact of Readiness on Efficiency:

Construction inefficiencies often stem from dependencies on other trades and delays in preceding tasks. All eight prerequisites must be in place to complete activities on time. In construction, delays in one task cascade through subsequent activities, affecting overall progress. While 100% readiness theoretically guarantees activity completion, achieving this consistently is rare in practice.

If each prerequisite is available with a 95% probability, the overall likelihood of completing the activity drops to 66% (calculated as 0.95^8). Even achieving 70–80% of planned weekly activities is considered excellent, provided teams learn from issues and prepare better for the following week. However, planning often proceeds despite low probabilities of material availability or work inspection, which commonly results in delays.

### 3.3.3 Focus on Prerequisites in CPS:

CPS emphasises ensuring near-perfect availability (99–100%) of prerequisites to achieve reliable and timely activity completion. Lean construction 4.0 achieves faster project completion by collaboratively addressing readiness issues and continuously learning from daily gaps between planned and actual performance. This proactive approach minimises delays and improves efficiency.

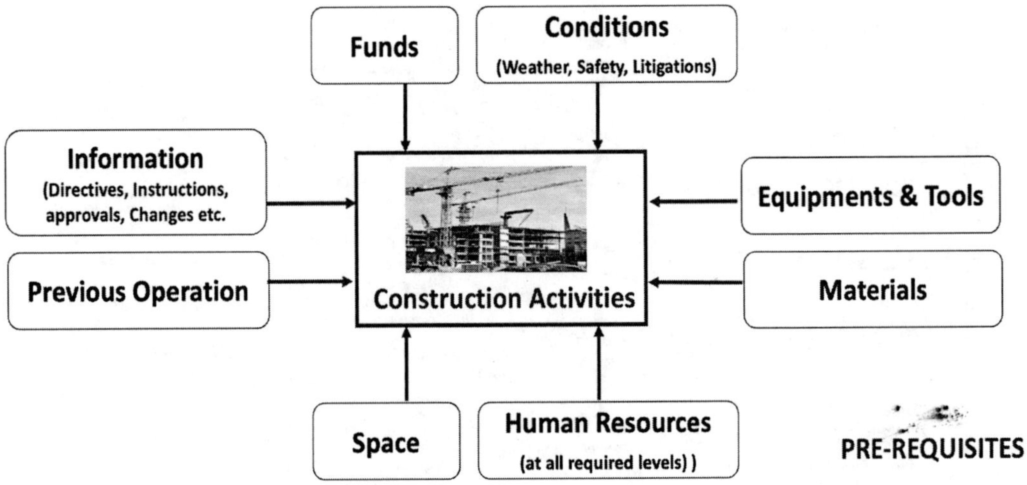


**Figure 3. 4 Pre-requisites for any Construction Activity to start/ finish**

### 3.3.4 Physical Focus in Lean Construction Planning:

Lean construction emphasises planning, scheduling, and monitoring the physical aspects of a project—those that are visible and provide a clear sense of progress. Major milestones (referred to as "boulders" in Lean), initially defined by the architect and design consultants for EPC contractors, may require refinement. These milestones should represent the completion of a distinct, visible physical unit of the project, encompassing all necessary trades such as civil, HVAC, fire-fighting (FF), MEP, and finishes, ensuring the unit is fully operational.

For example, completing a cafeteria in an office building requires finishing all civil, HVAC, electrical, and fire-fighting works, making it ready for kitchen equipment installation. This approach reflects the progressive elaboration of the master time schedule (MTS) approved by the client's top management.

### 3.3.5 Planning Steps Overview:

The overall planning process includes steps outlined in Figures. 3.5 and 3.6. These steps were briefly introduced in Chapter 2 (Section 2.2.4) and are explained in greater detail in Sections 3.3.3 to 3.3.6. This progressive planning method ensures alignment with project goals and enhances clarity on task completion.

**Visual representation:** The process of progressive elaboration is illustrated in fig. 3.5.

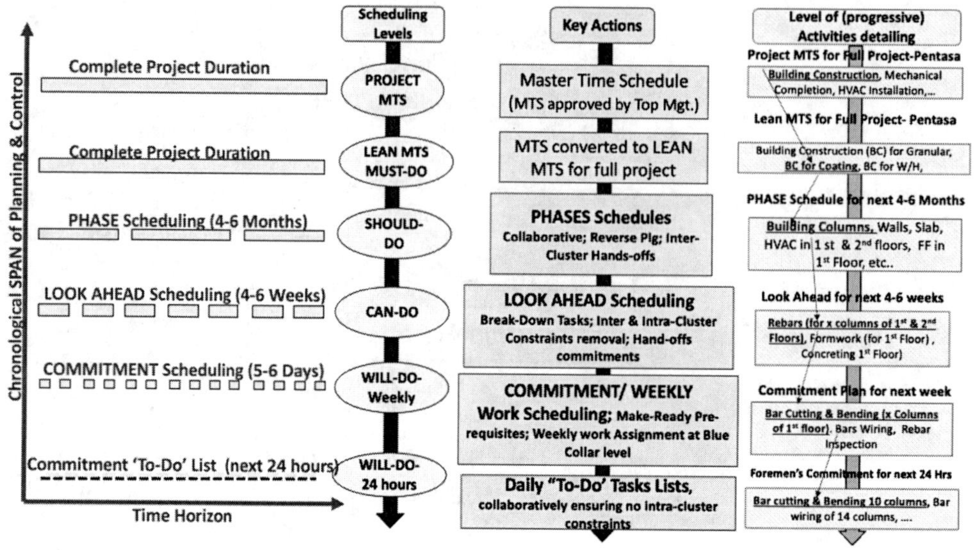


**Figure 3.5: Collaborative Production Planning & Control with added 5 levels**

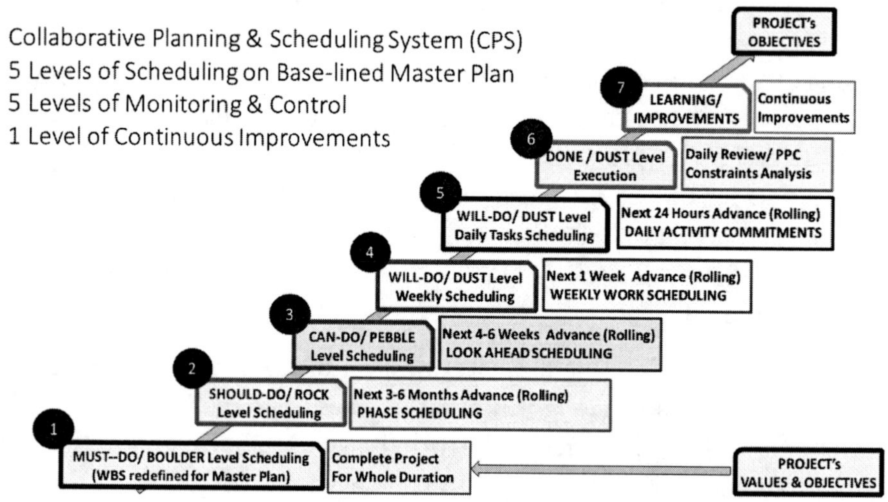


**Figure 3.6: Collaborative Production Planning & Control with added 5 levels**

## 3.3.6 Stakeholder Collaboration and Clustering

Construction projects involve numerous stakeholders, including designers, architects, consultants (project management, quality management, SMEs), engineering, procurement, and construction contractors, along with testing and validation experts. Their work is often interdependent, requiring extensive communication, cooperation, and collaboration. As outlined earlier in figure 3.4, the eight prerequisites provided by different stakeholders highlight the complexity of these interactions. Ensuring smooth collaboration among these diverse groups is challenging but critical.

A solution found in Lean construction is the concept of clusters. A cluster represents a group of interconnected tasks or activities that must be managed together for successful project execution. Clustering enhances communication and coordination among project teams and stakeholders. For instance, a trade cluster like detailed designers may include sub-clusters such as structural, HVAC, electrical, and interior designers, which need to collaborate internally and with other trade clusters like civil or HVAC installation. Poor communication and coordination within and between clusters often lead to errors, rework, delays, and cost overruns. Figure 3.7 & 3.8 provides an illustrative example of clusters, sub-clusters, and their communication needs.

The absence of one or more of the eight prerequisites (from figure 3.4) is termed a constraint, hindering task completion. Effective communication, coordination, and cooperation (abbreviated as 3C) are essential to ensure the timely availability of these prerequisites across clusters, sub-clusters, and sub-sub-clusters.

### 3.3.7 Structured Collaboration Through CPS:

To address the complexities of 3C, CPS offers a structured framework that ensures effective collaboration at various levels of planning. This includes:

- Monthly collaboration for the next 4–6 months,
- Weekly collaboration for the next 4–6 weeks,
- Daily collaboration for the next 24 hours,
- Ad hoc collaborations for immediate needs or issues.

This systematic approach helps clusters identify prerequisite needs in advance and collaboratively resolve constraints. Detailed explanations of these structured collaborations are provided in sections 3.3.3 to 3.3.6.

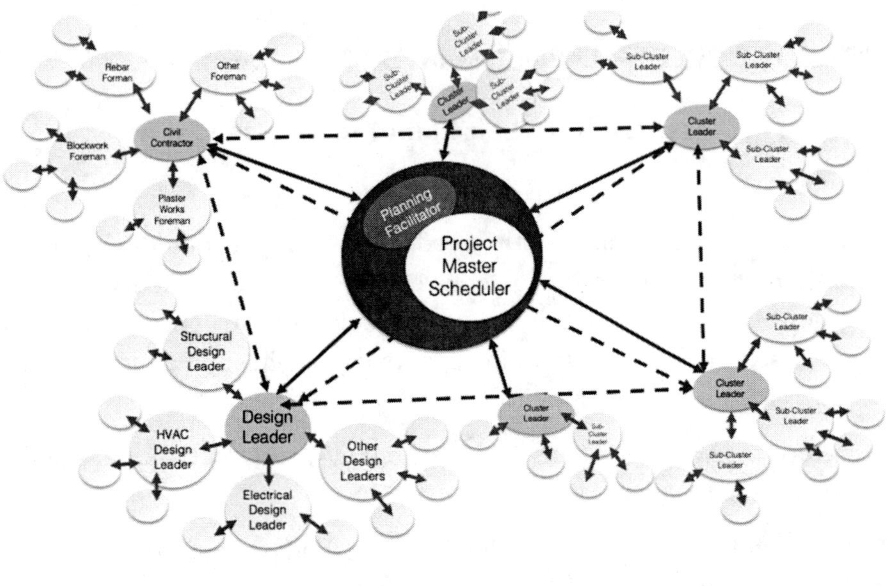


**Figure 3.7: Information Flow & 3C Collaboration in CPS among various Team Players**

| Allowed to say | Not allowed to say |
|---|---|
| I will do it | I will try my best |
| I will do it, provided... | Most probably I will complete it. |
| I will not do, unless..... | No, I will not do it |
| I am not capable of doing it | I will do it, if I can |
| You can bank on me | Surely, I will try to do it |
| Rest Assured | This is Impossible to do |
| I will verify and confirm after 3-3 hours | I challenge; it is just not possible |

As also explained briefly in section 3.4, CPS planning and scheduling is done in five levels mentioned below. The execution, monitoring, and control will happen in a separate stage, covered in section 3.4. The detailed steps involved in each of these five levels are given in the next sections.

In true collaboration, we should let people express their concerns and challenges in a transparent way. The earlier we engage the stakeholders, the more time we will have to listen to their requirements, whereas at the last minute, we will be forcing them and making them agree to confirm what we need. Developing the right mindset and a friendly environment is essential to get people to agree on what they can deliver—and to deliver what they agree.

### 3.3.8 MUST-DO / BOULDER Level: Lean WBS Milestones

#### Definitions and Process Steps for Their Planning and Scheduling

- **Guidelines:**

It was explained in chapter 1, the need for defining the boulder milestones, which are the most important milestones to achieve without failure. The following guidelines should be followed to define these Lean work breakdown structures. An example list of Lean boulder milestones for the building construction phase of an important project is also provided as an illustration in table 3.1.

Structure the whole project work in the way you would like to see the physical unit-wise progress of the project.

- Define high-level milestones, following a finished product structure (or locations) rather than trade-wise (civil, HVAC etc. operations/activities), as normally done.

- The boulder milestones should represent a completed physical unit of the project at a high level, which is visible. Completed means all the trades' work required, including civil, HVAC, FF, MEP, civil finishes etc., are completed and the unit is operational.
- It should cover the complete project plan from EPC start to finish, covering the complete lifecycle of the engineering and technical operations project.
- Lean WBS should follow the workflow requirements using an outside-in approach. This means all the utilities (water, power, steam etc.) that come from outside the building to inside the building for machineries and equipment should be ready and operational first, before machine qualification phase activities start.
- 'Stone-level' process activities, which are further detailing of activities to complete the boulder milestone, may not be defined at this stage.
- The target dates of completion of these boulder-level milestones should be captured from the base-lined plan and schedule approved by the steering committee, earlier prepared by the EBD consultant for EPC contracting. However, as stated in 4.3, the duration of the total project must be reduced by 10% for this scheduling. This 10% overall buffer time is kept for absorbing any real risks at the end of the project.
- As a thumb rule, the total number of boulder-level milestones should generally not exceed the manageable range of 60–80 for progress monitoring.

| DEFINING CLUSTERS of CONTRACTORS/ TEAMS, who have to COORDINATE/ COMMUNICATE and be in SYNC to achieve the VALUE of the WHOLE PROJECT | | | | | | | | | | | | | | | |
|---|---|---|---|---|---|---|---|---|---|---|---|---|---|---|---|
| sr # | MAIN CLUSTERS : Need INTER cluster Communication | SUB-CLUSTERS (within Main ) : Need INTRA Cluster communications) | | | | | | | | | | | | | |
| 1 | DETAILED DESIGN | Architect & Civil | HVAC & AHU | Fire Fighting | Electrical & LVS | Facilities Logistics | Waste Mgt | BMS/ EMS | Process Automation | Clean Room | Clean Utility | Black Room | | | |
| 2 | PROCUREMENT-FERRING | CSA | HVAC & AHU | Fire Fighting | Electrical & LVS | Logistics | ETP/ STP | BMS/ EMS | Process Automation | Clean Room | Clean Utility | Black Room | Clean Gases | Transfor mer | DG Set |
| 3 | PROCUREMENT-CONTRACTORS | CSA ??? | HVAC ?? | Fire Fighting | | | | | | | | | | | |
| 4 | BUILDING CONTRUCTION-MAIN | CSA | HVAC & AHU | Fire Fighting | Electrical & LVS | AS/RS | Electrical | BMS/ EMS | | Clean Room | Clean Utility | Black Room | | | |
| | (Sub-Sub-Clusters: Illustrative) | Rebar; Forming; Cementing; Joinery; Painting etc | Markers/ Cutter-Assemblers Installers/ Testers | Piping, Cabling, Sprinkling, Testing etc | Conduit ing, Cabling, Fixtures fixing etc | | | | | | | | | | |
| 5 | MACHINES INSTALLATION | | | | | | | | Process & Packing Equipments | | | | | Transfor mer | DG Set |
| 6 | MACHINES VALIDATION | | | | | | | | | | | | | | |
| 7 | MACHINES COMMISSIONING | | | | | | | | | | | | | | |
| | | | | | | | | | | | | | | | |
| Need to Sort-out: | INTER-CLUSTER CONTRAINTS & HANDOVERS | INTRA-CLUSTER "CONSTRAINTS" (non-availability of Eight Pre-requisites) & HANDOVERS | | | | | | | | | | | | | |

**Figure 3.8: Project Total work allocated to different Clusters of Contractors/ Consultants**

- **CPS Planning & Scheduling Process for at Boulder Milestones level:**
  - Understand clearly the big picture and all the values (including design, technology, duration, budget and quality etc.) the client expects from the project. Also understand any design, technology, procurement challenges and uniqueness of the project.
  - Get the detailed base-lined plan and schedule approved by the steering committee, for preparing Lean way of planning, scheduling and execution plan for the main contractor.
  - Understand and review the major (boulder) milestones defined there and refine those (by adding any missing, deleting any extra or modifying the milestones as per the guidelines defined for Lean WBS).
  - Elaborate further and break each boulder milestone into several boulder activities or processes and their expected (new and 10% reduced) duration for the entire project. The earlier base-lined master plan will be the basis of this.
  - Discuss boulder milestones of the complete project with relevant design and procurement teams for long lead-time items procurements and any complex or difficult design from an execution point of view. Modify the base-lined plan if any major problem arises and has no solution in sight at the moment.
  - Discuss the complete project boulder milestones plan for its completeness and get approval from the steering committee for execution and revising the baseline of the project.
  - On the basis of this (revised and Lean-oriented) base-lined project, validate the resources required (full-time people, the budget, construction and other machinery etc.). Call this agreed new plan the Lean MTS (master time schedule). Treat this as the basis for all future planning, scheduling, execution, monitoring and control.
  - Once the Lean MTS plan is agreed among the group, go to the next stage, "should-do" planning and scheduling in section 3.3.3.
  - Also update these on Lean MPS on Primavera or MS Project.

These steps are displayed in brief in the table below. The table also includes the recommended team involved in each step, the deliverables of each step and the possible risks and Lean construction solutions for those.

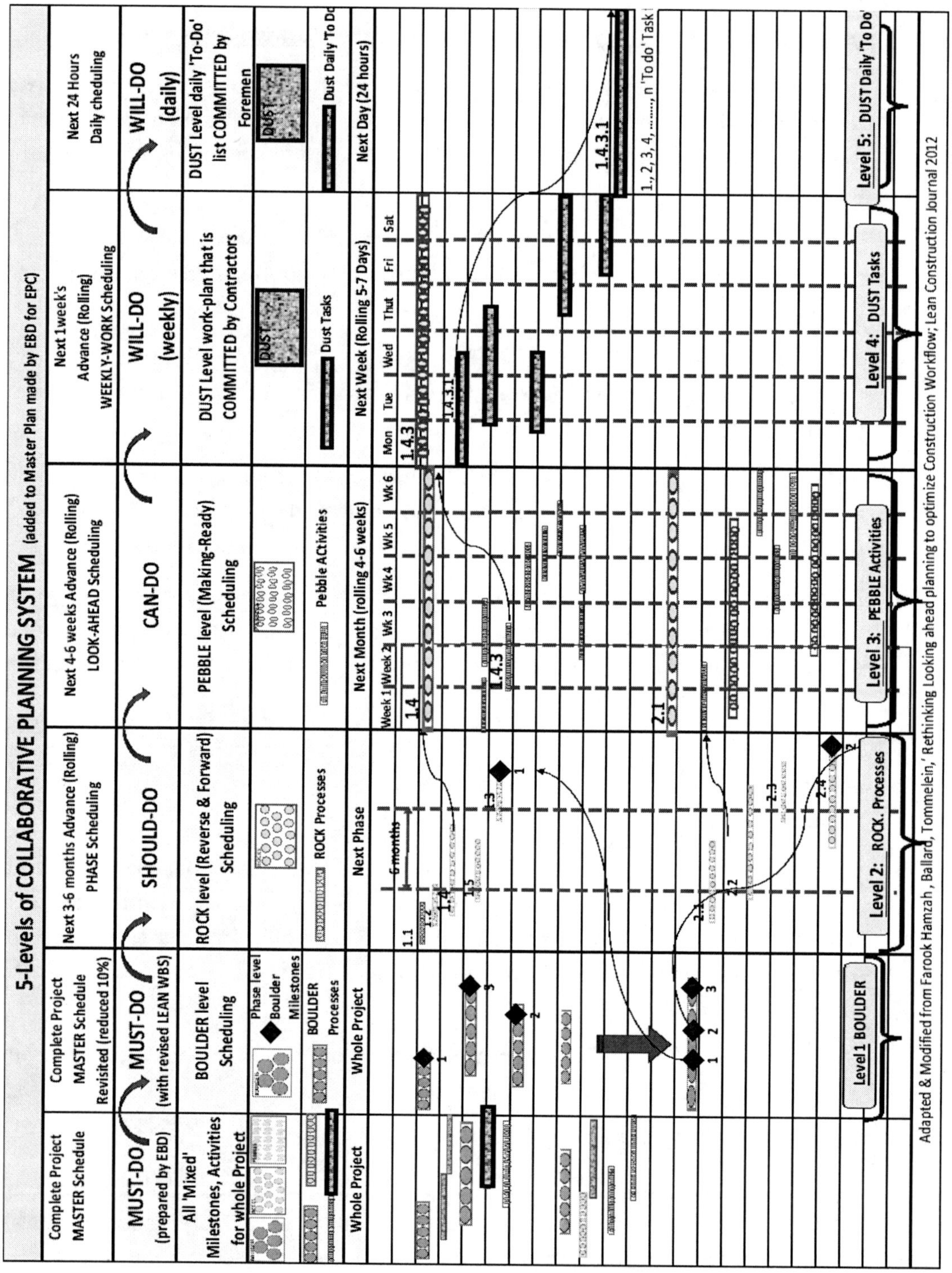


**Figure 3.9: Details of Collaborative Production Planning & Control at 5 levels**

The boulder milestones defined in 3.3.3 are expanded into greater detail by defining the rock processes to complete those boulder milestones. Please remember that each boulder milestone involves more than one trade or cluster (like civil, HVAC, electrical etc.), but rock processes generally involve only one main trade, like HVAC or civil. The following guidelines must be followed for this detailing:

(Also refer to red-colour boulder activities and black boulder milestones 1, 3, 3 in Figure 3.9 and five green-colour rock processes 1.1, 1.3, 1.4, 1.5, 1.3 above that)

- Each trade or cluster must review the boulder activities earlier defined in Lean MPS, for their correctness, and add, delete or modify the activities if necessary at this stage.
- Like Lean work structure (WBS) pattern, divide the boulder milestones and activities into greater detail (single) trade or cluster-wise rock activities for the next level of work structure.
- Rock activities must represent the completion of one trade's (or cluster's) work in the physical unit of the project, which is visible. It should also allow one to see the progress of the project unit or location (like first floor, third floor etc.) in a physical sense.
- While each boulder activity might involve many trades or clusters (perhaps 3–6), the rock processes generally involve only one trade or cluster (like CSA). These may have interdependent sub-clusters, however (like reinforcing bars, formwork, cementing, joinery etc.).
- As a thumb rule, the number of high-level rock processes should generally not exceed 4–5 per boulder, per trade or cluster. Further detailing of these rock processes is done in later phases at pebble and dust levels (which are at sub-cluster and sub-sub-cluster levels).
- Completeness: These rock activities must collectively represent 100% completion of the boulder milestones, as 95% completion is not acceptable in Lean. In Lean, completion of a process is either 100% or 0%.
- Collaborate and plan for all rock activities for their start and finish dates in a minimum of three iterations, collaboratively. This starts with reverse scheduling, and finally ends with forward scheduling. This is explained in greater detail below.

## Table 3.1- Refining & getting approval for earlier Base-lined MPS at "MUST-DO" BOULDER Milestones level.

**Table 3.1 Refining & getting approval for earlierBase-lined MPS at " MUST-DO " BOULDER Milestones level.**

| "MUST-DO, BOULDER MILESTONES" PLANNING for 30 Months Project (Planned for conpleting 27 months-3 Months Buffer in the end | | | | | | | | | | | | | | | | | | | | | |
|---|---|---|---|---|---|---|---|---|---|---|---|---|---|---|---|---|---|---|---|---|---|
| Mile Stone Number | WORK-STRUCTURE BOULDER MILESTONES | Typical Stages : Design, Procurment, Construction, Installation, Qualification, Commissioning: Showing : Consturction , Installation, Qualification, Commissioning | | | | | | | | | | | | | | | | | | | |
| | | Month 1 | Month 2 | Month 3 | Month 4 | Month 5 | Month 6 | Month 7 | Month 8 | Month 9 | Month 10 | Month 11 | Month 12 | Month 13 | Month 14 | Month 15 | Month 16 | Month 17 | Month 18 | Month 19 | Month 20 |
| 1 | Pre-Construction Startup Building Statutory Approvals Obtained | | | | | | | | | | | | | | | | | | | | |
| 2 | Key Equipments (Long Lead Time & Priority 1) Procured | | | | | | | | | | | | | | | | | | | | |
| 3 | Other Equipments & Items Procurement | | | | | | | | | | | | | | | | | | | | |
| 4 | All Construction Services Procured | | | | | | | | | | | | | | | | | | | | |
| 5 | All Management Services Procured | | | | | | | | | | | | | | | | | | | | |
| 6 | Final Building Statutory Approvals Obtained | | | | | | | | | | | | | | | | | | | | |
| 7 | Sub-Station Building Operational | | | | | | | | | | | | | | | | | | | | |
| 8 | Boiler House Building Operational | | | | | | | | | | | | | | | | | | | | |
| 9 | Solvant Storage Tank Farm Operational | | | | | | | | | | | | | | | | | | | | |
| 10 | Liquid Nitrogen Storage Tank Farm Operational | | | | | | | | | | | | | | | | | | | | |
| 11 | Extended ETP Plant Operational | | | | | | | | | | | | | | | | | | | | |
| 12 | Raw Water Storage & Supply Operational | | | | | | | | | | | | | | | | | | | | |
| 13 | Scrap Yard Operational | | | | | | | | | | | | | | | | | | | | |
| 14 | Potable Water System Operational | | | | | | | | | | | | | | | | | | | | |
| 15 | Fire Hydrant System Operational | | | | | | | | | | | | | | | | | | | | |
| 16 | PipeRack & Roads Operational | | | | | | | | | | | | | | | | | | | | |
| 17 | TOTAL BUILDING CONSTRUCTION | | | | | | | | | | | | | | | | | | | | |
| 18 | ALL EQUIPMENTS DELIVERY | | | | | | | | | | | | | | | | | | | | |
| 19 | ALL AUTOMATION COMPLETION | | | | | | | | | | | | | | | | | | | | |
| 20 | | | | | | | | | | | | | | | | | | | | | |
| 21 | ALL QUALIFICATIONS | | | | | | | | | | | | | | | | | | | | |
| 22 | GRANULAR PART of Building Construction | | | | | | | | | | | | | | | | | | | | |
| 23 | Granulation – Equipment Delivery | | | | | | | | | | | | | | | | | | | | |
| 24 | Granulation – Mechanical Completion | | | | | | | | | | | | | | | | | | | | |
| 25 | Granulation – Automation Completion | | | | | | | | | | | | | | | | | | | | |
| 26 | Granulation – Commissioning | | | | | | | | | | | | | | | | | | | | |
| 27 | Granulation – Qualification | | | | | | | | | | | | | | | | | | | | |
| 28 | COATING Part of Building Construction | | | | | | | | | | | | | | | | | | | | |
| 29 | Coating – Equipment Delivery | | | | | | | | | | | | | | | | | | | | |
| 30 | Coating – Mechanical Completion | | | | | | | | | | | | | | | | | | | | |
| 31 | Coating – Automation Completion | | | | | | | | | | | | | | | | | | | | |
| 32 | Coating – Commissioning | | | | | | | | | | | | | | | | | | | | |
| 33 | Coating – Qualification | | | | | | | | | | | | | | | | | | | | |
| 34 | BLISTER PACKING PART of Building Construction | | | | | | | | | | | | | | | | | | | | |
| 35 | Blister Packaging – Equipment Delivery | | | | | | | | | | | | | | | | | | | | |
| 36 | Blister Packaging - Mechanical Completion | | | | | | | | | | | | | | | | | | | | |
| 37 | Blister Packaging - Automation Completion | | | | | | | | | | | | | | | | | | | | |
| 38 | Blister Packaging - Commissioning | | | | | | | | | | | | | | | | | | | | |
| 39 | Blister Packaging - Qualification | | | | | | | | | | | | | | | | | | | | |
| 40 | STICK PACK PART of Building Construction | | | | | | | | | | | | | | | | | | | | |
| 41 | Stick pack packaging - Equipment Delivery | | | | | | | | | | | | | | | | | | | | |
| 42 | Stick pack packaging - Mechanical Completion | | | | | | | | | | | | | | | | | | | | |
| 43 | Stick pack packaging - Automation Completion | | | | | | | | | | | | | | | | | | | | |
| 44 | Stick pack packaging - Commissioning | | | | | | | | | | | | | | | | | | | | |
| 45 | Stick pack packaging - Qualification | | | | | | | | | | | | | | | | | | | | |
| 46 | Warehouse | | | | | | | | | | | | | | | | | | | | |
| 47 | Warehouse Equipment Delivery | | | | | | | | | | | | | | | | | | | | |
| 48 | Warehouse Mechanical Completion | | | | | | | | | | | | | | | | | | | | |
| 49 | Warehouse Commissioning | | | | | | | | | | | | | | | | | | | | |

### 3.3.9 SHOULD-DO / ROCK Level: Guidelines for ROCK Activities

**Definitions & Process Steps for 'Reverse Iterative' Planning & Scheduling :**

- **CPS Planning & Scheduling Process for at Rock Processes level:**
    - Agree on duration for the next phase for planning by selecting the next phase rock-level planning and scheduling duration (typically 4–6 months).
    - Each trade or cluster should elaborate further and break each boulder milestone into several rock activities or processes for their part of the trade, for the same boulder milestone during the chosen phase period.
    - Each trade team or cluster comes prepared with the sequence of activities, duration and target completion date as per the Lean MTS provided. Each sub-contractor or sub-cluster also comes prepared with resource (staff or labour), equipment, and material requirements at site with key dependencies and their main concerns and constraints regarding the eight pre-requisites, along with a temporary rough schedule for the phase duration.
    - Each contractor team or cluster prepares post-its with rock activities defined, along with their durations and critical dependencies on other sub-contractors, clusters and sub-clusters, and participates in the "should-do" collaborative planning in three iterations. This is done in the collaboration room (big room), Figure 3.10, as suggested below.

**Iteration 1 (reverse or backward planning):**

- Each contractor or cluster pastes their post-its on the board to schedule backwards for all their rock activities, taking the boulder milestone as the target date. Then they determine the start and finish time for those activities.

**Iteration 2 (resolving handing over conflicts and inconsistencies):**

- Each contractor or cluster studies the schedules of other contractors or clusters for mismatches or dependency conflicts and critically evaluates all such tasks (e.g. high-duration items) and challenges, to reduce cycle time and sequence processes to remove conflicts or mismatches. Perform inter-cluster constraints analysis for all the eight pre-requisites and discuss possible solutions to resolve these. This may involve:
    - Thinking of a different sequence of activities
    - Planning more rock activities in parallel (which may lead to more hand-offs with smaller batch sizes)
    - Enhancing the resources

- Engaging better skilled persons
- Low-cost automation of the process
- Removing all eight construction wastes (using VSM or 5W1H)
- Identifying if any rock activity is new to the cluster or if the team has limited experience in it

For this, advance piloting needs to be done to standardise the optimal process. Also, critically analyse the interdependencies between activities and understand their complexity. This process continues until all the trades or clusters agree on handovers for successor work. After all these discussions and problem resolutions, agree on the handover dates and sequence.

### Iteration 3 (forward scheduling):

- After removing conflicts, freeze the start and finish dates of the agreed activities and create the forward schedule for all these rock-level activities. There may still be some bottleneck rock activities during these three iterations. These need to be de-bottlenecked with improved processes or higher resource deployment. These must also be recorded so that they can be de-bottlenecked in advance. You may use VSM or waste elimination tools for this.
- All clusters, including engineering design and procurement, must commit to the sequence and the handover time to the next cluster, which is also an internal customer, based on the plan above.
- Also update these on the Lean MPS in Primavera or MS Project.
- Make plans for additional separate discussions and identify the relevant teams, clusters, and sub-clusters for resolving bottleneck activities and removing constraints.

This is also shown in the table below along with other information such as team involved, the deliverables of each step, possible risks, and Lean construction solutions for those. A sample rock process is provided in Annexure.

**Figure 3.10: The BIG ROOM for Phase Pull Planning (photo courtesy Ferring Pharmaceutical, Hyderabad)**

### 3.3.10 CAN-DO / PEBBLE Level:

**Guidelines for PEBBLE Activities definitions & Process Steps for Planning & Scheduling :**

The should-do phase schedules the rock processes for the next 4–6 months. These are then studied in greater detail at this level, for a look-ahead period of 4–6 weeks. Rock processes scheduled in the next 4–6 weeks only are further detailed into 5–6 sub-activities required to complete the parent rock processes. These sub-activities, named 'pebble' activities, generally belong to one single trade or cluster, such as civil or HVAC. However, there may be many sub-clusters involved. Follow the guidelines given below:

- Each trade, cluster and sub-cluster must review the rock activities defined in the phase plan for validation, and suggest any changes if required.
- Divide the rock activities into greater detail (single speciality trade-wise or sub-cluster-wise) pebble sub-activities or activities for the next level of work structure. Take help from the original MPS prepared for the client organisation's activities or processes, if required.
- As a thumb rule, the number of pebble activities should generally be no more than 4–6 per rock process. These will later be further expanded at dust level, by the sub-clusters and sub-sub-clusters.

- Completeness – these pebble activities must collectively represent 100% completion of the rock processes, as Lean does not accept 95% completion.

- **CPS Planning & Scheduling Process for at Pebble Activities level:**
  - Agree on duration for 'look-ahead' period planning for pebble-level planning and scheduling (typically 4–6 weeks).
  - Each trade should study the rock activities scheduled for the agreed look-ahead period in greater detail for the next 4–6 weeks. They must also study the status of the inter-cluster constraints listed for that period, related to the eight pre-requisites. Design and procurement clusters (and relevant sub-clusters) must be part of this, along with the client's project head. They should also elaborate further and break each rock process into several pebble activities or sub-activities for their part of the trade or sub-clusters.
  - Each trade team, cluster and relevant sub-cluster leaders must come prepared with further elaboration of these pebble activities, and clearly define the sequence, duration, dependencies, and key resource requirements.
  - Each trade vendor, cluster leader and sub-cluster leaders create the pebble activities collaborative plan (can-do) in the collaboration room (big room), using post-its of their activities. Each trade team studies the plans of other trade teams. They identify and resolve the constraints against these activities of the look-ahead period and add or update the constraint log. These can be resolved by some of the following:
    - Thinking of a different sequence of activities
    - Planning more rock activities in parallel (may lead to more hand-offs with smaller batch sizes)
    - Enhancing the resources
    - Engaging better skilled persons
    - Low-cost automation of the process
    - Removing all eight construction wastes (using 5W1H)
  - Also ask if any pebble activity is new to the cluster, or they have less experience in that. For this, advance piloting should be planned to standardise the process.
  - After resolving conflicts, finalise the schedule and exchange the details of handover points between trades, clusters, and sub-clusters. They also make plans for special, separate meetings for any process improvements or for resolving some constraints.

- All clusters, including engineering design and procurement clusters and sub-clusters, must commit the handover time to the next internal customer, based on the above plan.
- Update these on Lean MPS on Primavera or MS Project.

**Important note:** To ensure good utilisation of the human and machine capacities available, if there is still spare capacity, leaders can take up those activities or processes which have all the pre-requisites satisfied but are not yet required in the next 1–3 weeks as per the look-ahead or weekly work plan.

The magic to making the work ready lies in true collaboration. Collaboration means much more than mere attendance for many project leaders. Involving all stakeholder representatives and getting their signature on the attendance sheet is not the collaboration referred to here. The major prerequisite for effective communication is active listening. In a chain of events, if it happens only through instructions, many will agree in meetings to deliver materials, designs, and work, but the reality is different when the time comes. Negotiation and understanding the ground reality of both the requesting and performing party is crucial for establishing a firm make-ready plan. One magic statement used in collaborative and pull planning sessions is: 'I can do it, provided...' People involved in collaborative planning are encouraged to point the finger at management (e.g. "I can start the painting by the beginning of next month provided I get the approval within one week"), designers, procurement etc.

These steps are also shown in the table below, with additional information like teams involved, the deliverables of each step, and possible risks and Lean construction solutions.

**Important note:** This look-ahead planning and scheduling meeting is also used to review execution, progress, monitoring, and control of the project. These activities and templates are defined in section 4.4.3.

### 3.3.11 WILL-DO-Next Week/ Dust Level:

#### Guidelines for DUST Activities definitions and Process Steps for Planning & Scheduling :

The can-do phase schedule is studied further in greater detail for the next week on a rolling plan basis, for a period of 5–6 days, depending on the number of working days in the week. Pebble processes scheduled in the next 5–6 days are further detailed into 5–6 dust tasks, required to complete the parent pebble activities (please also refer to blue-coloured pebble task 1.4.3 in figure 3.9 and the corresponding 5 black-coloured dust tasks shown below that).

**The following are the guidelines for this:**

- Each trade, cluster or sub-cluster must review the pebble activities defined in the look-ahead plan for the immediate next week of 5–6 days, for their correctness, and indicate any missing ones to include after discussions.
- Divide the pebble activities scheduled during this period into greater detail, called dust tasks.
- As a thumb rule, the number of dust tasks should generally be no more than 4–5 per pebble activity per rock process.
- Completeness – these dust tasks must collectively represent 100% completion of the pebble activity, as Lean does not accept 95% completion.

- **CPS Planning & Scheduling Process for at Dust Tasks level:**
  - Agree on the number of working days in a week. This may be different for the client organisation, EPC contractors and consultants (typically, 5 or 6 days), but for execution at the worksite, it should be agreed by all.
  - Each trade or sub-cluster comes prepared with the dust-level tasks for the week in line with the rolling look-ahead plan and schedule for the next week, with practically no constraints or only a small number of manageable constraints. Relevant design and procurement sub-clusters have to be a part of these discussions.
  - Each trade team, cluster or sub-cluster team leader comes prepared with further elaboration of these dust tasks and clearly defines the sequence, duration, dependencies and all eight pre-requisites for these tasks.
  - Each trade team, cluster or sub-cluster team leader makes a tentative dust task will-do plan and studies the plans of all other trade teams for the next week in the big room. Then they discuss and ensure works are made ready for the week by identifying and resolving all the constraints related to the eight pre-requisites. Relevant design and procurement sub-cluster leaders must be part of these collaborative meetings.
  - They identify and resolve the constraints against these activities for the weekly work period and add or update the constraint log. They select the dust tasks that are well defined, constraint-free, properly sequenced and have all eight pre-requisites.
  - With those dust tasks, the whole team finalises the schedule and exchanges the details of handover points among the sub-clusters or trades. They also make plans for special extra meetings if required, using the 5W1H process improvement method to resolve some of the smaller constraints.

- All sub-clusters, including relevant engineering design and procurement sub-clusters, must commit to the handover time to the next internal customer, based on the above plan.

These steps are also shown in the table below with other information like the team involved, the deliverables of each step and possible risks and Lean Construction solutions for those.

**Important note:** This weekly make-ready meeting for weekly work planning and scheduling is also used to review the execution, progress, monitoring and control of the project. Those activities and templates are defined in section 4.4.3.

A 3-minute summary video for rolling planning at three levels will make it clear: https://www.youtube.com/watch?v=M8ODlDxpbOQ

The will-do weekly level scheduled the dust tasks for the next 5–6 days. These are studied in greater detail by foremen and supervisors (the sub-sub-cluster leaders), who actually allocate the work to the team working on those tasks. Then they derive their to-do tasks for the next 24 hours from there. They have to ensure that there are no constraints related to the eight pre-requisites such as manpower, work front (previous task completed), drawings and other approvals, equipment, etc., for the 24 hours. If, by chance, there is some minor constraint that can be resolved jointly with others before the task starts, they can consider those for the daily to-do list. These dust tasks are further detailed into 5–6 to-do list items of smaller tasks, required to complete the parent dust task. (Please refer to black dust task 1.4.3.1 in figure 3.9 and the list of to-do tasks below that.) Follow the guidelines given below:

- **<u>CPS Planning & Scheduling Process for at Daily 'To-Do' Tasks level:</u>**
    - Each trade shares the weekly work plan of will-do tasks with their foremen and supervisors (the sub-cluster or sub-sub-cluster leaders). It is assumed that the pre-requisites related to relevant design and procurement sub-clusters have already been cleared.
    - Each foreman or supervisor comes prepared with further elaboration of these dust tasks into their daily to-do tasks and clearly defines the sequence, duration, dependencies, and all resource requirements for the next 24 hours.
    - These teams study the plans of all other such trade teams for the next 24 hours and the week in the big room. They discuss and ensure work can be done uninterrupted during the next 24 hours.
    - In case there are minor hitches, they can resolve those constraints for the next 24 hours (such as handovers or sharing common equipment like cranes). The design and procurement sub-clusters may join this huddle only twice a week, not every day.
    - All teams, including engineering design and procurement, must ensure they stick to the agreed daily to-do list and achieve 100% of their commitments,

based on the above plan.

- Please remember that the ability to say no to certain requests is at the core of making reliable promises in terms of the to-do task lists.

These steps are also shown in the figure 3.11 below, along with additional information such as the team involved, the deliverables for each step, potential risks, and Lean construction solutions for those.

Important note: This daily 30-minute standing meeting for next 24-hour planning and scheduling is also used to review execution, progress, monitoring, and control of the project on a daily basis. The activities and templates involved in these are defined in section 3.4.4.

In Lean construction, although planning and scheduling go hand-in-hand with production monitoring and execution, they are explained separately in section 3.4 to help understand the details better.

## 3.4 Production Monitoring & Control at different levels of CPS

Although the production planning and control in CPS is done top-down (starting from the Lean MTS of Boulder Milestones to Rock Processes, Pebble Activities, Dust Tasks, and finally daily 'To-Do' lists), the execution actually happens bottom-up—from daily to weekly, to monthly, to phase-wise, and finally to the total project. But all of these are aligned to the Lean MTS. The monitoring and control are explained according to how they actually occur.

### 3.4.1 WILL-DO-Next Week/ Dust Level:

**Weekly & Daily Process of Production Monitoring & Control :**

These teams study the plans of all other such trade teams for the next 34 hours and the week in the big room. They discuss and ensure that work can proceed uninterrupted over the next 34 hours.

In case there are minor hitches, they can resolve those constraints for the 34-hour period (such as handovers or sharing common equipment like a crane). The design and procurement sub-clusters attend this huddle only twice a week, not every day.

All parties, including engineering design and procurement, must adhere to the agreed daily 'to-do' list and strive to achieve 100% of their commitments based on the plan.

Please remember that the ability to say 'no' to certain requests is central to making reliable promises, particularly with respect to daily task lists.

These steps are also shown in the table below, along with details such as the teams involved, deliverables for each step, possible risks, and Lean construction solutions.

**Important note:** This daily 30-minute standing meeting for planning and scheduling the next 34 hours is also used to review execution, progress, and the monitoring and control of the project on a daily basis. The activities and templates involved are defined in section 3.4.4. In Lean construction, although planning and scheduling go hand in hand with production monitoring and execution, they are explained separately in section 3.4 for clarity.

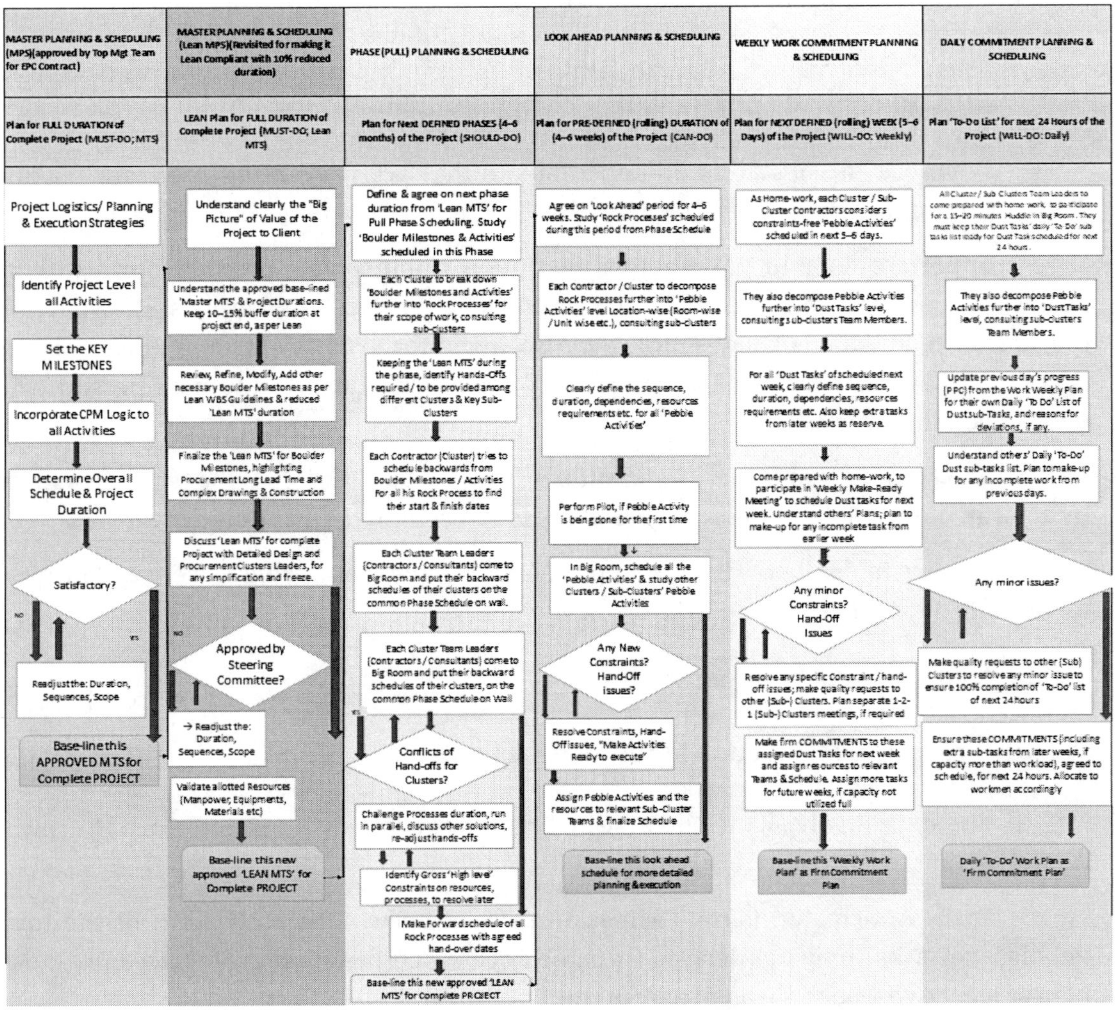

**Figure 3.11: Summary of Collaborative Planning System**

## 3.4.2 Production Monitoring & Control at different levels of CPS

Although production planning and control in CPS is done top-down (starting from the Lean MTS of boulder milestones to rock processes, pebble activities, dust tasks, and daily 'to-do' lists), the actual execution happens bottom-up—from daily to weekly to monthly, phase-wise, and then to the total project. However, all these are aligned with the Lean MTS. The monitoring and control are described in the order they occur in practice.

### 3.4.3 WILL-DO-Next Week/ Dust Level:

**Weekly & Daily Process of Production Monitoring & Control:**

The sub-cluster team leaders assemble in the big room each morning for a collaborative review of the last 24 hours. They update others and ensure commitments (including extra sub-tasks pulled from later weeks, if capacity exceeds workload) are aligned to the schedule for the next 24 hours, and allocate tasks to workmen accordingly.

- What did I accomplish since the last daily meeting?
- What do I plan to work on in the next 34 hours?
- What are the obstacles or impediments that are preventing me from making progress?

Important note: In this meeting, no time is spent solving problems or justifying delays. It is only for updates and listing problems or issues. You simply state what you did, and then do what you say—publicly. Also, avoid the "90% syndrome". Either a task is completed, or it is not; partial completion is not acceptable to the next person who depends on that work.

The reliability of these commitments (answer to question 1 above) is measured by a parameter called PPC (percentage of promises completed), as defined earlier.

The two tables below list the steps involved in the monitoring and control process for 'will-do' next 24 hours and 'will-do' next week.

### 3.4.4 CAN-DO /PEBBLE Level:

**Weekly Process of Production Monitoring & Control:**

The steps involved, the outcome and risks are listed in the table below. However, it is important to learn two terms here:

Tasks Anticipated (TA): TA measures % of tasks anticipated (out of the total Pebble level tasks/activities derived from the Rock Processes detailing for the look-ahead plan at least two weeks ahead of execution.

The Sub-Clusters Teams Leaders assemble in Big Room daily in morning, for updating others collaborative review of their last 24 hours Ensure these COMMITMENTS (including extra sub-tasks from later weeks, if capacity more than workload), agreed to schedule, for next 24 hours. Allocate to workmen accordingly.

i. What did I accomplish since the last daily Meeting?
ii. What do I plan to work on by the next 24 hours?
iii. What are the obstacles or impediments that are preventing me from making progress?

**Important Note**: In this meeting, NO Time is spent for problem solving, or justifying delays. It is ONLY updates and Problem/ Issues LISTING. You just say, what you did, and do what you say, publicly. Also avoid 90% Syndrome. Either you have completed, or not completed. There is nothing like % completion, acceptable to next person, for doing his/ her work on that.

The reliability of accomplishment (answer to question 1 above) of promises is measured in terms of a parameter called PPC (Percentage Promises Completed), defined earlier.

Below two tables lists the steps involved for monitoring and control process at Will-DO- next 24 hours and WILL-DO- next week, in brief.

The steps involved, the outcomes, and the risks are listed in the table below. However, it is important to understand two terms here:

- **Tasks anticipated (TA):** TA measures the percentage of tasks anticipated (out of the total pebble-level tasks/activities derived from the rock process detailing) for the look-ahead plan at least two weeks before execution.
- **Tasks made ready (TMR):** TMR measures the effectiveness of look-ahead planning in identifying and removing constraints to ensure tasks are ready for execution in the upcoming week.

Since the look-ahead plan and weekly work plan must relate to the master (phase) plan, it is important to include these two metrics and use them alongside PPC (percentage of promises completed) as performance parameters. In practical application, TA and TMR indicate the production team's capability to forecast and prepare tasks for execution. Figure 3.12 below provides a sample illustration of this.

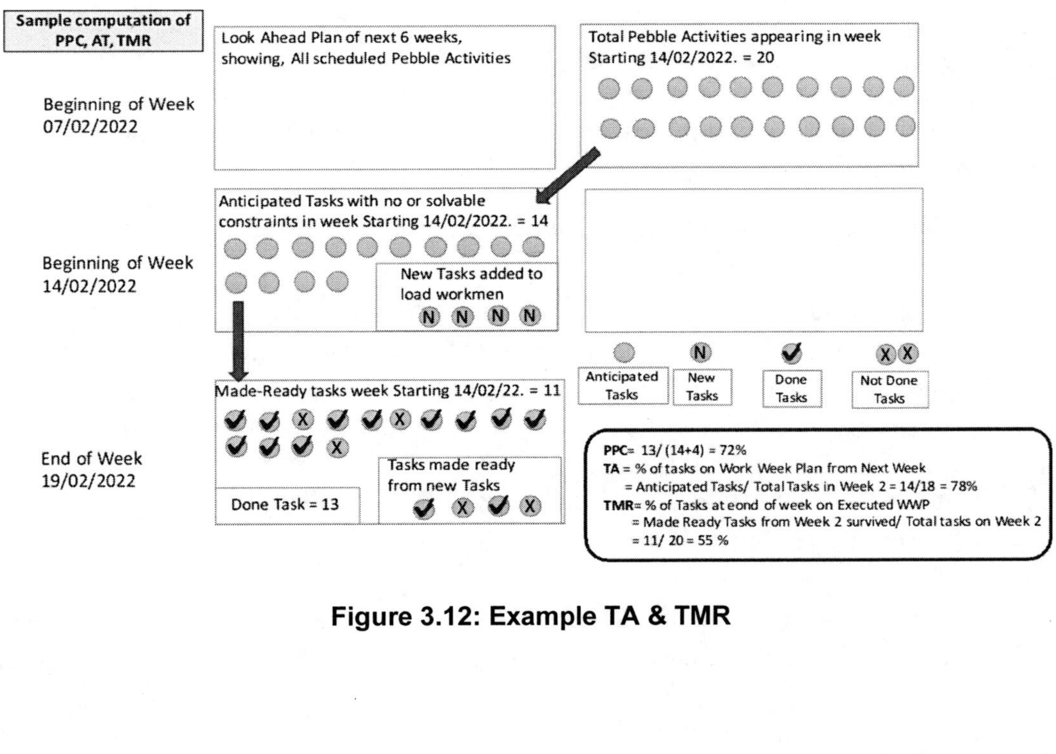


**Figure 3.12: Example TA & TMR**

**Important Note:** It is highly recommended that the CPS Scheduler must MEET each cluster team Leader separately 3 days before the 6 weekly rolling Plan to find & update the progress, constraints (including inter-cluster), to be more effective in Look Ahead and Work Week Planning.

### 3.4.5 SHOULD-DO / ROCK Level:

**Monthly Process of Production Monitoring & Control :**

This is very similar to section 3.4.3. The only difference is that it references the phase pull plan (should-do plan) instead of the pebble-level plan.

It also ensures that actions are taken to create the deliverables for the monthly progress report, as outlined in section 3.4.5. This requires each cluster to identify potential bottleneck operations and implement preventive and corrective actions using tools such as value stream mapping (VSM), kaizen, or the 5W1H method.

### 3.4.6 MUST-DO / BOULDER Level:

**Monthly Process of Production Monitoring & Control :**

Production planning and control in CPS is structured in a top-down manner, starting from the Lean MTS of boulder milestones to rock processes, pebble activities, dust tasks, and daily "to-do" lists. However, execution follows a bottom-up approach, progressing from daily tasks to weekly, monthly, phase-wise, and finally to overall project completion. All levels are aligned with the Lean MTS to ensure consistency and adherence to the original commitments made to the client. Monitoring and control processes are explained to reflect how implementation occurs in practice.

### 3.4.7 Key Risks and Lean Solutions

Anticipated risks and mitigation strategies are outlined in the respective tables. Critical points include:

- Training all team members on Lean construction practices.
- Mentoring to demonstrate tangible benefits.
- Incorporating relevant clauses in the contract documents.
- Holding regular HR counselling sessions for key stakeholders.
- Agreeing on a reward and penalty system to ensure compliance with Lean principles.

## 3.5 Summary of Five levels of Planning & Scheduling and Two Layers of Execution in CPS:

The Collaborative Planning System (CPS) is structured around five distinct levels of planning and scheduling, each with a progressively finer focus - from long-term milestones to daily work execution. These five levels - Must-Do, Should-Do, Can-Do, Will-Do Weekly, and Will-Do 24 Hours - are designed to ensure that planning is both comprehensive and adaptive across time scales. Complementing this, CPS incorporates two interconnected layers of execution: the Look-Ahead layer, which focuses on readiness and flow planning, and the Daily Execution layer, which drives task-level reliability and learning. The following summaries detail the purpose, process, key deliverables, typical risks, and Lean solutions associated with each level, providing a holistic view.

**Table 3.2: Brief Process at Must-Do level CPS Scheduling**

| Step | BRIEF PROCESS at MUST-DO level | DELIVERABLES | KEY RISKS / ISSUES | LEAN SOLUTIONS |
|---|---|---|---|---|
| 1 | Understand 'Big Picture' & base-lined complete Project MPS Schedule & Key Challenges | MUST-DO BOULDER lean MPS for complete Project | Earlier MPS is unrealistic; prepared without considering EPC 'real' Constraints | CPS MPS at "MUST-DO" level & BOULDER Process Scheduling |
| 2 | Review & refine Base-lined Boulder Milestones defined in MPS as per Lean WBS guidelines. | MOM for changes and any Action or major constraints | Difficult to monitor & control Trade-wise Project progress, Uncertainties of most activities completion in future | Better visualization and early alerts for focused efforts to complete the project faster |
| 3 | Discuss Lean MTS with Design & Procurement Teams. Resolve all issues Long Lead Time & Complex drawings. | MOM for changes and any Action or major constraints | Real constraints of some critical Designs or Procurement may force to change target dates approved Baselined MPS | Identify some critical/bottleneck Milestones/Activities, crash using VSM |
| 4 | Get Approval from the Steering Committee for execution of revised baselined lean MPS | MUST-DO BOULDER lean MPS for Project approved & baselined for execution | | |
| 5 | Validate the allocated Resources (Manpower, Budget, Equipments, Materials etc.) | MOM for changes and any Action or major constraints | EPC Contractors don't commit and allocate the resources in time for VSM, Design changes, Process improvements | Training, convincing for benefits and also contractual obligation |

**Table 3.3: CPS Process at Should-Do level Scheduling Rock Activities**

| Step | CPS PROCESS at WILL-DO- 24 Hours level Scheduling With To-Do Plan | DELIVERABLES | KEY RISKS / ISSUES | LEAN SOLUTIONS |
|---|---|---|---|---|
| 1 | Define & Agree on next phase duration from Lean MPS; Divides Boulder Activities into Rock Activities, for next level of work-structure. | Duration of Next Phase Details about Rock Activities | 'All' required relevant participants (including E, P, C) don't participate physically/virtually | "Should-Do" part of CPS/LPS: The ROCK level Planning. |
| 2 | agreed duration for next phase for Should-Do planning, each Trade Team divides Boulder's Activities of that phase, into Rock Activities. Define sequence, duration, dependencies, Key resources these Rock Activities. Do collaborative Iterative Planning in 3 steps, using Post-its. | Duration of Next Phase Details about Rock Activities | People play over-safe and often not willing to agree to crash duration required, despite knowing that big buffer is built-in there. | Training, convincing for benefits and also contractual obligation |
| 3 | Iteration 1 (Reverse or Backward planning): Each Contractor tries to schedule backwards for all his ROCK Activities, considering target date for Boulder MTS. Then finds start and finish time for those activities. | SHOULD-DO ROCK Processes Phase Plan- Iterations 1 | People are not willing to agree to changing the activities sequencing in the best interest of the project completion. Retain "my contract" syndrome. | Training, convincing for benefits and also contractual obligation |
| 4 | Iteration 2 (Resolving Hanging Over Conflicts & Inconsistencies): Each Team studies all other trades schedules for mismatch/dependency conflicts & critically evaluate all such tasks, challenge to reduce cycle time to remove conflicts, performs Constraint analysis; discusses & resolves conflicts regarding sequence & hand-over dates. | SHOULD-DO ROCK Processes Phase Plan- Iterations 2 | There may be uncertainties for some Rock Activities which need to be sorted out collaboratively. Solutions initially may look difficult. | Some HR inputs are required to emphasize Vision, Mission & Goals of Lean |
| 5 | Iteration 3 (Forward Scheduling): Make the forward schedule for all ROCK level activities & freeze the start & finish dates of agreed list, Longest & Bottleneck process where action is to be taken in future, using VSM, Waste Elimination tools etc. | SHOULD-DO ROCK Processes Phase Plan- Iterations 3 | There may be uncertainties for some Rock Activities which need to be sorted out collaboratively. Solutions initially may look difficult. | Use, VE, design simplification, and VSM, Kaizen, 5WLH, etc |
| 6 | All, including Engineering-Design and Procurements must connect the 'hand over' dates to next 'internal customer', based on above outcomes. Make Plans for additional discussions to resolve the Constraints | Handover Commitments and Constraints List | Participants don't commit and allocate resources (in time) for VSM, Waste elimination, Process improvements etc. | Training, convincing for benefits and also contractual obligation |

## Table 3.4: CPS Process at Can – Do level Scheduling Pebble Activities

| Step | CPS PROCESS at CAN-DO level Scheduling PEBBLE Activities | DELIVERABLES | KEY RISKS / ISSUES | LEAN SOLUTIONS |
|---|---|---|---|---|
| 1 | Agree on 'Look Ahead' period, take 'Rock Activities' schedule with greater details along with the Constraints list for that period. Design & Procurement agreement has to be a part of this. | Agreement on Look Ahead period | All' required relevant participants (including E, P, J) don't participate physically/virtually | Training, convincing for benefits and also stressing for contractual obligation |
| 2 | Each Trade team comes prepared with further elaboration of Rock processes into Pebble Activities (Room-wise, system-wise) for the Look Ahead period. Clearly define the sequence, duration, dependencies and key resources requirements for these Pebble Activities. | CAN-DO PEBBLE Activities Plan | People play over-safe and often not willing to agree to critical duration required, despite knowing that big buffer is built-in there | Training, convincing for benefits and also stressing for contractual obligation; HR intervention |
| 3 | Each Trade Vendor, using the list of Pebble Activities of 'all' the Trade teams collaboratively identifies and resolves the constraints against the activities of Look Ahead period and adds onto the constraints log. | CAN-DO PEBBLE Activities Plan & Updated Constraints List | There may be uncertainties for some Activities which need to be sorted out collaboratively. Solutions initially may look Difficult | Use VE, design simplification and VSM, Kaizen, 5W1H, etc. |
| 4 | After resolving conflicts, finalizes the schedule and exchange the detail of handover between trades. They also make commitments using VSM, process improvements, Waste elimination, HR, plans for special meeting, etc. for resolving some constraints. | Handover Commitments and Constraints List | Participants don't commit and allocate resources (in time) for VSM, Waste elimination, Process improvements etc. | Training, convincing for benefits and also stressing for contractual obligation; HR process intervention |

## Table 3.5: CPS Process at Will-Do-Weekly level Scheduling Dust Tasks

| Step | CPS PROCESS at WILL-DO-Weekly level Scheduling DUST Tasks | DELIVERABLES | KEY RISKS / ISSUES | LEAN SOLUTIONS |
|---|---|---|---|---|
| 1 | Each trade comes prepared for the Dust level activities for week in line with 'Look Ahead' Plan & Schedule, with no or manageable constraints | List of Comprehensive Dust Activities | 'All' required relevant participants (including E, P, C) don't participate physically/virtually | Training, convincing for benefits and also contractual obligation |
| 2 | Each trade writes the key Dust activities as 'Post it' under next 5/6 days in week. Details the duration, sequence, all resources required. Each Trade team make Dust Task plan and studies the plans of all other trade Teams, for next week. Discuss and ensure works are 'made ready' for week, by identifying and resolving constraints | CAN-DO Dust Task Plan | People play over-safe and often not willing to agree to crash duration required, despite knowing that big buffer is built-in there. | Training, convincing for benefits and also contractual obligation; HR Intervention |
| 3 | Discuss the inter trade dependencies, activity sequence, resource requirement. See Interruptions and Constraints and discuss possible corrective actions and agree on the chosen ones. | Updated activity sequence, Interruptions and Constraints List & solutions | Some Dust Tasks sequence need to be changed or improved. Solutions initially may look difficult. Participants don't commit improvements | Training, convincing for benefits and also contractual obligation; HR Intervention |

**Table 3.6: CPS Process at 'Will-Do-24 Hours' level Scheduling with 'To-Do' Plan**

| Step | CPS PROCESS at WILL-DO- 24 Hours level Scheduling With To-Do Plan | DELIVERABLES | KEY RISKS / ISSUES | LEAN SOLUTIONS |
|---|---|---|---|---|
| 1 | Vendors/Subcontractor's Foremen & Supervisor to refer the Level 4 'Will-Do-Weekly' Plan and 'pull' the activities for next day in 'daily To-Do list' template. | WILL-DO 24 hours 'To-Do' List by all Foremen-tentative | 'All' required relevant participants don't participate physically/virtually | Training, contractual obligation; HR Intervention |
| 2 | Project manager & engineers to ensure that the work planned for next 24 hours is 'constraint free' and as per the 'Will-Do-Weekly' plan and agreed hand-off. | WILL-DO 24 hours 'To-Do' List by all Foremen-tentative | People are not willing to agree to changing the activities sequencing. | Training, contractual obligation; HR Intervention |
| 3 | Take any reconfirmation/ follow-up to ensure no constraints for next 24 hours. Discuss activities of each trade. Resolve minor issues. | WILL-DO 24 hours 'To-Do' List by all Foremen-Final | Minor issues related with some Dust Tasks, need to be sorted out collaboratively. | Training, contractual obligation; HR Intervention |

**Table 3.7: CPS Monitoring & Controlling Processes at 'Will-Do-24 Hours' level 'To-Do' Dust Tasks & Subtasks**

| Step | CPS Monitoring & Control Process at Will- Do-24 Hours level To-Do DUST Task Sub- tasks | DELIVERABLES | KEY RISKS / ISSUES | LEAN SOLUTIONS |
|---|---|---|---|---|
| 1 | Complete "To-Do" Booklet & update the 100% completed Tasks.Also update reasons for incomplete. Record reasons Code, in the Booklet | List of Comprehensive Daily " To Do" DUST Tasks/Sub-Tasks | All' required relevant participants don't participate physically/virtually | raining, convincing for benefits and also contractual obligation; HR Counseling |
| 2 | Plot the 'daily' PPCs for all trades on Big Room Wall. Also input for making Weekly PPC Chart. | Daily PPC Chart for Foremen | All' required relevant participants don't participate physically/virtually | Training, convincing for benefits,HR Counseling |
| 3 | Record the causes for incomplete task, on Wall Chart. | Daily Chart for Causes of Variance for Foremen | All' required relevant participants don't participate physically/virtually | Training, convincing for benefits and also contractual obligation; HR Counseling |
| 4 | Study PPC trends & causes of Variance, and which Constraints were not resolved, despite promises. Discuss the inter trade dependencies, activity sequence, resource requirement etc. | Learning about why To-Do list not completed. | Foremen play over-safe and often not willing to agree to put efforts to improve sequence required , | Training, convincing for benefits, even contractually; HR Counseling/ Penalty? |

| | | | | |
|---|---|---|---|---|
| **5** | Discuss Interruptions and Constraints and agree on possible corrective actions to implement. Also see, how to make-up for the loss due to incomplete "To-Do" Tasks/Sub-tasks. Based on this, plan for the next 24 hours. | List of actions for the next 24 hours | Clusters/Sub-Clusters hide the problems | Training, convincing for benefits and also contractual obligation; HR Counseling. Any Penalty? |

**Table 3.8: CPS Monitoring & Controlling Processes at 'Will-Do-Weekly' level Pebble Activities/ Dust Tasks**

| Step | CPS Monitoring & Control Process at Will- Do-Weekly levrel PEBBLE Activities / DUST Task | DELIVERABLES | KEY RISKS / ISSUES | LEAN SOLUTIONS |
|---|---|---|---|---|
| 1 | Update the 100% completed Dust Tasks and/or Pebble Activities scheduled for the week, based on Foremen daily data. Find the Tasks/Activities missed. Compute the weekly PPC for Cluster/Sub-Cluster. Understand & validate the reasons for variance from Foremen. | List of Comprehensive Daily "To Do" Dust Tasks/Sub-Tasks | All' required relevant participants don't participate physically/virtually | Training, convincing for benefits and also contractual obligation; HR Counseling |
| 2 | Plot the Weekly PPCs for all trades on Big Room Wall. | Updated Weekly PPC Chart | All' required relevant participants don't participate physically/virtually | Training, convincing for benefits and also contractual obligation; HR Counseling |
| 3 | Review the Weekly Constraints and their resolution. Record the causes for incomplete tasks/schedule variance, on Weekly Wall Chart. | Updated Weekly Chart for Causes of Variance | All' required relevant participants don't participate physically/virtually | Training, convincing for benefits and also contractual obligation; HR Counseling |
| 4 | Study PPC trends & missed Dust Tasks and Pebble Activities from Committed Schedule and causes of Variance. Learn from these interruptions and Constraints, prioritize those to resolve and agree on possible corrective actions to implement. | List of Constraints to be acted upon and their target dates to resolve | Clusters/Sub-Clusters play over-safe and often not willing to agree to put efforts to improve sequence required etc., despite knowing possibilities. | Training, convincing for benefits and also contractual obligation; HR Counseling. Any Penalty? |
| 5 | Spend time, in a separate meeting with its Team of relevant people, about how to make-up for the time lost due to incomplete Dust Tasks, by improving process. | Productivity improvement in the immediate future for the chosen Tasks | Clusters/Sub-Clusters play over-safe and often not willing to agree to put efforts to improve sequence required etc., despite knowing possibilities. | Training, convincing for benefits and also contractual obligation; HR Counseling. Any Penalty? |

| | | | | |
|---|---|---|---|---|
| 6 | Based on this, prepare inputs for action points for the next Make-Ready Meeting for better planning. Share inputs for the Weekly Project Progress Report. Also identify any help required for Referring management for resolution. | List of actions & other inputs for the next Make-Ready Meeting & Weekly Project Review | Clusters/Sub-Clusters hide the problems | Training, convincing for benefits and also contractual obligation; HR Counseling. Any Penalty? |

**Table 3.9: CPS Monitoring & Control Processes at 'Can-Do-6 W' level Pebble Level Activities**

| Step | CPS Monitoring & Control Process at Will- Do-Weekly levrel PEBBLE Activities / DUST Task | DELIVERABLES | KEY RISKS / ISSUES | LEAN SOLUTIONS |
|---|---|---|---|---|
| 1 | Update the 100% completed Dust Tasks and/or Pebble Activities scheduled for the week, based on Foremen daily data. Find the Tasks/Activities missed. Compute the weekly PPC for Cluster/Sub-Cluster. Understand & validate the reasons for variance from Foremen | List of Comprehensive Daily "To-Do" Dust Tasks/Sub-Tasks | All' required relevant participants don't participate physically/virtually | Training, convincing for benefits and also contractual obligation; HR counseling |
| 2 | Plot the Weekly PPCs for all trades on Big Room Wall. | Updated Weekly PPC Chart | All' required relevant participants don't participate physically/virtually | Training, convincing for benefits and also contractual obligation; HR counseling |
| 3 | Review the Weekly Constraints and their resolution. Record the causes for incomplete tasks/schedule variance on Weekly Wall Chart. | Updated Weekly Chart for Causes of Variance | All' required relevant participants don't participate physically/virtually | Training, convincing for benefits and also contractual obligation; HR counseling |
| 4 | Study PPC trends & missed Dust Tasks and Pebble Activities from Committed Schedule and causes of Variance. Learn from these interruptions and Constraints, prioritize those to resolve and agree on possible corrective actions to implement. | List of Constraints to be acted upon and their target dates to resolve | Clusters/Sub-Clusters play over-safe and often not willing to agree to put efforts to improve sequence required etc., despite knowing possibilities. | Training, convincing for benefits and also contractual obligation; HR counseling. Any Penalty? |
| 5 | Spend time, in a separate meeting with its Team of relevant people, about how to 'make-up' for the time lost due to incomplete Dust Tasks, by improving process. | Productivity improvement in the immediate future for the chosen Tasks | Clusters/Sub-Clusters play over-safe and often not willing to agree to put efforts to improve sequence required etc., despite knowing possibilities. | Training, convincing for benefits and also contractual obligation; HR counseling. Any Penalty? |

| 6 | Based on this, prepare inputs for action points for the next Make-Ready Meeting for the better planning.For the next week, prepare inputs for the Weekly Project Progress Report. Also identify any help required for Referring management for resolution. | List of actions & other inputs for the next Make-Ready Meeting & Weekly Project Review | Clusters/Sub-Clusters hide the problems | Training, convincing for benefits and also contractual obligation; HR counseling. Any Penalty? |
|---|---|---|---|---|

## 3.6 ANNEXURE I Case Study- Real Estate Project: 650 Bed Teaching Hospital in Odisha[1]

### The Project & the Need:

This case study focuses on constructing a 650-bedded teaching hospital in Odisha, one of four identical projects in the state. Table 3.10 provides more details. These hospital projects were delayed by over 15 months, prompting the state government to explore Lean construction concepts to accelerate progress within the existing budget. One project was selected as a pilot to implement and evaluate Lean construction principles. A Lean expert was engaged for eight months to mentor the teams, with the goal of reducing the remaining 15 months of work to 10 months.

[1]*The authors wish to thank Er Narendra Ghadai, Superintending Engineer, and Er Ashis Kara, Head, PMO, Iron Triangle Ltd, for enabling the implementation of these concepts and providing the relevant data for this case study.*

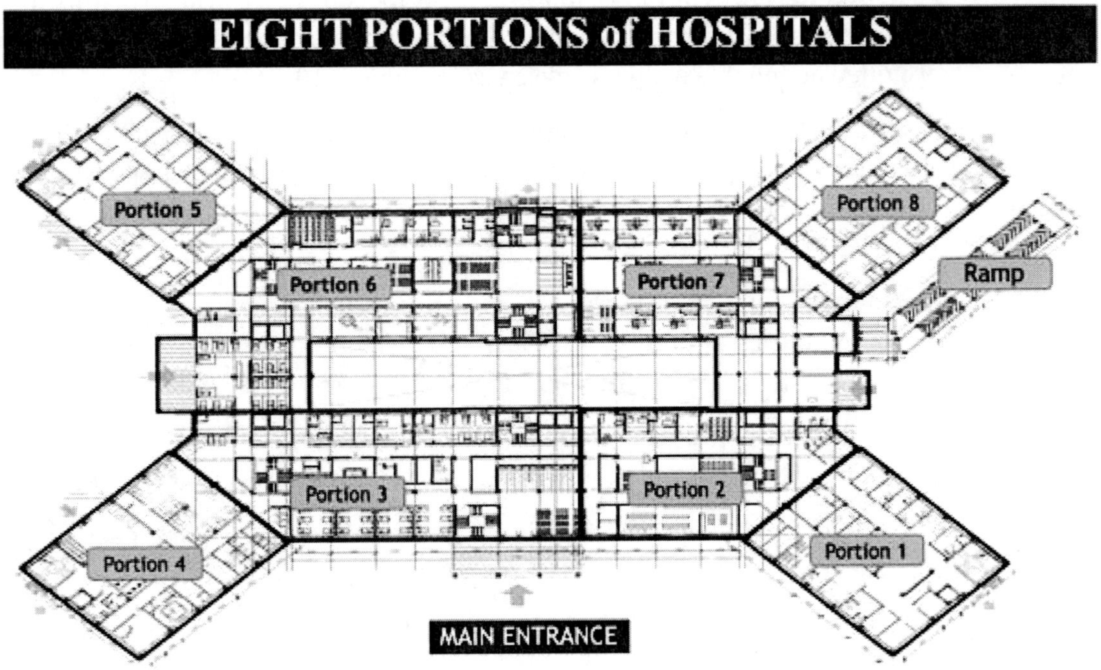


*Key Information about the Project:*

*Number of Beds = 650*
*Revised Budget = 256 Crores INR*
*Total Baseline Duration Promised = 30 Months*
*Total Balance Work after 30 M = 15 M*
*Max. Time approved to complete = 11 M*
*Number of Main Contractors involved = 8*
*(One each for 8 almost similar blocks)*
*Centralised PMO Agency for whole Project = 1;*
*Number of Floors = Ground + 5*
*Total Built-up Area = 53040 SQM*
*Each of Eight Blocks & Each Floor have similar facilities:*
*30 Beds Patient Ward; 9 Rooms for Doctors, Nurses, Students, Duty Room, Demo, Medicine etc; Utilities like Male, Female & Staff Wash Rooms, Pantry, Lifts, Operations Theatre*

**Figure 3.13 The basic details about the Hospital Project**

## The Methodology Used:

This challenge required an initial assessment of existing project planning, execution, and review methods to identify weaknesses. The main issues identified are summarized below:

- **Lack of tangible progress tracking:** the project primarily relied on quantity-based plans and reviews, which neither reflected physical progress nor provided actionable insights (please refer table 3.10).
- **Disconnection from dependencies:** plans overlooked interdependencies among activities, such as delays in blockwork affecting plastering or flooring, leading to unclear priorities for corrective actions (no Gantt chart, like in Table 3.12).
- **Inadequate quality assessment:** quantity-focused metrics ignored quality issues, resulting in defects, rework, and resource wastage, such as higher cement consumption due to poor blockwork or uneven beams.
- **Limited integration and coordination:** challenges related to coordination between contractors and their work packages were discussed in section 3.3.4.
- **Overlooking site-specific factors:** factors like limited workspace, safety requirements, and site-specific constraints added unforeseen delays.

After evaluating these issues, discussions with the PMO team and four contractors revealed that frequent design changes, such as increasing the height of nurses' residences, contributed significantly to delays.

The layout of a single floor was largely repetitive across eight sections and six floors, with identical facilities for 30 patient beds in each section. Recognizing this repetition allowed us to focus on resolving key bottlenecks in constructing a single floor, enabling efficient implementation of improvements.

**Key Actions Taken:**

- Detailed Gantt chart preparation: a comprehensive Gantt chart was created for one section and floor, as the remaining sections and floors were nearly identical. This included physical location-based timelines with clear start and end dates. Dependencies among subcontractors were highlighted, identifying bottlenecks in long-duration or critical-path activities.
- Risk identification and mitigation: anticipated risks related to design approvals, equipment, manpower, materials, weather, absenteeism, funding, and inter-departmental coordination were addressed proactively. Solutions were planned in advance, with senior-level support sought when necessary.

**Observations and Suggestions**

Efforts were initially restricted to addressing the top 4–5 issues to ensure manageable implementation. These focused on improving clarity, resolving bottlenecks, and enabling contractors to meet tight deadlines effectively.

## Table 3.10: A Project Construction Monthly Plan & Review (Sample: 650 Bedded)

Monthly Project Work plan "650 beded Hospital"

| Activity name | Unit | Total Quantity | Quantity Completed | Quantity /Balance | STATUS (31st December) | | | | | | | |
|---|---|---|---|---|---|---|---|---|---|---|---|---|
| | | | | | Nov | | Dec | | Jan | | Feb | |
| CONCRETE WORK | | | | | Target | Achive | Target | Achive | Target | Achive | Target | Achive |
| Ground Floor | | | | | | | | | | | | |
| Hospital Section 1 | Cum | 120 | 120 | 0 | 21.62 | 21.62 | 172.92 | 145.35 | 21.62 | | | |
| Hospital Section 2 | Cum | 216 | 0 | 216.00 | 214.71 | 158.23 | 53.68 | 62.76 | 0 | | | |
| Hospital Section 3 | Cum | 268 | 0 | 268.00 | 0 | | | | | | | |
| Hospital Section 4 | Cum | 140 | 140 | 0 | 0 | | | | | | | |
| Hospital Section 5 | Cum | 110 | 110 | 0 | 404.85 | 276.33 | | 123.33 | | | | |
| Hospital Section 6 | Cum | 405 | 0 | 405 | 299.94 | 213.42 | | 86.52 | | | | |
| Hospital Section 7 | Cum | 300 | 0 | 300 | 188.8 | 0 | | 153.32 | | | | |
| Hospital Section 8 | Cum | 189 | 0 | 189 | | | | | | | | |
| First Floor | | | | | | | | | | | | |
| Hospital Section 1 | Cum | 171 | 171 | 0 | | | | | | | | |
| Hospital Section 2 | Cum | 153 | 0 | 153.00 | | | | | | | | |
| Hospital Section 3 | Cum | 256 | 0 | 256 | | | | | 179.23 | | 76.81 | |
| Hospital Section 4 | Cum | 154 | 0 | 154 | | | | | 154.13 | | | |
| Hospital Section 5 | Cum | 154 | 154 | 0 | | | | | | | | |
| Hospital Section 6 | Cum | 352 | 352 | 0 | | | | | | | | |
| Hospital Section 7 | Cum | 232 | 232 | 0 | | | | | | | | |
| Hospital Section 8 | Cum | 159 | 159 | 0 | | | | | | | | |
| Second Floor | | | | | | | | | | | | |
| Hospital Section 1 | Cum | 165 | 0 | 165 | | | | | 164.98 | | | |
| Hospital Section 2 | Cum | 143 | 0 | 143.00 | | | | | | | | |
| Hospital Section 3 | Cum | 253 | 0 | 253 | | | | | | | | |
| Hospital Section 4 | Cum | 127 | 0 | 127 | | | | | | | | |
| Hospital Section 5 | Cum | 154 | 154 | 0 | | | | | | | | |
| Hospital Section 6 | Cum | 352 | 352 | 0 | | | | | | | | |
| Hospital Section 7 | Cum | 234 | 234 | 0 | | | | | | | | |
| Hospital Section 8 | Cum | 159 | 159 | 0 | | | | | | | | |

Monthly Project Work plan "650 beded Hospital"

| Activity name | Unit | Total Quantity | Quantity Completed | Quantity /Balance | STATUS (31st December) | | | | | | | |
|---|---|---|---|---|---|---|---|---|---|---|---|---|
| | | | | | Nov | | Dec | | Jan | | Feb | |
| | | | | | Target | Achive | Target | Achive | Target | Achive | Target | Achive |
| Ground Floor | | | | | | | | | | | | |
| Part-1 | | | | | | | | | | | | |
| Granite Flooring | Sqm | 14,552.24 | | 14,552.24 | 1455.22 | 933.14 | 11641.79 | 7674.81 | 1455.22 | | 0 | |
| Vitrified Flooring | Sqm | 2,541.04 | | 2,541.04 | 2032.83 | 700.23 | 508.21 | 773.22 | 0 | | | |
| Kota Stone | Sqm | 347.89 | | 347.89 | 347.89 | 315.87 | | 32.02 | | | | |
| Vinyl Flooring | Sqm | 2195.7 | | 2195.7 | 2195.7 | 2195.7 | | | | | | |
| Alkali Tile | Sqm | 92.67 | | 92.67 | 92.67 | 92.67 | | | | | | |
| Chequered Tile | Sqm | 144.14 | | 144.14 | 144.14 | 0 | | 144.14 | | | | |
| Ceramic Floor Tile | Sqm | 327.52 | | 327.52 | 327.52 | 103.67 | | 223.85 | | | | |
| Ceramic Wall Tile | Sqm | 956.73 | | 956.73 | 956.73 | 752.55 | | 204.18 | | | | |
| | | | | | | | | | | | | |
| Part-2 | | | | | | | | | | | | |
| Granite Flooring | Sqm | 11,888.19 | | 11,888.19 | | | | | 8321.74 | | 3566.46 | |
| Vitrified Flooring | Sqm | 2069.94 | | 2069.94 | | | | | 2069.94 | | | |
| Kota Stone | Sqm | 284.64 | | 284.64 | | | | | 284.64 | | | |
| Vinyl Flooring | Sqm | 1837.39 | | 1837.39 | | | | | 1837.39 | | | |
| Alkali Tile | Sqm | 75.82 | | 75.82 | | | | | 75.82 | | | |
| Chequered Tile | Sqm | 117.94 | | 117.94 | | | | | 117.94 | | | |
| Ceramic Floor Tile | Sqm | 267.97 | | 267.97 | | | | | 267.97 | | | |
| Ceramic Wall Tile | Sqm | 782.78 | | 782.78 | | | | | 782.78 | | | |

## Table 3.11 Gantt chart for Section 1 of the Hospital, to complement the Quantity-based Progress Plan & review

**Table 3.11: Gantt Chart for Section 1 of the Hospital to Complement the Quantity based Progress Plan & review**

| S No. | Section 1 | Activity | W1 | W2 | W3 | W4 | W1 | W2 | W3 | W4 | W1 | W2 | W3 | W4 | W1 | W2 | W3 |
|---|---|---|---|---|---|---|---|---|---|---|---|---|---|---|---|---|---|
| A | STRUCTURE | | | | | | | | | | | | | | | | |
| B | FINISHES | Brick Work Shaft Side Front Elevation (Office Side) | | | | | | | ■ | ■ | | | | | | | |
| | | Brick Work Shaft Side Right Side Elevation (Stock Yard) | | | | | | | | | | | | | | | |
| | | Brick Work Shaft Side Rear Elevation (Steel Yard) | | | | | | | | | | | | | | | |
| | | Plaster Shaft Side Front Elevation (Office Side) | | | | | | | | ■ | | | | | | | |
| | | Plaster Shaft Side Right Side Elevation (Stock Yard) | | | | | ■ | | | | | | | | | | |
| | | Plaster Shaft Side Rear Elevation (Steel Yard) | | | | | ■ | | | | | | | | | | |
| | | Final Plaster Front Elevation (Office Side) | | | | | | | | | | ■ | ■ | | | | |
| | | Final Plaster Right Side Elevation (Stock Yard) | | | | | | | ■ | ■ | | | | | | | |
| | | Final Plaster Rear Elevation (Steel Yard) | | | | | | | | ■ | ■ | | | | | | |
| | | Primer Front Elevation (Office Side) | | | | | | | | | | | | | ■ | | |
| | | Primer Right Side Elevation (Stock Yard) | | | | | | | | | | ■ | | | | | |
| | | Primer Rear Elevation (Steel Yard) | | | | | | | | | | | ■ | | | | |
| | | Trap Door Front Elevation (Office Side) | | | | | | | | | | | | | ■ | | |
| | | Trap Door Right Side Elevation (Stock Yard) | | | | | | | | | | ■ | | | | | |
| | | Trap Door Rear Elevation (Steel Yard) | | | | | | | | | | | ■ | | | | |
| | | Paint 1st Coat Front Elevation (Office Side) | | | | | | | | | | | | | | ■ | |
| | | Paint 1st Coat Right Side Elevation (Stock Yard) | | | | | | | | | | | ■ | | | | |
| | | Paint 1st Coat Rear Elevation (Steel Yard) | | | | | | | | | | | | ■ | | | |
| | | Paint Final Coat Front Elevation (Office Side) | | | | | | | | | | | | | | | ■ |
| | | Paint Final Coat Right Side Elevation (Stock Yard) | | | | | | | | | | | | ■ | ■ | | |
| | | Paint Final Coat Rear Elevation (Steel Yard) | | | | | | | | | | | | | ■ | | |
| | | STRUCTURAL GLAZING / LOUVERS | | | | | | | | | | | ■ | ■ | ■ | ■ | |
| c.1 | FIRE Fighting -Internal | Sprinkler Line | | | | | | | | | | | | | | | |
| | | Hydrant Line | | | | | | | | | | | | | | | |
| | | Valve Fixing | | | | | | | | | | | | | | | |
| | | Hose Cabinet / Reel Fixing | | | | | | | | | | | | | | | |
| | | Fire Extinguisher Fixing | | | | | | | | | | | | | | | |
| d.1 | EELECTRICAL | | | | | | | | | | | | | | | | |
| | Cable Tray | Tray Fixing | | | | | | | | | | | | | | | |
| | | Cable Laying | | | | | | | | | | | | | | | |
| E | INTERNAL PLUMBING | Water Supply Down take Front Elevation | | | | | | | | | ■ | ■ | | | | | |
| | | Water Supply Down take Right Side Elevation | | | | | | ■ | ■ | | | | | | | | |
| | | Water Supply Down take Rear Side Elevation | | | | | | ■ | ■ | | | | | | | | |
| | | Drainage Down take Front Elevation | | | | | | | | | ■ | ■ | | | | | |
| | | Drainage Down take Right Side Elevation | | | | | | ■ | ■ | | | | | | | | |
| | | Drainage Down take Rear Side Elevation | | | | | | ■ | ■ | | | | | | | | |

Among the several regular improvement measures, the following five key actions were undertaken for this project, as detailed below:

- **Addressing delays in approvals and deviations**

This project faced numerous changes, with the budget estimated to deviate by over 35–40% due to client-driven adjustments in facilities, basic specifications, and BoQ calculations. Approval delays from the PWD and the client had already caused significant setbacks and risked further delays.

To resolve this, the focus shifted to securing approvals for all changes, including drawings and material costs. Collaborative planning system (CPS) activities were paused, as it was futile to plan under uncertain circumstances. A senior-level team comprising representatives from the client, PWD, and PMO was formed. Three meetings were held to finalise and approve deviations. This process took about a month but resolved most bottlenecks, reassuring contractors that future payments would not face disruptions due to pending approvals or changes.

- **Ensuring material availability**

Following the approval of changes, the next priority was finalising orders and ensuring the supply of long lead-time and bulk items. A review team was formed, including PMO representatives, the project planner, the procurement team, a Lean expert, and key suppliers. A master list of critical materials, categorised by location, was prepared. Weekly reviews were conducted to monitor procurement and supply status, and any constraints were promptly assigned for resolution.

This process significantly improved material availability, which had previously been a major factor in project delays. Table 3.12 illustrates such a planner for better understanding.

**Table 3.12: The status of the long Lean Time Items and Bulk Items supply.**

**650 Beded TeachingHospital, Balangir, Odisha**

| Sl no | Package | UOM | Scope Quantity | Scope of work | | | | Drawing & Spec Tender | Finalized supplier & PO | Delivery time | Installation | T&C | Total Days | Suply | Supply | Installation | Installation | Remarks |
|---|---|---|---|---|---|---|---|---|---|---|---|---|---|---|---|---|---|---|
| | Long lead items | | | ITL | R&B | OSMCL | Others | | | | | | | Start | Ending | Start | Ending | |
| C-A | **Civil** | | | | | | | | | | | | | | | | | |
| 1 | Door Frame | | | ✓ | | | | | ✓ | 30 | 30 | | | 25th MaY | 08-Jun | 25th June | 30th June | |
| 2 | Doorn Shutter | | | ✓ | | | | | 10th May | 45 | 30 | | | 10TH June | 15th June | 10th July | 15th July | |
| 3 | Windows | | | ✓ | | | | | ✓ | 45 | 30 | | | 15th May | 15th June | 15th June | 15th July | |
| 4 | False Ceiling | | | ✓ | | | | | 10th May | 45 | 45 | | | 10th June | 25th June | 25th July | 10th Aug | |
| EL-A | **ESS Building** | | | | | | | | | | | | | | | | | |
| 1 | LT Panel | Nos | 64 | ✓ | | | ✓ | ✓ | ✓ | 45 | 15 | 15 | 75 | 25th MaY | 25th MaY | 10th June | 25th June | |
| 2 | DG 1500 KVA | Nos | 3 | ✓ | | | | | 10th May | 60 | 30 | 10 | 10 | | 10th July | 10th Aug | 15th | |
| EL-B | **Illuminatition system** | | | | | | | | | | | | | | | | | |
| 1 | PDB | Nos | 281 | ✓ | | | ✓ | ✓ | ✓ | 10 | 15 | 5 | 30 | 5th Feb | 10th Feb | 3rd March | 15th May | |
| 2 | Light Fixtures | Nos | 13014 | ✓ | | | ✓ | ✓ | ✓ | 35 | 30 | 10 | 75 | 5th June | 5th June | 6th June | 20th June | |
| 3 | Nurse call system | Nos | 170 | ✓ | | | x | | | | | | | | | | | |
| 4 | Rising mains | Set | 60 | ✓ | | | ✓ | ✓ | ✓ | 70 | 15 | 5 | 90 | 5th June | 5th June | 8th June | 18th June | |
| 5 | UPS System | Set | 7 | ✓ | | | ✓ | ✓ | ✓ | 35 | 20 | 5 | 60 | 30th May | 30th May | 2nd June | 12TH June | |
| EL-C | **Solar Power Paint System** | | | | | | | | | | | | | | | | | |
| 1 | Solar Power Paint System | Lot | 1 | ✓ | | | | | | | | | | | | | | |
| El-D | **Solar Water Heating system** | | | | | | | | | | | | | | | | | |
| 1 | Hot water pump | Set | 4 | ✓ | | | | | | | | | | | | | | |
| EL-E | **ELV System** | | | | | | | | | | | | | | | | | |
| 1 | Data 24 port switches/ Accessories | Nos | 1786 | ✓ | | | ✓ | ✓ | ✓ | 30 | 10 | 5 | 45 | 30th May | 30th May | 1st June | 15st June | |
| 2 | CCTV | Nos | 1857 | ✓ | | | ✓ | ✓ | ✓ | 35 | 15 | 10 | 60 | 5th April | | 1st June | 20st June | |
| 3 | CCTV Devices (rack and switches) | Lot | 59 | ✓ | | | ✓ | ✓ | ✓ | 35 | 10 | 10 | 55 | 30th May | 30th May | 1st June | 15st June | |
| 4 | Projectors for Demo room & Lecture Theatre | Lot | | | | ✓ | | | | | | | | | | | | |
| 5 | Internet Connection | Lot | | | | ✓ | ✓ | ✓ | ✓ | 30 | 15 | 10 | 55 | | | | | |
| 6 | Telephone Connection | Lot | | | | ✓ | ✓ | ✓ | ✓ | 30 | 15 | 10 | 55 | 5th April | 5th April | 1st June | 15th June | |
| EL-F | **EXTERNAL ELECTRICAL POWER** | | | | | | | | | | | | | | | | | |
| 1 | External HT Electrical Power | Lot | 1 | | | ✓ | | | | | | | | | | | | |
| 2 | Transformer | Nos | 4 | | | ✓ | | | | | | | | | | | | |
| ME-A | **Fire Fighting system** | | | | | | | | | | | | | | | | | |
| 1 | Fire motor driven main pump | Nos | 2 | ✓ | | | ✓ | ✓ | ✓ | 25 | 10 | 25 | 60 | 30th May | 30th May | 1st June | 15th June | |
| 2 | Fire diesel engine driven main pump | Nos | 1 | ✓ | | | ✓ | ✓ | ✓ | 25 | 10 | 25 | 60 | 30th May | 30th May | 1st June | 15th June | |
| 3 | Jockey pump | Nos | 1 | ✓ | | | ✓ | ✓ | ✓ | 25 | 10 | 25 | 60 | 30th May | 30th May | 1st June | 15th June | |
| 4 | Gas Suppression system | Lot | 1 | ✓ | | | ✓ | ✓ | ✓ | 15 | 20 | 10 | 45 | 30th May | 30th May | 1st June | 25th June | |
| ME-B | **HVACS System** | | | | | | | | | | | | | | | | | |
| 1 | Axial Fan | Nos | 127 | ✓ | | | ✓ | ✓ | ✓ | 30 | 25 | 15 | 70 | 2nd May | 2rd May | 3rd May | 28th May | |

- **Single and similar collaborative plan for each of the 48 floors**

Since one floor of one section of the hospital was the repetitive unit—repeated 48 times across the entire project—a master Gantt chart was prepared for this, showing the floor-wise handover of various work-package activities to the next subcontractor. This master plan was discussed with the contractors' team and then replicated for all six floors of each of the eight sections of the hospital.

A monthly plan was created for each section, and floor-wise handover schedules were agreed among the subcontractors. The monthly review also recorded the progress made

and constraints encountered in following the plan. The list of constraints was reviewed each month, and appropriate actions were taken to remove them. A sample floor-wise plan for one of the sections is shown in figure 3.13. Sample tables for constraints are also presented in Tables 3.14 and 3.15. This structured planning significantly helped the team accelerate project progress.

- **Debottlenecking the main bottleneck operation**

The fourth major bottleneck was in the final finishing stages—specifically for false ceiling installation and subsequent lighting fixture installation. A single subcontractor was hired for this critical operation, responsible for executing the false ceiling work across all eight sections of the hospital.

Engaging an alternate contractor at this stage was not feasible, as the false ceiling forms a highly visible element of the final finish. Poor workmanship would have been immediately noticeable and could lead to client dissatisfaction. To ensure consistent manpower capacity, minimise errors, and avoid delays due to rework, the following three steps were taken:

## Table 3.13 Project Construction Monthly Plan and Review (Sample for a Section 2)

**PROGRAM For Hospital, Balangir**

Legend: Completed / WIP / Not Started

Section 2 — Jan,23 | Feb,23 | Mar,23 | Apr,23 | May,23 | June,23 | July,23 | Aug,23 (W1 W2 W3 W4 for each month)

| S no. | | Activity |
|---|---|---|
| A | STRUCTURE 5TH FLOOR | Gallary Slab |
| | | ROOM |
| | | LMR |
| B | FINISHES | |
| B.1 | Brickwork | GROUND FLOOR |
| | | FIRST FLOOR |
| | | SECOND FLOOR |
| | | THRID FLOOR |
| | | FOURTH FLOOR |
| | | FITH FLOOR |
| | | ABOVE TERREACE |
| B.2 | Internal Plaster | GROUND FLOOR |
| | | FIRST FLOOR |
| | | SECOND FLOOR |
| | | THRID FLOOR |
| | | FOURTH FLOOR |
| | | FITH FLOOR |
| | | ABOVE TERREACE |
| B.3 | Puttty - 1st coat | GROUND FLOOR |
| | | FIRST FLOOR |
| | | SECOND FLOOR |
| | | THRID FLOOR |
| | | FOURTH FLOOR |
| | | FITH FLOOR |
| | | ABOVE TERREACE |
| B.4 | External Plaster | GROUND FLOOR |
| | | FIRST FLOOR |
| | | SECOND FLOOR |
| | | THRID FLOOR |
| | | FOURTH FLOOR |
| | | FITH FLOOR |
| | | ABOVE TERREACE |

## Table 3.14 Monthly Constraints and Review (Sample for a Building Section-7)

**NINE Weeks Look Ahead Plan -Construction : date 7th Feb 2023**

| Location Section | Trade | Main Activity | Milestone (Stone)- Next 2 months | Start | End | Constraints: FINAL DRG/ APPROVAL | Constraints: WORK FRONT (from PREVIOUS Contractors); Date | Constraints: ALL MATERIALS (as per BoQ for work) | Constraints: TOTAL Required MANPOWER | Constraints: Safety/ Work-space/ Holidays | Constraint Details - Action to be Taken | Action by Whom | Target to Remove Constraint |
|---|---|---|---|---|---|---|---|---|---|---|---|---|---|
| NINE Weeks Look Ahead Plan -Construction : date 7th Feb 2023 | | | | | | | | | | | | | |
| 1 | Civil | TILES WORK | TOILETS OF FLOOR - 1ST,2ND,3RD,4TH,5TH | | | | x | | | | WATER PROOFING AGENCY TO BE FINALISED | ASHISH KARA | 1ST WEEK OF MARCH |
| 2 | Civil | DOOR, WINDOW | ALL ROOMS DOOR & WINDOW | | | | x | | | | AGENCY NOT FINALISED | ASHISH KARA | 15TH OF FEB |
| 3 | Civil | FALSE CEILING | ALL ROOMS AND CORRIDOR | | | | x | | | | AGENCY NOT FINALISED, MATERIAL NOT AVAILABLE | ASHISH KARA | 1ST MARCH |
| 4 | Civil | GRANITE AND KOTA | STORE ROOM, AHU, STAIRCASE | | | | | x | | | MATERIAL NOT AVAILABLE | ASHISH KARA | 10TH FEB |
| 5 | HVAC | LOUVRES | LOUVRES FIXING | | | | | x | | | LOUVRES TO BE SUPPLY | EKAGO | 15.3.2023 |
| 6 | HVAC | INLINE FAN | INLINE FAN FIXING | | | | x | | | | INLINE FAN TDS TO BE APPROVE | MR. NARENDRA SHARMA | 10.3.2023 |
| 7 | HVAC | DUCT FIXING | | | | | x | | | | WALL OPENING TO BE REQUIRED | ANJAN & LILIT | 8.2.2023 |
| 8 | Electrical | NURSE CALL | DRAWING APPROVAL PENDING | | | | x | | | | DRAWING APPROVAL PENDING BY CONSULTANT | MR. NARENDRA SHARMA | 10.2.2023 |
| 9 | PH | DRAWING | GROUND FLOOR TO TERRACE FLOOR | | | | | | | | WATER PROOFING AGENCY TO BE FINALISED | ASHISH KARA | 15.2.2023 |

- **The total operation of false ceiling was broken into four activities:**

Grouting, framing, tile fixing, and finishing. A floor-wise, detailed daily plan (in the form of a Gantt chart) was prepared and displayed on each floor. A daily round by one of the planners from the PMO was conducted to ensure that the required materials, manpower, and tools were available at the planned floors. A sample plan is shown in Table 3.15 below.

**Table 3.15 The False Ceiling schedule for all the six floors of all the Eight Segments (also called Pours)**

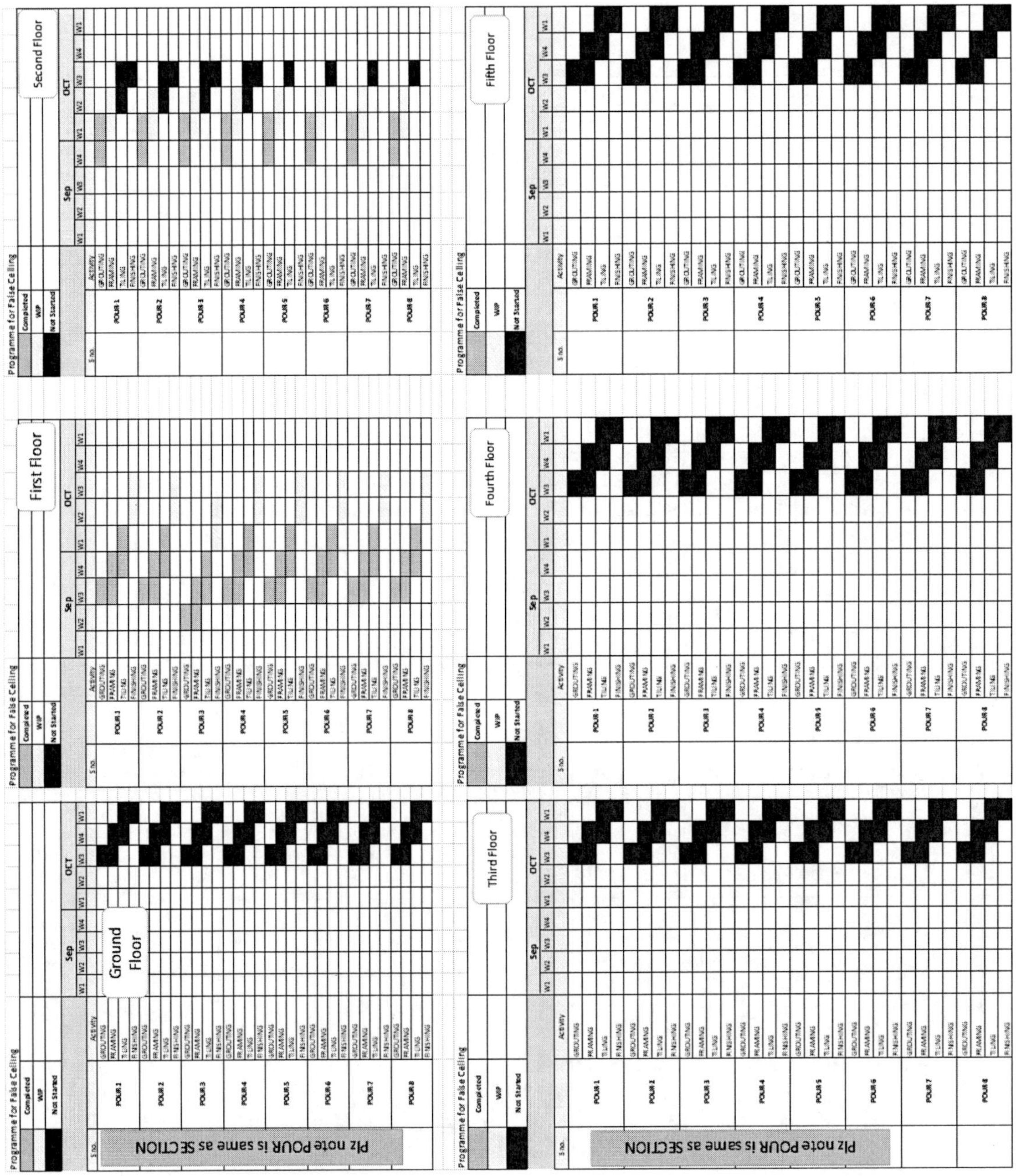

## Table 3.16 Template used for Boulder level Project Review & Monitoring

| Construction of Proposed Multipurposev Hall(MPH) at XYZ Location | | | | | | | | | | | | | | | | | | | | | | | | | | | | | | | | |
|---|---|---|---|---|---|---|---|---|---|---|---|---|---|---|---|---|---|---|---|---|---|---|---|---|---|---|---|---|---|---|---|---|
| | Name of the Main Contractor/Agency | ABC | | | | | | | | | | | | | | | | | | | | | | | | | | | | | | |
| | Agreement Value: | Rs.6,62,88,419,490 | | | | | | | | | | | | | | | | | | | | | | | | | | | | | | |
| Sl No | Description of items | Quantity As Per BOQ | Amount in Rs. as per | May-22 | Jun-22 | | | | Jul-22 | | | | Aug-22 | | | | Sep-22 | | | | Oct-22 | | | | Nov-22 | | | | Dec-22 | | | |
| 1 | CIVIL WORK (BUILDING) | | | WK4 | WK1 | WK2 | WK3 | WK4 | WK1 | WK2 | WK3 | WK4 | WK1 | WK2 | WK3 | WK4 | WK1 | WK2 | WK3 | WK4 | WK1 | WK2 | WK3 | WK4 | WK1 | WK2 | WK3 | WK4 | WK1 | WK2 | WK3 | WK4 |
| A) | Site Clearing | 3660.11 SQM | XYZ | | | | | | | | | | | | | | | | | | | | | | | | | | | | | |
| B) | Earth work in Excavation | 2727.87 CUM | XYZ | | | | | | | | | | | | | | | | | | | | | | | | | | | | | |
| C) | PCC Work | 280.47 CUM | XYZ | | | | | | | | | | | | | | | | | | | | | | | | | | | | | |
| D) | RCC WORK | | XYZ | | | | | | | | | | | | | | | | | | | | | | | | | | | | | |
| i) | GROUND FLOOR (Including Footing, Plinth Beam) | 471.76 CUM | XYZ | | | | | | | | | | | | | | | | | | | | | | | | | | | | | |
| ii) | GROUND FLOOR (Col, Beam, Slab, Lintel, Chajja) | 280.22 CUM | XYZ | | | | | | | | | | | | | | | | | | | | | | | | | | | | | |
| iii) | FIRST FLOOR (Col, Beam, Slab, Lintel, Chajja) | 153.21 CUM | XYZ | | | | | | | | | | | | | | | | | | | | | | | | | | | | | |
| iv) | SECOND FLOOR (Col, Beam, Slab, Lintel, Chajja) | 133.28 CUM | XYZ | | | | | | | | | | | | | | | | | | | | | | | | | | | | | |
| v) | TERRACE FLOOR (Col, Beam, Slab, Lintel, Chajja) | 99.57 CUM | XYZ | | | | | | | | | | | | | | | | | | | | | | | | | | | | | |
| E) | BRICK WORK | | XYZ | | | | | | | | | | | | | | | | | | | | | | | | | | | | | |
| i) | FOUNDATION & PLINTH | 21.52 CUM | XYZ | | | | | | | | | | | | | | | | | | | | | | | | | | | | | |
| ii) | GROUND FLOOR | 1827.48 CUM | XYZ | | | | | | | | | | | | | | | | | | | | | | | | | | | | | |
| iii) | FIRST FLOOR | 173.33 CUM | XYZ | | | | | | | | | | | | | | | | | | | | | | | | | | | | | |
| iv) | SECOND FLOOR | 117.38 CUM | XYZ | | | | | | | | | | | | | | | | | | | | | | | | | | | | | |
| v) | TERRACE FLOOR | 71.98 CUM | XYZ | | | | | | | | | | | | | | | | | | | | | | | | | | | | | |
| F) | PLINTH PROTECTION | 97.50 SQM | XYZ | | | | | | | | | | | | | | | | | | | | | | | | | | | | | |
| G) | ROOF TREATMENT | | XYZ | | | | | | | | | | | | | | | | | | | | | | | | | | | | | |
| H) | TERRACE WATER PROOFING | 603.11 SQM | XYZ | | | | | | | | | | | | | | | | | | | | | | | | | | | | | |
| I) | 6mm Thick Plaster (Ceiling) | | XYZ | | | | | | | | | | | | | | | | | | | | | | | | | | | | | |
| i) | Ground Floor | 730.31 SQM | XYZ | | | | | | | | | | | | | | | | | | | | | | | | | | | | | |
| ii) | First Floor | 610.99 SQM | XYZ | | | | | | | | | | | | | | | | | | | | | | | | | | | | | |
| iii) | Second Floor | 621.84 SQM | XYZ | | | | | | | | | | | | | | | | | | | | | | | | | | | | | |
| iv) | Terrace Floor | 432.39 SQM | XYZ | | | | | | | | | | | | | | | | | | | | | | | | | | | | | |
| J) | 12mm Thick Plaster (Inside) | | XYZ | | | | | | | | | | | | | | | | | | | | | | | | | | | | | |
| i) | Ground Floor | 274.02 SQM | XYZ | | | | | | | | | | | | | | | | | | | | | | | | | | | | | |
| ii) | First Floor | 68.48 SQM | XYZ | | | | | | | | | | | | | | | | | | | | | | | | | | | | | |
| iii) | Second Floor | 58.45 SQM | XYZ | | | | | | | | | | | | | | | | | | | | | | | | | | | | | |
| iv) | Terrace Floor | 98.85 SQM | XYZ | | | | | | | | | | | | | | | | | | | | | | | | | | | | | |
| K) | 18mm Thick Plaster (Outside) | | XYZ | | | | | | | | | | | | | | | | | | | | | | | | | | | | | |
| i) | Ground Floor | 1982.89 SQM | XYZ | | | | | | | | | | | | | | | | | | | | | | | | | | | | | |
| ii) | First Floor | 638.80 SQM | XYZ | | | | | | | | | | | | | | | | | | | | | | | | | | | | | |
| iii) | Second Floor | 451.87 SQM | XYZ | | | | | | | | | | | | | | | | | | | | | | | | | | | | | |
| iv) | Terrace Floor | 970.37 SQM | XYZ | | | | | | | | | | | | | | | | | | | | | | | | | | | | | |
| L) | Putty With Paint (Interior) | | XYZ | | | | | | | | | | | | | | | | | | | | | | | | | | | | | |
| i) | Ground Floor | 3427.45 SQM | XYZ | | | | | | | | | | | | | | | | | | | | | | | | | | | | | |
| ii) | First Floor | 2298.37 SQM | XYZ | | | | | | | | | | | | | | | | | | | | | | | | | | | | | |
| iii) | Second Floor | 2054.93 SQM | XYZ | | | | | | | | | | | | | | | | | | | | | | | | | | | | | |
| iv) | Terrace Floor | 1501.61 SQM | XYZ | | | | | | | | | | | | | | | | | | | | | | | | | | | | | |
| M) | Exterior Paint | | XYZ | | | | | | | | | | | | | | | | | | | | | | | | | | | | | |
| i) | Ground Floor | 1910.26 SQM | XYZ | | | | | | | | | | | | | | | | | | | | | | | | | | | | | |
| ii) | First Floor | 638.80 SQM | XYZ | | | | | | | | | | | | | | | | | | | | | | | | | | | | | |
| iii) | Second Floor | 451.87 SQM | XYZ | | | | | | | | | | | | | | | | | | | | | | | | | | | | | |
| iv) | Terrace Floor | 1316.62 SQM | XYZ | | | | | | | | | | | | | | | | | | | | | | | | | | | | | |
| N) | False Ceiling | 21.27 SQM | XYZ | | | | | | | | | | | | | | | | | | | | | | | | | | | | | |
| O) | Flooring | | XYZ | | | | | | | | | | | | | | | | | | | | | | | | | | | | | |
| P) | (Floor Tiles) (600x600) & Wall Tiles | 1081.31 SQM | XYZ | | | | | | | | | | | | | | | | | | | | | | | | | | | | | |
| Q) | Granite Work | 227.93 SQM | XYZ | | | | | | | | | | | | | | | | | | | | | | | | | | | | | |
| R) | Wooden Flooring | 898.80 SQM | XYZ | | | | | | | | | | | | | | | | | | | | | | | | | | | | | |
| S) | Supplying and installation of 4.5mm thick (BWF Certi | 1089.28 SQM | XYZ | | | | | | | | | | | | | | | | | | | | | | | | | | | | | |
| T) | Supply of 7.5 mm thick vinyl sports flooring | 344.65 SQM | XYZ | | | | | | | | | | | | | | | | | | | | | | | | | | | | | |
| U) | Providing & installation of Safelock 25 mm interlocki | 91.65 SQM | XYZ | | | | | | | | | | | | | | | | | | | | | | | | | | | | | |
| V) | Supply and Installation of 80m thick weight lifting | 2 Sets | XYZ | | | | | | | | | | | | | | | | | | | | | | | | | | | | | |
| W) | VDF Flooring | 898.80 SQM | XYZ | | | | | | | | | | | | | | | | | | | | | | | | | | | | | |
| X) | AS Flooring | 190.48 SQM | XYZ | | | | | | | | | | | | | | | | | | | | | | | | | | | | | |

**Manage the risk better at an early stage:** A major bottleneck occurred during the final finishing stage, particularly in the false ceiling operations. Certain slow activities were identified and improved through joint discussions with the site team. The changes made and the resulting reduced durations were incorporated into the schedule shown above. One such brainstorming session is illustrated in Figure 3.14.

**Figure 3.14 Brainstorming to reduce the activity duration and making collaborative plan for achieving that**

The above measures, among others, ensured that the project was completed on time. In fact, out of the four identical 650-bed hospitals, this project—by embracing these Lean principles—was completed almost three months ahead of the others.

# CHAPTER 4

# ENHANCING VALUE AND PROFIT THROUGH WASTE ELIMINATION

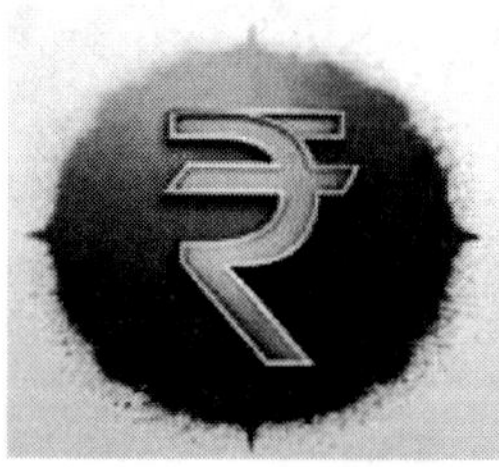

# 4. Enhancing 'Value & Profits'

Construction projects succeed when they deliver measurable 'value' to every stakeholder. Lean construction achieves this by aligning objectives with waste elimination, ensuring 'profits' and performance go hand in hand.

## 4.1 Need and benefits of Construction Waste Hunting & Elimination

In Lean construction, identifying stakeholder 'value' and eliminating wastes is paramount. Projects generate large amounts of tangible and intangible waste across labour, machinery, materials, and money. Minimising these wastes is essential for faster progress, cost-effectiveness, higher quality, improved safety, sustainability, and greater stakeholder value.

Identifying such wastes presents several challenges due to the complex and dynamic nature of the construction industry. Some of the key difficulties include:

- **Inability to define value clearly for various stakeholders, especially customers:** Construction projects often fail to articulate the value customers expect from their investments. Some values are tangible (e.g. cost or time savings), while others are intangible (e.g. user convenience).
- **Variability**: Construction projects typically involve architects and several engineering, procurement and construction consultants, contractors, sub-contractors, and suppliers. Changing client requirements, along with external influences such as design trends, new technologies, and emerging materials, result in high variability across construction and business processes. This variability often causes wastage in terms of delays, cost overruns, unsatisfied customers, and material losses. Identifying such variability is essential to reducing waste and improving value.
- **Non-standardisation**: Unlike manufacturing processes, construction work is rarely standardised. Projects vary by design, site conditions, and client preferences. This lack of uniformity makes it difficult to establish universal waste categories or measurement criteria. However, some attempts to categorise wastes are explained in the next section.
- **Invisible waste**: Some types of waste—such as waiting time for resources, excessive worker movement, or over-processing of business procedures—are not immediately visible like material waste. 'Hunting' these hidden inefficiencies requires a deeper understanding of project operations.
- **Data availability**: Many construction projects lack reliable data collection mechanisms, making it difficult to analyse waste accurately and in a timely manner.

- **Multi-layered activities**: Numerous concurrent activities make it hard to isolate individual sources of waste. For example, plastering a room, installing concealed wiring, pulling wires, and sealing conduits may happen simultaneously—often by different subcontractors—making waste identification complex.
- **Perception of waste**: Stakeholders often differ in their understanding of what constitutes waste. This variation makes it difficult to build consensus around what needs to be eliminated. The term 'waste hunting' is used deliberately here, as many intangible wastes go unnoticed.
- **Time constraints**: Tight project schedules often limit the time available for properly identifying and analysing waste.

Despite these challenges, the benefits of identifying and eliminating waste in construction projects through Lean principles make the effort worthwhile. This investment can lead to the following advantages:

- **Increased productivity**: Lean construction principles of waste hunting and elimination promote better utilisation of workmen and machines by reducing idle time. By eliminating waste, teams can focus more on value-adding activities, thereby improving overall productivity.
- **Improved resource management**: Lean construction methodologies emphasise efficient allocation and synchronisation of resources. Eliminating waste allows for better management of materials, labour, and equipment, ensuring optimal utilisation.
- **Enhanced quality**: Poor design or workmanship often leads to errors, rework, and defects. By addressing these issues early, the quality of construction improves, reducing the need for corrective action and saving both time and cost.
- **Cost savings**: Waste in construction—such as underutilised labour or equipment—can significantly inflate project costs. Lean methods help identify inefficiencies and introduce corrective measures, resulting in substantial cost savings and improved profitability for all parties.
- **Safer work environment**: A cluttered or disorganised site increases the risk of accidents. Eliminating waste contributes to a cleaner, more orderly workspace, reducing safety hazards and promoting a healthier work environment. It also supports better workflow and fewer unnecessary delays, leading to improved timelines and reduced costs.
- **Environmental sustainability**: Construction waste frequently ends up in landfills, contributing to pollution. Eliminating avoidable waste minimises environmental impact and supports sustainable practices. For example, in one large gated community project with apartment towers and villas, all debris was reused within the site—for landscaping and other purposes—with nothing transported outside the project boundary.

- **Stakeholder satisfaction**: Clients and stakeholders value projects that are efficient, cost-effective, and environmentally responsible. By adopting waste elimination practices, construction firms can build trust, enhance their brand, and strengthen their market position.
- **Long-term business success**: Implementing Lean practices fosters a culture of continuous improvement. Organisations that prioritise waste reduction not only improve their current project outcomes but also position themselves for sustained success in a competitive industry.

## 4.2 Eight Type of Common Construction Wastes:

Lean construction broadly categorises waste into three major types:

- **Muda (waste):** This represents all forms of non-value-adding activities in the construction process that consume effort and money but provide no benefit to the customer. These are further classified into eight types, which are explained in the next section.
- **Mura (unevenness):** This refers to irregularities and variations in construction workflows, leading to inconsistent performance, imbalanced resource utilisation, and uneven workloads. Such variability causes inefficiencies and reduced productivity.
- **Muri (overburden)**: This describes excessive strain placed on workers, equipment, or processes due to unrealistic expectations or poor resource allocation. Overburdened resources are more prone to errors, fatigue, and performance breakdowns. Eliminating muri is essential for maintaining smooth workflows and preventing worker burnout.

Among these three, the most well-known and widely discussed is Muda, which is explained in greater detail below. Mura and muri are addressed in Chapter 3 on the collaborative planning system and Chapter 6 on Value Stream Mapping. Some examples of these are also illustrated in Figure 4.1.

1. **Waiting / Delay**: This involves time lost due to waiting for various inputs such as the work-front from the previous contractor, drawing clarifications or RFIs, instructions, approvals, materials, skilled labour, equipment, tools, or even payments. Waiting can also arise due to unfavourable weather, unavailable workspace, or local cultural practices (e.g. prayer times or rituals).
2. **Rework / Defect**: Refers to repeated efforts and material waste caused by poor-quality workmanship or material defects. This includes functional or aesthetic defects, rejected work following inspections, and any client-requested changes that lead to scrap or rework.

3. **Inventory**: Unnecessary or excessive inventory causes avoidable storage costs and logistical issues. This includes early material deliveries, rejected materials held onsite, unused tools or surplus items, packaging waste, and damaged or expired stock.
4. **People Motion**: Refers to unnecessary movement of personnel, such as searching for team members, engineers, or tools; walking long distances within large campuses or high-rise buildings to access drawings or equipment.
5. **Transportation**: Any redundant movement of materials, including double handling, shifting items unnecessarily, transporting bulk materials inefficiently, or excessive repositioning within storage areas, constitutes this waste.
6. **Overproduction**: This difficult-to-detect waste includes constructing or ordering more than required at a given time. Examples include producing work not yet needed by downstream stakeholders, surplus material orders, or activities for which the client does not recognise value or payment.
7. **Overprocessing**: Non-value-adding activities such as redundant supervision, frequent design iterations, or inefficient approvals fall into this category. It also includes under-capacity equipment usage, duplicated reporting, or bureaucratic delays caused by excessive sign-offs.
8. **Unutilised Capability**: This waste refers to not leveraging the experience, creativity, or decision-making abilities of available professionals—engineers, designers, foremen, or tradespeople—leading to missed opportunities for improvement.

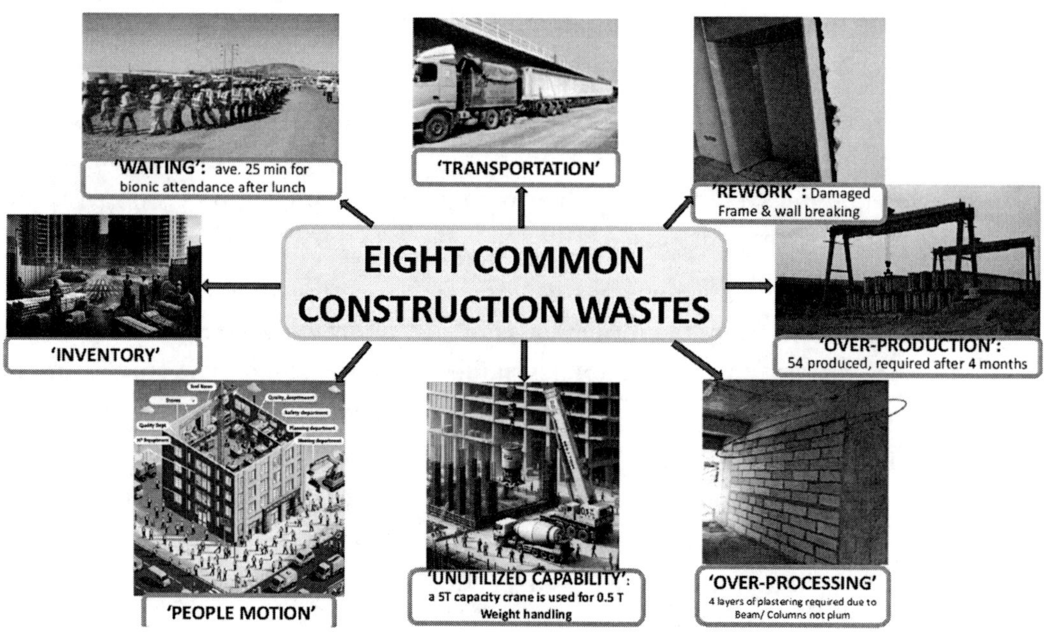


**Figure 4.1 Examples of Eight Common Wastes in Construction**

### 4.2.1 Need for Waste Elimination

Various studies (by the Construction Industry Institute, Lean Construction Institute, IIT Madras, and McGraw Hill Research) estimate that 10–30% of construction costs and 20–50% of project durations are waste, adding no real value to customers. At least 30–80% of these wastes can be prevented and converted into value for stakeholders, from clients to consultants and contractors. A conscious effort is therefore needed to identify and minimise wasteful costs and efforts. This chapter explains, beyond CPS, what else can be done to save project costs and shorten completion times significantly.

### 4.2.2 Process for Waste Hunting or Waste Identification:

For developing the culture of Lean ways of working, it is better to act proactively on reducing the wastes of project delays, project costs, and prevent rework. Some of these triggers can come from CPS. It should be developed as a routine matter, like quality checks. However, it can also be done when future bottlenecks are expected or there are some frequent causes of execution variance or low PPC. Following steps are recommended to assure you that your team's initiatives have been successful.

- Conduct an internal survey to identify the value improvement and waste identification and elimination opportunities. A survey template is given in Annexure-I (Chapter 4) as an illustration. The annexure also depicts the results of such a survey.
- This survey is helpful in identifying the opportunities to target high-value wastes, leveraging the working people's opinion and their past experiences. On such high-opportunity areas, you should try to improve the construction and business processes. This should be done, preferably in the 'big room' workshop, with advance notice to the project teams. A full Chapter 5 is written on the concept and benefits of a 'big room', because this is one of the most important recommendations of Lean philosophy for better cooperation, collaboration, and communication. The following steps are recommended for this:
    - Finalise the team to be invited for this survey to identify the opportunities with significant wastes in the project. Some people can also be surveyed via email or other virtual means to cover a wider segment of the organisation.
    - Review, customise, and finalise the survey templates if required.
    - Explain the purpose and the template, and conduct the survey.
    - Analyse results, and deliberate on the wastes and issues identified.
    - Prioritise the top 10 wastes or project issues and make plans to eliminate these wastes as the project progresses.

- Form separate teams for resource-wise different wastes (like material waste, manpower waiting waste etc.). This focused effort will provide better benefits.
- It is important to involve many members in this, as everyone's contribution would help to cover eight different types of construction wastes, as also shown in Fig 4.1. Therefore, choose one leader for each type of the potential waste. Some possible team compositions, with suggested leaders in brackets, are recommended below:
    - **Team 1 –** Focus to minimise 'delays' in construction processes (site manager)
    - **Team 2 –** Focus to minimise 'overproduction' and 'overprocessing' (planning manager)
    - **Team 3 –** Focus to minimise 'rework' (quality manager)
    - **Team 4 –** Focus to minimise 'procurement' delays and internal transportation (procurement manager)
    - **Team 5 –** Focus to utilise unutilised 'capabilities' of the people and equipment (HR manager)
- Get the concurrence of the teams formed and the team leaders. Some possible team members are suggested below:
    - **Team 1:** project manager, site manager, admin manager, main contractor's senior representative, safety manager, estate manager (if any)
    - **Team 2:** site manager, shift engineer, site engineers, planning manager
    - **Team 3:** QC manager, QC engineers, site manager
    - **Team 4:** procurement manager, site manager, site engineers
    - **Team 5:** HR manager, project manager, site manager
- Conduct the 'waste observation walks' or 'Gemba walks' with the team with a very critical eye. During Gemba walk, ensure all the safety precautions are followed. Keep all the facilities required (like template, pen, paper, mobile phone etc.) for recording observations, texting, taking pictures, videos, audios etc. Follow the steps mentioned below:
    - Visit the workplace site for the observations
    - Explain the purpose of the visit to the people working there
    - Do everything necessary to gain their confidence, that you are there for their betterment and not for pointing out their weaknesses
    - Ask, observe, and listen to them and share the team's theme or goals
    - Ask the right questions based on your theme
    - Listen to their views, problems, and suggestions with empathy

- Listen more than you talk. Appreciate them for their contributions wherever possible
- Observe the workplace site yourself and ensure the accuracy of the information given
- Hunt, with a critical eye, the wastes relevant to your theme
- Record, take pictures, audio, video on your mobile, with their help, especially related to work interruptions due to materials and equipment availability and quality.

A recommended template, which can be studied in advance and used for Gemba walk, is provided in Table 4.1 on the next page.

## Table 4.1: Gemba Walk Template for Identifying Construction Wastes and Improvement Opportunities

| WASTE OBSERVATION Or GEMBA WALK: Some POINTERS | | | |
|---|---|---|---|
| **Questions (for identifying 8 Wastes)** | **Is the MAJOR PROBLEM: yes/no** | **If Yes- What is the Priority: High/Med/Low** | **Remarks/Special Observation** |
| For OVER PRODUCTION waste ASK QUESTIONS like: | | | |
| Its some work being done much ahead of schedule, compared to its succeeding work? | | | |
| Are there too much of not-immediately-required Raw Material/ Semi-finished Inventory? | | | |
| Is the Floor very cluttered with material? | | | |
| Is next day's/ week's demand for their planned work known to people? | | | |
| Are various Plans & Progress information available visually? | | | |
| Is Planning done collaboratively, involving the relevant people? | | | |
| Is Waste handling defined & streamlined? | | | |
| For WAITING waste ASK QUESTIONS like: | | | |
| Are people idle & waiting for Drawings/ Specs/ Instruction? | | | |
| Are people idle & waiting for Materials? | | | |
| Are people idle & waiting for Equipments & Tools? | | | |
| Are people idle & waiting for previous work completions? | | | |
| Are Equipments idle often? (Crane/ Hoist/ Forklifts) | | | |
| Is there list of tasks available to people, which can be completed now? | | | |
| Are there too many testing & Inspections activities involved? | | | |
| For PEOPLE MOTION waste ASK QUESTIONS like: | | | |
| Are people often busy in searching Tools, Equipments, Drawings? | | | |
| Is the path to move around smooth and uncluttered for easy movement? | | | |
| Do people have to wait a lot for morning for Attendance before work starts? | | | |
| Do people have to walk a lot for personal needs like Lunch, Tea, Toilets? | | | |
| Do people have required tools, information nearby? | | | |
| Is there any active formal Five S program running? | | | |
| Can people connect to internal/ external people outside easily (over phones)? | | | |
| Are there frequent unplanned face-to-face meetings? | | | |
| For TRANSPORTATION waste ASK QUESTIONS like: | | | |
| Do you see lots of equipments movement at site? | | | |
| Are Handling equipments enough, as per the people working there? | | | |
| Are Trucks/ Cranes waiting for long for being unloaded? | | | |
| Are too many people involved in movement of materials? | | | |
| Is storage area is quite big? | | | |
| Is material delivered at point of use, in the first instance? | | | |
| Are there too many hands-offs (materials moving to different people)? | | | |
| Is waste disposal well planned and smooth? | | | |
| For OVERPROCESSING waste ASK QUESTIONS like: | | | |
| Are there frequent delays due to several stages of approvals? | | | |
| Is delay in processing felts very often? | | | |
| Are too many clarifications (RFIs) required? | | | |
| Are sorting, testing, inspection of materials done very often? | | | |
| Are resources being asked to/ extra resources deployed but are often idle? | | | |
| Is the work been done first time? | | | |
| Are there many complaints about equipments' capacity? | | | |
| Is materials consumption far more than the BoQ standards? | | | |
| For INVENTORY waste ASK QUESTIONS like: | | | |
| Are there items, which have not moved for more than 1 week/ 1 month? | | | |
| Are the floors and racks fully occupied? | | | |
| Is the WIP Inventory high, occupying lots of space? | | | |
| Are there excess/ surplus materials, lying for more than 1 week? | | | |
| Are there material which are obsolete now? | | | |
| Is there demand for more storage area/ racks/ equipments? | | | |
| For DEFECTS waste ASK QUESTIONS like: | | | |
| Are standard checklists available for all trades? | | | |
| Are standard checklists updated during last one year? | | | |
| Does the equipments have a preventive maintenance schedule? | | | |
| Are there quality measurements for each part of the construction process? | | | |
| Does each trade know their quality rating? | | | |
| Do employees generally have sufficient time to complete the task? | | | |
| Are their any cost effective training programs? | | | |
| For UNUTILIZED Employees' Capability ASK QUESTIONS like: | | | |
| Are you spending most of your time only in ROUTINE activities? | | | |
| Have you given at least one suggestion for improvement during last one week? | | | |
| Did you spend at least one hour during last 1 week in thinking how your work can be done more smartly? | | | |
| Do you write or get more than 30 m | | | |

## 4.3 Process for Waste Elimination or Reduction:

After identifying and hunting the wastes, the collected information must be critically analysed to find opportunities to eliminate or at least reduce the wastes. The following steps are recommended for this process of improvement.

Keep your critical and fundamental questioning capabilities—5W1H and 5 Whys—sharp. 5W1H involves asking why, what, where, who, when, and how for every action. This popular, simple, and very effective questioning technique is depicted in Table 4.4. 5 Whys refers to conducting root cause analysis by asking why at least five times to reach the core issue.

Use these two methods extensively during analysis. Challenge the current state—both the process and the outcome—at a fundamental level. The present situation may be the result of an outdated practice developed under very different conditions, and may no longer be optimal for the current context.

You may refer to the diagram in Figure 4.2 for this analysis approach.

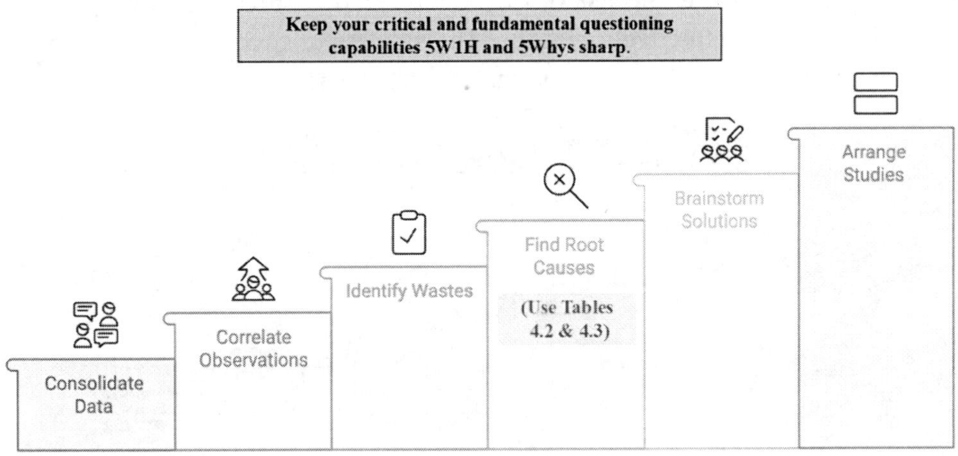


**Figure 4.2: High-level Steps involved – Elimination of Waste**

### 1. Conduct the process improvement study and get the end-users' acceptance for problems and the improvement changes

Ensure the involvement of the site people who perform the work, to study and understand their problems and suggestions. Be open to innovative ideas from the site team as well for the improvements. Sharpen your critical questioning and innovation skills. You can use a common and effective 'A3' one-pager methodology for identifying the improvement opportunities and proposed changes. A simplified version of this A3 methodology is provided in Table 4.5 for quick reference.

- Conduct a detailed kaizen study in the area suggested by the core team
- Propose improvements in the process to reduce duration, cost, or defects, as required
- Discuss the suggestions with the people performing the activities and obtain their concurrence for the changes. Incorporate their suggestions where relevant
- Complete the templates for the study and submit them to the core team members

**2. Get management's approval and budgets to implement the changes. Also, follow up periodically to ensure the change has delivered the intended benefits**

Prepare a cost–benefit report comparing current and proposed practices, focusing on improved value to the customer and other stakeholders. Quantify the tangible benefits and document any possible intangible benefits. Present this to the decision-making and fund-approval teams, and ensure proper implementation.

## 4.4 Key Measures of success for greater Value for all

Such initiatives are successful only when people invest quality time in these activities. Therefore, to effectively measure the success of such initiatives, the following KPIs are recommended, as shown in the diagram below (Figure 4.3):

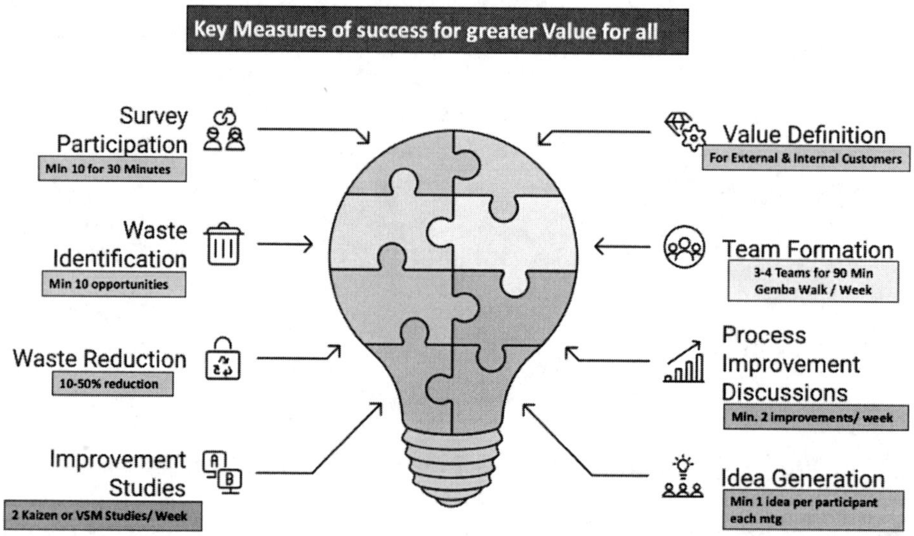


**Figure 4. 3. Key Measures of success of the Value Creation Meetings for all Stakeholders**

During these meetings and discussions, the templates for work sampling, kaizen, and A3 should be used every week. The planned savings and improvements in efficiency must also be discussed and documented by core team members. It is important to note that, during these discussions, the workers are speaking about the management's interest in driving improvements. This reflects a positive sign of trust between both parties.

A summary of the waste identification and elimination process is given below in Tables 4.2 to 4.6 for reference.

**Table 4. 2: Waste Identification and Elimination Process.**

| Step | Waste Identification & Elimination Process | Teams Involved | Deliverables | Key Risks / Issues | Lean Solutions |
|---|---|---|---|---|---|
| 1 | Conduct manual survey to identify value improvement areas and eight types of waste | All relevant teams (physical/virtual participation) | High-potential zones for waste reduction | Incomplete/ Poor team participation | Training, awareness, contractual obligation, HR facilitation |
| 2 | Identify elimination opportunities across teams (e.g., rework, delays, resource misuse) | 2–4 willing teams for value stream mapping | Mapped value streams for key waste areas | Lack of interest; excuses of unavailability | Training, benefit orientation, contractual obligation |
| 3 | Apply 4-R approach: Just Do, Scale Up, Needs-Based, and Preventive actions | Volunteers observing waste hotspots | Documented hotspots of 8 wastes | Uneven participation; inconsistent weightage of similar wastes | Training, contractual obligation, HR support |
| 4 | Conduct structured waste observation using standard templates | Minimum 6 team visits (1.5 hrs each) with observation templates on prediction of waste | Completed observation rounds with predicted waste patterns | Inconsistent team registration; varied interpretation | Training, contractual obligation, HR facilitation |
| 5 | Quantify value improvement opportunities collaboratively | Open team discussions and idea sharing | Consolidated team ideas for waste reduction | Low participation or engagement | Training, contractual obligation, HR support |
| 6 | Conduct process improvement study; identify root causes and corrective actions | Conduct 5W1H sessions; root cause analysis and solution mapping | Improvement proposals with impact assessment | Incomplete participation; unclear problem ownership | Training, contractual obligation, HR facilitation |
| 7 | Obtain management approval and monitor implementation effectiveness | Project team with management endorsement | Documented change implementation and tracking | Lack of awareness , so may throw the templates | Training, contractual obligation, HR support |

**Table 4.3: Symptoms of Few Construction Wastes**

| OVER-PRODUCTION Waste | Root Causes of OVER-PRODUCTION Waste |
|---|---|
| Excessive raw materials | Missing Just-in-case logic in Materials Planning |
| Extra inventory | Misuse of automation at few places |
| Unnecessary work | People staying busy most of the time in ROUTINES |
| Excessive floor space utilized | Long process set-up times |
| Unbalanced material flow | Unleveled scheduling among dependent trades |
| Backups between departments | Unbalanced workload distribution among dependent trades |
| Complex information management | Poor Workmanship having frequent defects |
| Disposal charges | Very high Safety factors considered in equipment/capability selection |
| Extra waste handling & treatment | Redundant inspections |
| High utility costs | Lack of reuse & recycling |
| **TRANSPORTATION Waste** | **Root Causes of TRANSPORTATION Waste** |
| Extra handling | Mis-located materials |
| Surplus handing equipments | Unleveled scheduling |
| Extra inventory | Unfavorable facility layout |
| Extra storage areas | Redundant inspections |
| Over-staffing | Poor work placeorganization & House keepinmg |
| Reduced quality | Poor Managed waste stearms |
| Extra paperwork | Unleveled scheduling among dependent trades |
| Extra hand-offs | Material stored away from point of use |
| Transport of disposal | Poor supply chain management |
| Excessive energy consumption | |
| **OVER-PROCESSING Waste** | **Root Causes of OVER-PROCESSING Waste** |
| Unnecessary work being done | Product changes without process changes |
| Extra equipments used | Practicing Old Processes (it's always been done) |
| Longer lead time | Lack of communication |
| Reduced productivity | Redundant approvals |
| Extra material movement | Extra copies, excess information |
| Frequent sorting, testing, inspection | Undefined customer requirements |
| Inappropriate use of resources | "Temporary" measures that become part of the process |
| Excess energy consumption | Lack of reuse / recycling |
| Processing by-products | |
| **INVENTORY Waste** | **Some Root Causes of INVENTORY Waste** |
| Complex tracking systems | Product complexity |
| Extra storage & handling | Incapable processes |

| | |
|---|---|
| Extra rework / hidden problems | Unleveled scheduling |
| Paperwork / documents | Poor market forecast |
| Stagnated information flow | Unbalanced workload |
| High disposal costs | "Just-in-case" |
| Obsolete material | Unreliable supplier shipments |
| In-process packaging | Inadequate measurement |
| High utility costs | Misunderstood communications |
| | Reward system (wrong metric) |
| Waiting as Waste | Root Causes of Waiting |
| Under-utilization of resources | Unbalanced work load |
| Reduced productivity | Late start of meetings |
| Increase in investment | Unplanned maintenance |
| Idle equipment | Long process set-up times |
| Large waste handling rooms | Misuse of automation |
| Equipment running, not producing | Quality problems |
| Unnecessary testing | Unleveled scheduling |
| | Ineffective layout |
| | Specialization |
| | Batch processing |
| | Things Not at right location |
| Motion of People as Waste | Root Causes of Motion of People |
| Reduced productivity | Poor ergonomics / layout |
| Large reach / walk distances | Machine / process design |
| Longer lead times | Inconsistent work methods |
| Excess handling | Poor workplace organization & housekeeping |
| Reduced quality | Extra "busy" movements while waiting |
| People / machines waiting | |
| High energy cost | |
| Inappropriate use of resources | |

## Table 4.4: Causes of Construction Waste

| CAUSES OF CONSTRUCTION WASTE | | | |
|---|---|---|---|
| GROUP | CAUSES OF CONSTRUCTION WASTE | Is THIS MAJOR CAUSE Yes/No | Priority for SOLVING High/Medium/Low |
| DESIGN | Frequent design changes (including Client's Requirements Change) | | |
| | Frequent Design errors (by designer, or lack of info from dependent Designer) | | |
| | Lack of Design information | | |
| | Poor design quality (Missing Dimensions, Tolerances, Clashes etc.) | | |
| | Slow drawing distribution (different versions floating) | | |
| | Complex design (with poor Constructability) | | |
| | Inexperienced designer (leading to error, missing info etc.) | | |
| | Low interaction among various specialists (like among MEP) | | |
| MATERIALS HANDLING | Shifting of material (not stored at right place first time) | | |
| | Complex material handling (wrong method of handling) | | |
| | Damage during transportation | | |
| | Poor quality of material, needing replacement | | |
| | Delay during delivery | | |
| | Equipment & Tools capacity not appropriate | | |
| WORKER | Worker's mistakes/ Human Errors | | |
| | Incompetent worker/ Lack of Skills & Training | | |
| | Poor attitudes of workers | | |
| | Demotivated Workmen/ Deliberate Mistakes | | |
| | Lack of experience | | |
| | Shortage of skilled worker | | |
| | Inappropriate use of material | | |
| | Poor workmanship | | |
| | Too much overtime for workers | | |
| MANAGEMENT | Poor Planning (not taking care of all dependencies) | | |
| | Poor site management (Supervision, Clarity of Instruction etc.) | | |
| | Inappropriate construction methods | | |
| | Lack of coordination among parties | | |
| | Poor information quality | | |
| | Late information flow among parties | | |
| | Scarcity of equipment | | |
| | Resources adequacy problem | | |
| | Frequent Rework | | |
| | High and frequent waiting time | | |
| | Communication problem | | |
| | Outdated equipments | | |
| | Lack of waste management | | |
| | Lack of environmental and safety awareness | | |
| SITE CONDITION | Excess/ Left over/ Obsolete/ Rejected material on site | | |
| | Poor site condition; Cluttered and Congested | | |
| | High amount of waste generated from packaging (Wood, Packaging papers) | | |
| | No proper space & facilities for Lunch, Tea, Wash-Rooms etc. | | |
| | Poorly Lit and prone to accidents | | |
| | Congested space for working causing Crews interference | | |
| PROCUREMENT | Ordering error (Specs, Quantities, delivery etc.) | | |
| | Errors and delays in shipping | | |
| | Mistakes in quality surveys | | |
| | Ignorance of specifications | | |
| | Rejection after testing, lack of testing facilities, 3rd Party testing | | |
| EXTERNAL | Effect of Weather (Heat, Cold, rains, Storms, Dust etc.) | | |
| | Accidents | | |
| | Pilferage | | |
| | Vandalism | | |
| | Damages caused by third parties | | |
| | Festivities | | |
| | Unpredictable local conditions including political | | |
| | Lack of Law enforcement | | |

**Table 4.5: Eliminating Wastes using 5W1H to improve Value for Stakeholder**

| ***ADDING GREATER VALUE TO CUSTOMERS using 5W1H*** | | | | |
|---|---|---|---|---|
| **5W1H** | **BASIC ISSUE** | **PRIMARY QUESTIONS** | **SECONDARY QUESTIONS** | **BASIC FOCUS TO IMPROVE** |
| **WHAT** | **Purpose ?** | **WHAT** is being done | What **ELSE** can it be done? | |
| **WHY** | Essential? | **WHY** is it being done ? (value to customer/ user) | What **SHOULD** this be done? | |
| | | What implication, if one does not do at all ? | | |
| **WHERE** | **Place ?** | **WHERE** is it being done | Where **ELSE** can it be done? | |
| **WHY** | (dept./Section/ Building/Machine | **WHY** at that place only ? | Where **SHOULD** this be done? | **In order to ELIMINATE/SIMPLIFY/MERGE/AUTOMATE some of the activities/tasks ??????** |
| | | What implication, if one does not do there ? | | |
| **WHEN** | **Sequence?Time?** | **WHEN** is being done | When **ELSE** (or what other sequence) this be done | |
| **WHY** | | **WHY** at that time/ sequence only ? | When **SHOULD** (or which sequence) this be done? | |
| | | What implication, if that person does not do ? | | |
| **WHO** | **Person ?** | **WHO** is doing it? | Hho **ELSE** can do it? | |
| **WHY** | | **WHY** only that person ? | Hho **SHOULD** do it? | |
| | | What implication, if that person does not do ? | | |
| **HOW** | **Method ?** | **HOW** is it being done | How **ELSE** can it be done? | |
| **HHY** | | **WHY** only that way ? | How **SHOULD** can it be done? | Reduce Cost |
| | | What implication, if one does not do that way ? | | Reduce Processing Time |

**Table 4.6: Simplified A3 Template for Construction process Improvement**

| Section | Guiding Prompts | Details |
|---|---|---|
| **Title / Theme** | Clearly state the problem to be solved or the improvement goal to be achieved using this A3. | |
| **Organization Details** | - Organization name | |
| | - Department | |
| | - Date | |
| | - Place | |
| **1. Background (Importance & Context)** | • Identify key stakeholders affected by the problem | |
| | • Describe the value this solution will bring (quantitative in rupee terms and qualitative) | |
| | • Explain the impact if the problem is not addressed | |
| | • Clarify why this issue is important now | |
| **2. Current Situation (Problem Statement & Process Map)** | • Define the specific problem based on stakeholder needs | |
| | • Collect and verify data (not perception) | |
| | • Use facts, visuals, and process maps | |
| | • Note conflicts of interest among stakeholders | |
| **3. Set Target / Goals (Desired Outcome & Success Matrix)** | • Define the desired future state | |
| | • Establish SMART goals (Specific, Measurable, Achievable, Relevant, Time-bound) | |
| | • Benchmark against internal or external standards | |
| | • State clear measures of success | |
| **4. Root Cause Analysis** | • Apply 5 Whys or Fishbone diagrams | |
| | • Identify high-priority constraints (manpower, machines, methods, measurements, collaboration, budget, vendors) | |
| | • Highlight performance gaps between current and desired KPIs | |
| | • Support findings with data or visuals | |
| **5. Countermeasures (Solutions to Meet Goals)** | • Propose actionable solutions (e.g., VSM, Collaborative Planning, Waste Elimination, 5S) | |
| | • Link solutions to goals and root causes | |
| | • Explain how measures reduce waste and improve value | |
| | • Revise goals if needed | |
| **6. Implementation Plan** | • List required activities and timelines | |
| | • Identify resources (manpower, budget, approvals, equipment) | |

| | | |
|---|---|---|
| | • Anticipate risks and mitigation strategies | |
| | • Assign responsibilities | |
| **7. Follow-Up (Monitoring & Control)** | • Define monitoring activities and review frequency | |
| | • Outline reporting mechanisms and success documentation | |
| | • Specify standardization or process modifications | |
| | • Ensure PDCA cycle completion and capture lessons learned | |
| **Team Members (Credits & History)** | • List contributors to the A3 process | |
| | • Include preparation timeline | |
| | • Note roles and responsibilities | |
| | • Capture acknowledgements | |
| | | |

## 4.5 References


1. *Rastogi, S., & Laghate, A. (2021). Quality management of construction in the 21st century: Leveraging Lean philosophy and digital technologies. Innovative Lean Mentoring Academy.*
2. *Rastogi, S. (2015, August). Construction 4.0: New generation construction in a hyper-connected world. The Master Builder.*
3. *Engineers Australia, Western Australia Division. (2012). Recommended practices for the application of Lean construction methods to building new Australian LNG capacity. Perth, Australia: Engineers Australia.*
4. *Rastogi, S. (2015, February 5–6). Elimination of wastes in construction: A case study. In Proceedings of the Indian Lean Construction Conference (ILCC 2015). Mumbai, India.*
5. *Rastogi, S. C. (2015, February). DTBB: Designer test-based build – A new approach to quality assurance. In Proceedings of the National Conference ILCE 2015, Indian Lean Construction Conference (pp. 224–230). Mumbai, India.*

## 4.6 Annexure I: Case Study I Assessment of Construction Wastes & Ways to Eliminate

**The Project:**

SM is a leading global design and build company that specialises in creating innovative and transformative workplaces. With a strong presence in multiple countries, SM offers end-to-end solutions, combining cutting-edge design concepts with seamless construction execution. Their diverse team of architects, interior designers, and project managers collaborates closely with clients to understand their unique needs and vision, ensuring the delivery of bespoke and functional spaces that inspire productivity and foster creativity. They have an impressive portfolio of projects across various industries, including corporate offices, retail spaces, hospitality, and healthcare facilities. While they have been great designers and architects since the beginning, they have improved their build capability to its current level after focusing on Lean Construction 4.0 a few years back.

**The Objectives of this project:**

- The company was globally renowned in design and architecture and had won several global awards in those disciplines till then. However, they also wanted the build capability to be world-class and very client-friendly, so they wanted to find out if Lean Construction could bring that change.
- They wanted to assess the current state of issues related to construction waste, and see the potential for improvement in that area.
- They also wanted to see the impact of eliminating some of those wastes through a three-month pilot project to determine whether the benefits of Lean Construction could actually be achieved.

**Outline of the Methodology followed:**

The following methodology was followed for identifying the top six wastes and the main causes for those top six wastes. A survey of 30 relevant people in SM, representing 15 senior and 15 middle management level people, was conducted through a questionnaire. This questionnaire was preceded by a 60-minute webinar about the Lean Construction concept and the typical wastes. The steps used are briefed below:

- Design a waste assessment survey to cover 15 pre-identified senior executives (heading the regions or business) and 15 middle-level project managers, including a few other relevant people in workplace design and build.
- Conduct a webinar for all of them, explaining the overall Lean Construction concept and the purpose of taking inputs from them. This was related to identifying the top six wastes in workplace build, and also the main causes of

the top wastes. This would enable focused future actions to minimise those, for faster and better completion of projects.

- Request them to complete the survey and give their inputs, based on their long experience, which varied from 8 to 20 years in their domain, and mostly in this company.
- Define the methodology (Fuzzy Function, not explained in this book) to reflect the importance of relative ranking from 1 to 6, in terms of representative weightages.
- Conduct a Pareto analysis of the inputs received from them based on the following parameters:
  - Top executives' inputs
  - Operating-level executives' inputs
- Identify the top 6 wastes and their causes, based on this Pareto analysis.
- Compare the causes from both groups, and also perform some sensitivity analysis by changing the weightages, to ensure correct identification.
- Finalise the top wastes in SM workplace build and how these could be eliminated proactively in their next projects.
- Discuss these top wastes, their causes, and recommended solutions with SM senior leadership and also the core team for the next pilot project for Lean Construction.
- Provide final short-term and long-term recommendations for eliminating or reducing the top wastes.

This is also summarised below in a simple diagram, Figure 4.5, for quick understanding:

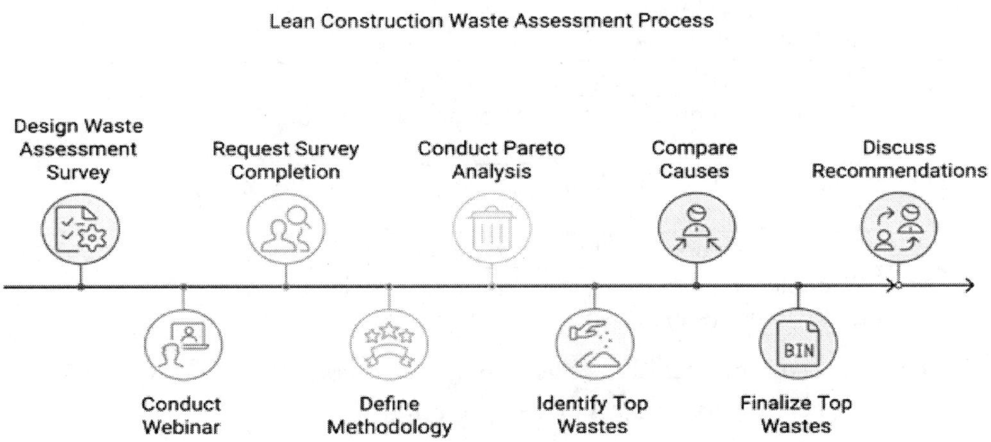


**Figure 4.4: Methodology for Identifying and Analysing Top Wastes in Workplace Build Projects**

## Identification of Top Six Wastes and Their Causes

The Pareto analysis of the top executive group inputs is provided below in Table 4.7.

**Table 4.7: Results of Pareto analysis**

| | Top Leadership Teams Inputs | Middle Level Executives Inputs |
|---|---|---|
| 1 | Rework required due to Poor quality of work (on account of functional or aesthetic reasons) | Delay/ Waiting/ Interruptions (of any kind) |
| 2 | Delay/ Waiting/ Interruptions (of any kind) | Rework required due to Poor quality of work (on account of functional or aesthetic reasons) |
| 3 | Mismatch in interdisciplinary drawing/ specifications/ customer needs | Frequent Design Changes (for any reason)/ Incorrect BoQ |
| 4 | Incomplete work (stopped for > 1 week) | Waiting for clearance from other vendor/ contractor (The Work Front) |
| 5 | Waiting for clearance from other vendor/ contractor (The Work Front) | Incomplete work (stopped for > 1 week) |
| 6 | Frequent Design Changes (for any reason)/ Incorrect BoQ | Error in design/ drawing/ specifications |

If we consider the combined inputs from all 30 participants, along with their rankings, the top six wastes are listed below in Table 4.8.

**Table 4.8: Combined inputs of all Participants**

| Rank | Waste Type | Total Fuzzy Weightage |
|---|---|---|
| 1 | Rework due to poor quality of work (functional or aesthetic deficiencies) | 24.8 |
| 2 | Delays, waiting, or interruptions of any kind | 22.6 |
| 3 | Frequent design changes or incorrect Bill of Quantities (BoQ) | 14.4 |
| 4 | Waiting for clearance from other vendors or contractors (no work front available) | 14.4 |
| 5 | Incomplete work halted for more than one week | 12.6 |
| 6 | Mismatch in interdisciplinary drawings, specifications, or customer requirements | 11.4 |

Hence, it is recommended that SM should focus on eliminating or minimising these top construction-related wastes. It may be noted that:

i. The top two issues—rework and delay—together account for approximately 39% of total problems related to construction waste.

**ii.** The top six wastes account for 82% of the total waste-related issues.

The aggregated responses from all 30 participants revealed the primary causes of the leading construction wastes, detailed in Table 4.9.

**Table 4.9: Top Construction Wastes**

| Rank | Cause of Waste | Score |
|---|---|---|
| 1 | Poor coordination and communication among team members in daily operations | 71 |
| 2 | Inadequate planning and scheduling | 63 |
| 3 | Incorrect or incomplete information | 57 |
| 4 | Low worker skill levels and poor workmanship | 55 |
| 5 | Weak information sharing and internal departmental communication | 55 |
| 6 | Shortage of manpower | 43 |
| 7 | Delayed information flow | 42 |
| 8 | Excessive or ineffective project management control | 42 |
| 9 | Poor material quality and high rejection rates | 29 |
| 10 | Material shortages | 29 |
| 11 | Delayed payments to subcontractors or suppliers (based on participant comments) | 27 |

Based on this analysis, the following recommendations were given to SM and subsequently implemented. These proved highly successful in the full project, as documented in Annexure to Chapter 2.

## Preventive actions implemented in the project to eliminate the top six wastes:

Most of these preventive action recommendations (Table 4.10) were actually implemented in the next full project from the beginning, resulting in excellent outcomes that had never been achieved in SM's history until then. The outcomes and benefits are already documented in the case study in the annexure to Chapter 2. A standard template has also been prepared for future projects to conduct waste assessment (Table 4.11).

**Table 4.10: Recommended Preventive Actions**

| COMBINED INPUTS OF All PARTICIPATED | | | |
|---|---|---|---|
| **Rank** | **WASTE TYPE** | **Top Causes of Wastes** | **PREVENTIVE ACTIONS** |
| 1 | Rework required due to Poor quality of work (on account of functional or aesthetic reasons) | Poor Skill of workers; Poor Co-ordination & Communications; Poor Planning; Shortage of People; Incorrect / incomplete info etc. | Specially UAT driven development & long-term partnership with key Sub-contractors, using Educate-Reward-Punish model;<br>Use of Collaborative Planning System, Weekly Lean & Make-Ready and Daily Stand-up meeting;<br>Visual control thru' Dash Board;<br>Review of Site manpower requirements |
| 2 | Delay/Waiting /Interruptions (of any kind) | Poor Skill of workers; Delay in Payments to sub-contractors; Shortage of People; Shortage of Materials; Poor Coordination & Communications; Poor Planning; Incorrect / incomplete info etc. | Specially UAT driven development & long-term partnership with key Sub-contractors, using Educate-Reward-Punish model;<br>Use of Collaborative Planning System, Weekly Lean & Make-Ready and Daily Stand-up meeting;<br>Visual control thru' Dash Board;<br>Review of Site manpower requirements<br>Review of Sub-contractors payments terms. |
| 3 | Frequent Design Changes (for any reason)/ Incorrect BoQ | Poor Co-ordination & Communications; Poor Planning; Late information; Incorrect / incomplete info; Poor communication among internal team etc. | Use of Collaborative Planning System, Weekly Lean & Make-Ready and Daily Stand-up meeting;<br>Communication thru' integrated one-page-document for various most problematic defects<br>Visual control thru' Dash Board |
| 4 | Waiting for clearance from other vendor/ contractor<br>(No Work Front) | Shortage of People ; Shortage of Materials; Poor Co-ordination & Communications; Poor PM control; Delay in Payment to Sub-contractors etc. | Mandatory participation in Weekly Lean & Make-Ready & Daily Stand-up meeting;<br>Communication thru' integrated one-page-document for various most problematic defects.<br>Visual control thru' Dash Board;<br>Review of Sub-contractors payments terms. |

| | | | |
|---|---|---|---|
| **5** | Incomplete work (stopped for > 1 week) | Shortage of People (for balancing various Capacities); Shortage of Materials; Poor Co-ordination & Communications; Poor PM control; Delay in Payment to Sub-contractors etc. | Mandatory participation in Weekly Lean & Make-Ready & Daily Stand-up meeting;<br>Communication thru' integrated one-page-document for various most problematic defects.<br>Visual control thru' Dash Board;<br>Review of Sub-contractors payments terms. |
| **6** | Mismatch in drawing/specification/customer needs | Poor Co-ordination & Communications; Incorrect / incomplete info; Poor communication among internal team; Poor Planning; Late information; Poor Skill of Designers; etc. | Mandatory participation in Weekly Lean & Make-Ready Planner;<br>Daily Stand-up meeting;<br>Communication thru' integrated one-page-document for various most problematic defects<br>Visual control thru' Dash Board;<br>More & better communications & Co-ordination among MEP & Civil Designers and Contractors. |

## Table 4.11: Template for Construction Waste Assessment

Table 4.11: Template for Construction Waste Assessment

**SM Inc** — CONSTRUCTION WASTE ASSESSMENT (INPUTS FROM MANAGEMENT)

GUIDELINES for FILLING during the CLASS-ROOM WORKSHOP;
1.Out of 17 wastes listed below,plz first rank top waste types from 16;(1-highest;6-lowest),accord to your experience in SM Projects
2.For those 'chosen top six wastes out of 17,assig your reasons/causes for wastes based your past Experience in SM.Put 'X' ,In six rows,chosen by yo (you may mark 'X',in more than one box of Reasons) 3.Plz mail it to your Global head Of work place Design,in next 30 min time,before webinar is over

| | | | CAUSES OF WASTES | | | | | | | | | | | | | | | | | | | | | | | |
|---|---|---|---|---|---|---|---|---|---|---|---|---|---|---|---|---|---|---|---|---|---|---|---|---|---|---|
| | | | MANAGEMENT RELATED | | | | | | | RESOURCES RELATED | | | | | | | INFORMATION | | | | | OTHERS | | | | |
| sr no | WASTE TYPE | TOP SIX WASTES RANKING | Poor planning & Scheduling | Poor co-ordination & Communications | Excessive /Poor Project Management Control | Too much of paper work/forms to be filled | Delay in Payment to Sub Sontractors/Suppliers | Not much focus on Processes improvements | Other,plz specify | Shortage of people | Shortage of Material | Poor Skills of workers | Bad quality of material/heavy rejections | Excessive quantity of material | Lack of Workfront Availability from previous Trade | Other,plz specify | Incorrect/incomplete information | Unnecessary information | Late information | Poor communication | Others | Abnormal Weather | Frequent Equiment Breakdowns | Accidents | Other plz specify | Remarks if any |
| | | 1 TO 6 | Plz fill 'X',In appropriate 'cause' colums below agaist the top six wastes,according to you | | | | | | | | | | | | | | | | | | | | | | | |
| 1 | Delay/Waiting/Interuptions(of any kind) | | | | | | | | | | | | | | | | | | | | | | | | | |
| 2 | Rework required due to poor quality of wo (on account of functional or aesthetic reasons) | | | | | | | | | | | | | | | | | | | | | | | | | |
| 3 | Material Wastage of any type at site | | | | | | | | | | | | | | | | | | | | | | | | | |
| 4 | Damaged material received | | | | | | | | | | | | | | | | | | | | | | | | | |
| 5 | Incomplete work(stopped for>1week) | | | | | | | | | | | | | | | | | | | | | | | | | |
| 6 | Excessive suoervision required for progres | | | | | | | | | | | | | | | | | | | | | | | | | |
| 7 | Frequent Design changes(for any reason) | | | | | | | | | | | | | | | | | | | | | | | | | |
| 8 | Error in design/drawing/specificatios | | | | | | | | | | | | | | | | | | | | | | | | | |
| 9 | Mismatch in interdisciplinary | | | | | | | | | | | | | | | | | | | | | | | | | |
| 10 | incorrect installation (deviation from drawing) | | | | | | | | | | | | | | | | | | | | | | | | | |
| 11 | Redundant reporting/multiple approvals | | | | | | | | | | | | | | | | | | | | | | | | | |
| 12 | Duplication of efforts for Drawing or report | | | | | | | | | | | | | | | | | | | | | | | | | |
| 13 | Excess material handling /searching materials/Tools | | | | | | | | | | | | | | | | | | | | | | | | | |
| 14 | Unnecessary Labour movements a site | | | | | | | | | | | | | | | | | | | | | | | | | |
| 15 | Unnecessary materials on site(causing congestion) | | | | | | | | | | | | | | | | | | | | | | | | | |
| 16 | Waiting for dwaing /specification /RFI repl | | | | | | | | | | | | | | | | | | | | | | | | | |
| 17 | Waiting for clearance from other vendor/contractor (The work front) | | | | | | | | | | | | | | | | | | | | | | | | | |

This standard template can be used for future projects to conduct waste assessment. This template enables systematic identification and ranking of construction wastes, ensuring transparency and accountability. It helps teams prioritize corrective actions, reduce costs, and improve project efficiency across all stakeholders.

## 4.7 Annexure II: Case Study II Finding hidden waste of Over-Production to improve Profitability

### The Project:

PPI Bangalore, India, is a premier real estate company offering innovative and sustainable solutions in luxurious residential projects to cutting-edge commercial spaces. They are among the most trusted property developers in Bangalore, India. They are committed to excellence, integrity, and customer satisfaction. However, in order to improve their brand image further and also improve their profitability, they wanted to embrace some of the Lean Construction practices to see whether those are useful to them. For this, the company wanted to try pilot Lean Construction in one of their real estate projects. The project was already delayed significantly, and also created a hole in their pocket because of cash-flow problems. Experience of Lean Construction with that project would enable them to take an educated decision whether they should adopt these new practices.

This project is spread over 2.5 acres and consisted of three towers of 12 floors each, having a total of 144 premium apartments. The project is situated in a prime location in Bangalore, offering a convenient and well-connected address for its residents. The project also has a distinctive and contemporary architectural design, aiming to create a harmonious blend of aesthetics and functionality. It also offers a wide range of modern amenities like landscaped gardens, a well-equipped fitness centre, children's play area, swimming pool, clubhouse, and 24/7 security services.

### The Problems to be addressed:

This project, although very prestigious, was not selling fast and created a huge cash-flow issue for the company. There was significant expenditure for materials and labour payments to the contractors and suppliers, but revenue was not being received from customers as initially forecast. The following three key problems were specifically stated:

- **Delayed construction**: Many clients who had already made part-payments were experiencing huge delays in the construction of their apartments. These delays caused inconvenience and financial strain on the buyers, impacting their plans for moving in or investing in the property.
- **Quality concerns**: Some clients had raised issues regarding the quality of construction and also the poor-quality materials used in their apartments. This was leading to dissatisfaction and the need for repairs and renovations.
- **Cash-flow**: The builders were facing severe cash-flow problems, as payments from the customers were slow, and apartments were not selling fast due to slow progress of the project. At the same time, builders were not able to delay the

payments for the bulk materials suppliers (steel and cement) and the labour contractors.

## The Key Construction Wastes Identified:

Two areas were identified for construction waste reduction and also for resolving part of the cash-flow problem. This was done after a few discussions with the client, the main contractor and the project management and quality management consultant teams. One of the two key reasons for the huge project delay was identified as the bottleneck at the cutting and bar bending shop. The other reason was heavy material expenses, which were far more than the BoQ standards estimated till that stage of completion. The two solutions for both these reasons are briefly mentioned below.

## The Solutions Implemented:

- **Bottleneck Bar Bending & Cutting:**

One of the main bottlenecks for slow progress was found to be the steel bar cutting and bending shop. The workshop at the site had one bending machine and one cutting machine as per the contract terms. They were supplying the steel cut and bent bars to all the three towers under structural construction. One contractor was making two towers while another was making one tower. A lot of workers waiting time was observed due to lack of bars for column and beam construction in the three towers. The figure 4.5 below gives the rough layout of the towers and the bar bending workshop.

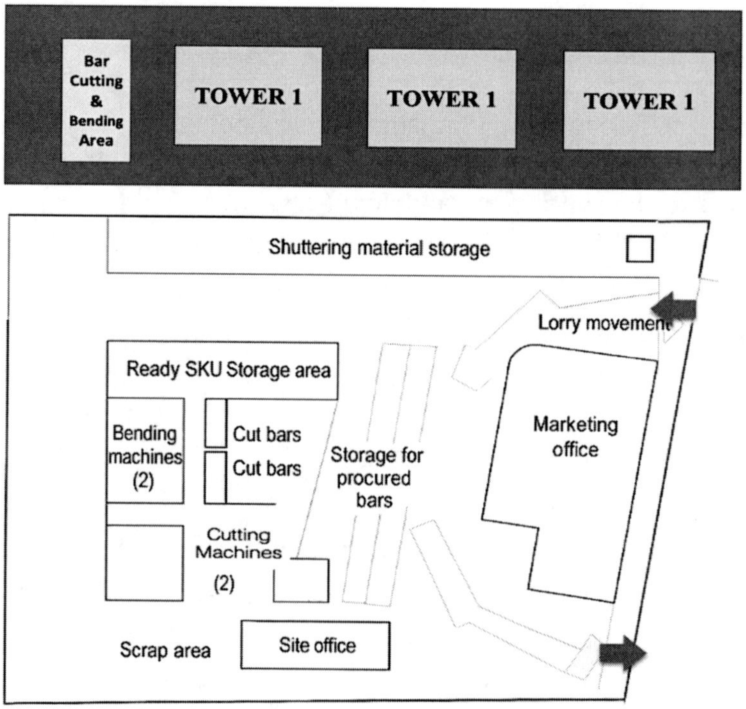


**Figure 4.5: Layout of Three Towers & Bar Bending Workshop (not to scale)**

**Following were the main constraints due to this:**

- There was very limited space, as the project site was very narrow. The plots on both sides of the project site were under legal disputes, and did not allow even an inch of space for material storage or movement. This also made the layout very congested.
- There was a high amount of material movement for supply of steel to all the three different towers in a row, with several hindrances.
- The slab-to-slab cycle time was very high—as much as 32 to 35 days, requiring four pours due to complex apartment design, compared to normally only three pours. Still, it was difficult to meet the requirement of the three towers. The total cutting time for one floor of steel requirement was 6–7 days, followed by a 14–15-day cycle of bending.
- There was a lack of proper bar cutting and bending schedules.
- The problem was going to be increased as there was already a Lean Expert Team working to reduce the slab-to-slab cycle time to 24 days, despite having four pours.

This was leading to a large number of carpenters and fitters waiting at the site and also slow progress of the project.

After identifying this requirement of shorter slab-to-slab cycle time, a 10-day time and method study was conducted by two construction management post-graduate interns, to see how to improve the productivity and output of this shop. Their recommendations included the following:

- Prepare a scientific bar bending and cutting schedule, as the requirements for 12 floors in each of the three towers were almost identical.
- The schedule should be based on the overlap of the different towers' requirements, in terms of daily bar diameters, cut length, shape, and numbers required for the next three days' requirements of each tower.
- Improve the site layout and proper storage system of the cut and bent bar.
- Fabricate a lever attached to the cutting machine to prevent marking and measuring frequently.

This led to about a 35% increase in the output of this shop, but still was not adequate to meet the revised improved slab-to-slab cycle time. Therefore, the only option was to add one more bending machine, as that was the bigger bottleneck. But this suggestion was not accepted as the bar cutting and bending contractor's contract included only one machine each of bar cutting and bar bending, despite these machines being low-cost machines.

Thus, the slow progress of the project continued for some time, till there was a severe cash-flow problem with the builders. On one side, the apartments were not selling due to slow progress of the project, and on the other hand, there were huge outstanding payments to be made to the contractors and bulk materials suppliers. Finally, they had to raise almost ₹30 crores at 18% per annum interest rate, as the buyers were pressuring for faster progress of the project. And this made the Lean Expert think of a new way of selling the idea of adding one bar bending machine to the CFO, even though it was not included in the original contract of the steel contractor. We recommended investing in an additional bending machine. It was estimated by Lean Experts that the project would be delayed by more than two months due to this bottleneck. This meant they would have 'additional' interest liability of ₹90 lakhs due to delay in clearing payments caused by this bottleneck. While the additional bending machine cost just ₹1.25 lakhs, and could be bought immediately—even from the builder's side—to save his own money. This justification was agreed, and finally the delays were reduced tremendously, and this section was able to meet the slab-to-slab cycle time requirement of 24 days per floor for each of the three towers.

There were two other alternatives also considered for the solution. One was to buy Smart Steel (pre-cut and bent) and another was to have a central workshop for bending and cutting for various projects of the same builders in the Bangalore city. These alternatives were in place of retaining the bar cutting and bending activities at each site. The central facility would have far better machines and skilled workers, and therefore far higher productivity and quality, and also cheaper, despite adding the logistics costs of those bars from central place to different sites. However, these two alternatives were deferred for future consideration.

- **Extra Material and Labour being wasted in Plastering:**

The third-party quality management consultant often mentioned in their monthly quality reports that the column and beam casts were not plumb, and so was the case with blockwork, which might also be a reason for extra cement consumption and slow progress.

Therefore, two completed apartments of 3 bedrooms were taken up to actually measure the amount of extra plastering done. Because the columns and blockwork were not plumb, extra cement coats were required to keep the external finished plastered wall plumb, for aesthetic reasons. The details of the results are provided below in the pictures and the tables. Plastering variation was the result of non-compliance with quality specifications for structural concrete work, which was 13 mm thick plaster. The measurement was done in each room of the two apartments chosen as a sample. The total actual plastering thickness was measured by the size of the plastering buttons made to get a final smooth and plumb wall. A table of such button sizes for one of the two flats is given below in figure 4.6 as a sample. Two flats were chosen for this purpose to get an average of additional plastering required.

| Flat- 01 | Plastering thickness (mm), derived from Plastering Button put at different locations | | | | |
|---|---|---|---|---|---|
| | | | Beam | Column | Wall |
| | C1 | Column projected 10mm more than beam | 30 | 15 | |
| | C2 | Column projected 14mm more than beam | 30 | 15 | |
| | B1 | Beam is projected 15 mm more than wall | 10 | | 25 |
| | B2 | Beam is projected 10 mm more than wall | 15 | | 20 |
| | B3 | Beam is projected 13 mm more than wall | 12 | | 25 |
| | B4 | Beam is projected 15 mm more than wall | 10 | | 25 |
| | B5 | Beam is projected 10 mm more than wall | 15 | | 20 |
| | B6 | Beam is projected 5 mm more than wall | 15 | | 20 |
| | B7 | Beam is projected 8 mm more than wall | 15 | | 23 |
| | B8 | Beam is projected 10 mm more than wall | 15 | | 23 |
| | W1 | Wall is projected 12 mm more than beam | 27 | | 15 |
| | W2 | Wall is proje cted 15 mm more than beam | 30 | | 15 |

**Figure 4.6: Samples of Actual Plastering Thickness for Column & Beams**

The same was done in all the rooms of the apartments. Similar images (figure 4.7) of the living room and master bedroom are shown below as example.

| LIVING ROOM |
|---|
| Plastering Button 1 = 28 mm |
| Plastering Button 2= 24 mm |

| MASTER BED ROOM |
|---|
| Plastering Button 3 = 32 mm |
| Plastering Button 4= 26 mm |

**Figure 4.7: Samples of Actual Plastering Thickness for Column & Beams (Living & Master Bed Room)**

Based on all such data collected, the additional material consumed and its cost was computed, with respect to the standard BoQ specified. Since at some places the thickness of the plaster was up to 32 mm as well, the plaster had to be completed in 3 to 4 iterations, in order to prevent any crack formation later. This not only added labour efforts and cost but also delayed the completion, because of the curing time required between two plaster coats. These were all calculated and the following additional costs and efforts were computed per apartment (Table 4.12).

**Table 4.12: Additional Cost & Efforts due to Poor Blockwork Quality**

| Sl No. | Description | Quantity | Cost (Rs) | | Time (Hrs) |
|---|---|---|---|---|---|
| | | | Per Bag | Total | |
| Total internal plastering area of mock-up flat: 4120 Sqft | | | | | |
| At mock-up flat ACE ready made mixture used as plastering mortar, which 30 kg bag can do 15 sqft of 13 | | | | | |
| 1 | With the above productivity, the mock-up flat required plastering mortar bags. | 275 bags | 360 | 99000 | 192 |
| 2 | Actual consumption of readymade plastering mortar bags. | 384 bags | 360 | 138240 | 441 |
| **Additional bags consumed due to structural column, beam, and block wall not being in** | | **114 bags** | **360** | **39240** | **249** |

This meant approximately 40% more consumption of cement, and 120% extra consumption of labour efforts and costs. This obviously will increase the cash-out flow unnecessarily, and impact the profits as well.

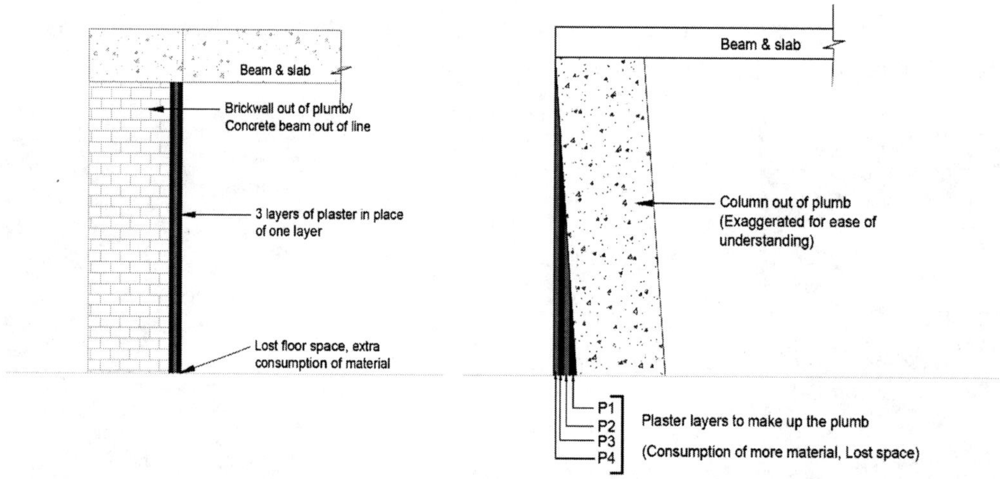


**Figure 4.8: Depiction of Plastering Thicknesses at Different Plaster Coats**

Besides this, one problem, which most builders and even the buyers ignore is that because of extra thickness of plastering on all four sides of the room, it will also reduce the carpet area available to the buyer (Figure 4.8). Since, as per Indian government regulations, the rate for the house should be based on carpet area (not built-up or super built-up area), the buyer unknowingly pays more for receiving less actual area. Someday, buyers will become wiser, and this might become an issue in future. Thus, elimination of construction wastes benefits everyone: the customer, the builder, and the contractors.

*Construction Waste isn't just debris—it's lost value. Every discarded material reflects inefficiency, cost, and environmental harm. By reducing waste, we build smarter, save resources, and create sustainable spaces for future generations.*

# CHAPTER 5

# CREATING THE BIG ROOM : ENHANCING COLLABORATION

# 5 Creating 'THE BIG ROOM'

The concept of the Big Room draws inspiration from high-performance team environments, where interdisciplinary collaboration is enabled through physical or virtual proximity. In construction projects, often riddled with fragmented workflows and communication delays, creating a Big Room helps bridge the divide between planning and execution. The Big Room is not just a meeting space; it's a dynamic coordination hub designed to foster transparency, fast decision-making, and mutual accountability among all stakeholders.

## 5.1 Need and Benefits of the Big Room

Since it is essential to bring together all key stakeholders, including designers, domain experts, contractors, and client representatives into a shared environment, the need for a Big Room emerges. This integrated approach fosters transparency, faster problem-solving, and mutual accountability, thereby enhancing project efficiency and quality outcomes for all involved.

### 5.1.1 Need for a Big Room

In construction projects, a large number of people are required to be involved to complete them successfully. These include Designers, Architects, Technical Consultants, Project Management Consultants, Quality Management Consultants, Domain Experts, etc. In addition, there are Engineering, Procurement and Construction Contractors, Specialty Contractors, Testing & Validation Consultants, etc. Key personnel are also needed from several other departments, including experts from the Client's side. All these individuals are expected to work independently in their respective areas of expertise, while also collaborating closely with others, as their activities are highly interdependent.

As mentioned earlier, there are prerequisites involving eight types of resources and other essential elements that must be in place before the next construction activity can begin. These eight prerequisites are provided by different sets of the above-listed individuals. Therefore, unless all eight prerequisites are made available continuously, the project cannot be completed on time. This requires a high degree of collaboration, cooperation, and communication among all stakeholders throughout the life cycle of the project.

To ensure the required Communications, Coordination, and Cooperation (termed '3C' for simplicity) among these stakeholders—so that the eight prerequisites are reliably available during project execution—the CPS framework was explained in Chapter 3. To enable 3C, people were grouped into Clusters, Sub-clusters, and Sub-sub-clusters, as described previously, to foster productive collaboration. For successful CPS implementation, a well-structured 3C system is essential across five different levels of planning defined within CPS. This includes quarterly 3C for 'Rock' process scheduling, monthly 3C for 'Pebble' activity scheduling, weekly 3C for Dust task scheduling, and daily 3C for Dust sub-task scheduling.

The Big Room, also called the 'War Room' sometimes, is a concept that provides both infrastructure and a collaborative platform. It supports cross-functional team collaboration by jointly anticipating future constraints and workflow impediments in the short term (e.g., daily and weekly) or medium term (e.g., monthly and quarterly), and resolving them collectively in a timely manner so that issues are addressed proactively.

This is accomplished by bringing the larger team together to help accelerate the activities of other groups or individuals. It also enables teams to understand their impact across various clusters or work groups. Additionally, it provides dedicated time for project-wide discussions on concerns such as budgets, hot issues, or major changes required in the baseline scope, time, or expense schedules.

The main objectives for the Big Room can be summarized as:

- **Collaboration**: The Big Room encourages open communication, shared decision-making, and collaborative problem-solving among project team members. This leads to better coordination among different project participants and clusters, minimizing conflicts and delays. It adds value by identifying and resolving constraints and impediments, thereby enabling all the eight prerequisites (like Manpower, Machines, Money, Materials, Approvals/Drawings, Work-front) in time. It also facilitates rapid advancement of work within a short time frame.
- **Integration**: It facilitates the integration of design, construction, and operation processes, ensuring that all aspects of the project are considered from the early stages. Real-time communication supports quick decision-making, enabling timely progress and adherence to project schedules. This adds value by preserving the 'Big Picture' or the 'Value' of the project for all decision-making.
- **Efficiency**: By fostering real-time collaboration, it streamlines project timelines, reduces rework, and improves overall project efficiency. Efficient collaboration helps identify and resolve issues early, minimizing costly rework and change orders. The use of the Visual Management System also enables faster and better decision-making. This adds value by reducing overall project costs through improved methods and activity sequencing.
- **Knowledge Sharing**: The Big Room creates an environment for the exchange of expertise, best practices, and lessons learned, enabling continuous learning and improvement. It encourages innovative ideas and creative solutions to complex problems, resulting in better-informed decisions and stronger commitments.

In short, the BIG ROOM refers to a *physical* workspace for *collaboration, communication, and coordination* among various stakeholders involved in a construction project. It uses visual displays to present all the information needed to guide the team in making timely, proactive decisions.

It is typically a physical 'big room' where professionals from different disciplines—such as architects, engineers, contractors, and clients — gather to collaborate, plan, and coordinate their efforts to achieve better results and commitments.

However, it is more than a physical facility. It is also a mindset that addresses several issues commonly found in regular meetings. Typical issues observed in collaboration during routine meetings include:

- Absence of TRUST among team members
- Fear of CONFLICT with other Cluster groups, stemming from their 'Siloed' approach
- Lack of COMMITMENT from Team Members, especially due to lack of transparency
- Avoidance of ACCOUNTABILITY, with a tendency to blame others for problems
- Lack of FOCUS on RESULTS, the Big Picture, or the PROJECT VALUE for the Client
- Cultural Issues, such as resistance from individuals or organizations accustomed to traditional hierarchical structures and isolated decision-making

The BIG ROOM seeks to address these issues and may, at times, require support from the HR departments of key Consultants, Contractors, Specialty Contractors, and even the Client organization — particularly due to the nature of these soft issues.

### 5.1.2 Benefits of an effective Big Room:

We discussed above about the objectives of the Big Room; the key benefits of this are enhancing Project Success, as shown in the Figure 5.1 below. Besides CPS, the Big Room can be used for Construction & Business Processes improvements leveraging Value Stream Mapping, 5W1H, etc.

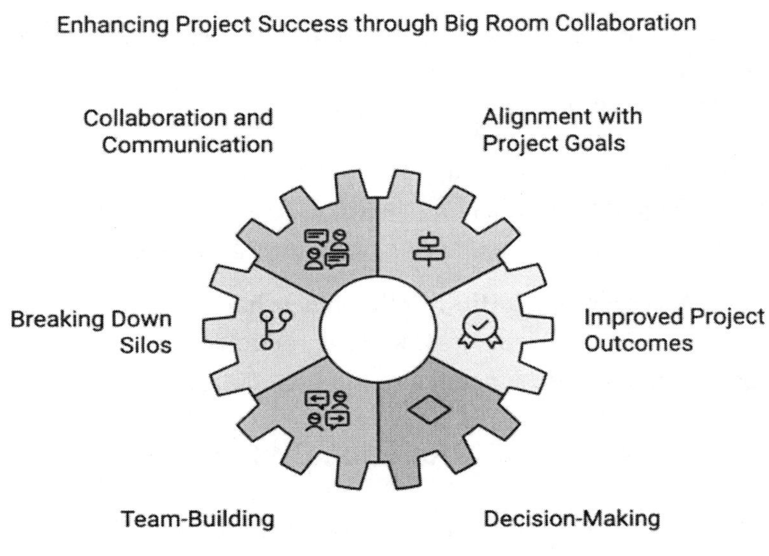


**Figure 5.1 Benefits of Big Room Collaboration**

### 5.1.3 What is an Effective BIG ROOM?

The Lean Construction Institute (USA) describes the BIG ROOM as a project approach that gathers key stakeholders in one space to accelerate communication, decision-making, and reduce siloed thinking. Held regularly from design through handover, it fosters cross-functional collaboration and coordination. In these sessions, stakeholders plan, update, share resources, invite feedback, demonstrate accountability, make decisions, and compare current progress against agreed goals — all aimed at fulfilling base project commitments.

### 5.1.4 Main Challenges for developing an effective BIG Room for Lean Construction:

While the concept and benefits of the BIG ROOM are excellent, there are several challenges in making it operational. Some of these were also mentioned above as softer issues in Section 5.1.1. As a brief, comprehensive summary, the following four challenges make it difficult to implement and derive full benefits from the concept:

- **Cultural Resistance**: Adopting the BIG ROOM may face resistance from individuals or organizations accustomed to traditional hierarchical structures and isolated decision-making. Many such organizations, especially in the Construction Sector, have remained conventional in their approach—unlike the Manufacturing or IT sectors. In the BIG ROOM, there is no hierarchy, and all participants are treated as equals.
- **Logistical Challenges**: Coordinating schedules, ensuring the availability of all relevant personnel for scheduled meetings, and securing physical space to accommodate all stakeholders productively and comfortably can be difficult—especially for large-scale projects.
- **Information Management**: Ensuring effective sharing, storage, and access to project-related data and documents for everyone in the room is challenging without adequate IT infrastructure, software systems, and tools.
- **Initial Investment**: Setting up a dedicated physical space or implementing suitable technology solutions for the BIG ROOM can involve significant costs, and many organizations may be unwilling to invest in anticipation of future benefits.

Therefore, in developing a BIG ROOM, these challenges must be addressed; only then can its full benefits be realized. The following section outlines the key principles that must be incorporated into its design, development, and usage.

## 5.2 Key principles of BIG Room for Lean Construction:

You must appreciate that the BIG ROOM is more of a mindset of working, and requires an attitudinal change among the people. Therefore, it is a good idea to involve the Human Resource (HR) departments of the key participants to emphasize this point.

The HR team should ensure that their key employees involved in the project truly appreciate the underlying philosophy.

**Below are the key principles to be kept in mind to make the BIG ROOM really effective:**

1. First and foremost is "Respect for the People" — their opinions and their time. Every person holds unique knowledge and both good and bad experiences, which must be leveraged for the benefit of the project.
2. Focus on optimizing the "Whole" — that is, the overall "Big Picture & Value Defined" for the project, rather than each individual's activity and cost, merely based on their 'Contract'.
3. Always be open and ready for improvements. People should not hold the mindset that "Only I have all the knowledge."
4. Encourage team members to support each other and align themselves with the overall goals and value identified for key stakeholders of the project.
5. Leaders must be Bold, Observant, and capable of creating Positive Team Energy, with a mindset of "Yes, we can." This also requires a proper understanding of team members' strengths and weaknesses, and their action-orientation.
6. The work environment must promote mentoring, coaching, and mutual support.
7. There must be a high level of 'Transparency' among participants — such as not hiding problems or bad news, and visually displaying all plans, achievements, failures, and areas or actions for improvement.

Since most of these requirements involve softer skills, HR professionals are best positioned to bring alignment and motivation among their team members. A Project Director may not always be the ideal person to ensure this part.

### 5.2.1 The Venue for the potential Big Room:

The venue for the BIG ROOM should preferably be physically co-located on a recurring basis, so that participants can interact face-to-face — the most effective form of interaction and communication. However, some participants may be located remotely or across different regions. In such cases, the venue can be designed as hybrid — a combination of physical co-location and virtual participation via platforms like Zoom, MS Teams, or video conferencing. Fortunately, this hybrid model has become widely accepted and prevalent since the COVID-19 pandemic.

One fundamental requirement for both physical and hybrid formats is that it must be a recurring activity, with all meetings and sessions pre-scheduled until the project is completed. The following points must be considered when choosing the venue:

1. The main room (even in a hybrid setup) must be large enough to accommodate the required number of participants comfortably and should include breakout areas.
2. There must be space for all members to sit comfortably, including provisions for working in smaller teams, if needed.
3. Sufficient wall space should be available for displaying all relevant charts and visuals.
4. A dedicated space is preferable, ensuring that other unrelated activities do not interfere.
5. Access to basic amenities like coffee, water, snacks, and restrooms must be ensured.
6. Available technology must fully support the team's activities — including a good PA system that enables clear communication for both physical and virtual attendees.
7. The venue should have connectivity to company servers, internet, email, and other systems to access necessary project information in real time.
8. Essential materials for training and discussions — such as whiteboards, sticky notes, markers, and flip charts — should be readily available.

Few pictures of Big Room are provided below in Figures 5.2 and 5.3. The detailed process for setting up a BIG ROOM, along with additional requirements, is described in the next section.

**Figure 5.2: Few pictures from an Effective Big-Room discussions and display**

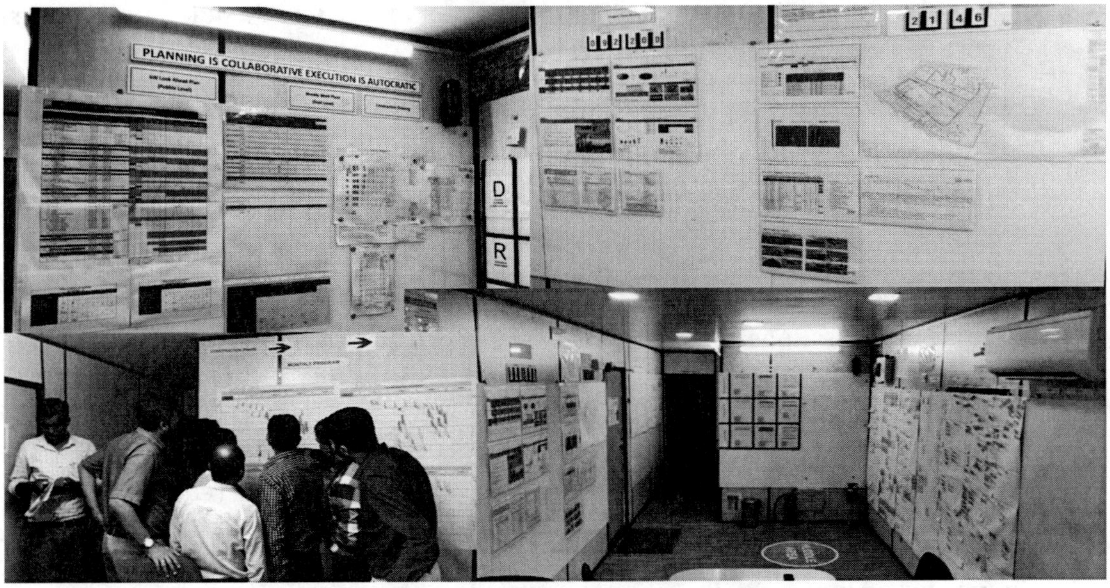


(Pics courtesy: Ferring Pharmaceutical Ltd.)

**Figure 5.3: Few pictures from an Effective Big-Room discussions and display**

## 5.3 Steps involved for developing an effective BIG ROOM:

It was highlighted earlier that BIG ROOM is more of a mind-set of working, not just a physically coming together. Therefore, the following points are important for this, which HR can enable to develop, through their interactions and discussions.

**Ensuring Desired Behaviour from participants**

- **Positive Leadership:** Develop a "yes we can" type of attitude, and avoid the mindset of "no, it won't work."
- **Effective Facilitation Practices:** The expected outcome should be *clearly communicated to all members*, preferably in advance.
- **Limit Cell Phone Usage:** Avoid using cell phones during discussions, except for occasional relevant pictures. No multitasking. Phones may be used only during scheduled breaks.
- **Trust and Respect for Others' Opinions & Time:** Cultivate this trait actively. Speak up with your own thoughts and ideas, but also listen attentively to others.
- **Maintain Enthusiasm:** Sustain energy throughout discussions. Include few breaks or ice-breakers to make the sessions fun and engaging..
- **Overcome 'Silos' in Discussions**: If needed, intervene to encourage broader participation across clusters.
- **Willingness to Learn:** Be open to others' perspectives. Avoid a rigid mindset like "I know everything."

- **Regularity and Structure:** Maintain consistent meeting schedules and establish core working hours for the project team.
- **Combat Virtual Meeting Fatigue:** Use virtual "breakout rooms" for smaller group discussions to maintain engagement.
- **Document Key Issues:** Record all problems and challenges that emerge during discussions.
- **Emphasise Visual Controls:** Refer frequently to displayed charts and visuals to guide conversations and decisions.
- **Promote Trust and Friendship:** Foster mutual respect and camaraderie among participants who spend extended time working together.

Table 5.1 provides a summary of the key points to be considered for developing an effective Big Room.

**Table 5.1: Key points for consideration for designing Effective big room**

| Commonly Occuring Risks/Bad Practices | Importance of Learning |
|---|---|
| The team spirit & enthusiasm wanes and people gravitate to their silos and "us-them" type of thinking violating the team-work spirit. | The participating teams should 'hunt' the repetitive or other major errors made in the planning, executing or monitoring & controlling the project and continuously learn together from those for resolution and future prevention. This will enable them to resolve the real issues of project impediment. |
| Thoughts like, "we don't have time to do our regular work because we're always in meeting." | The team should have the Quest for "Continuous improvement". They should not be complacement after reaching the targeted goals, but always think how to do that even better next time. |
| ▪ Meeting start late because of late arrivals of some people.<br>▪ Leadership members stop listening and becomes very directive to the team-members. | ▪ There should be focused knowledge transfer, whenever required<br>▪ Any Releveant short videos should be viewed and discussed so that all are in sync with the knowledge regarding better practices. |
| Planners don't follow the plan for the scheduled meeting or value stream mapping or waste elimination workshop, and go deviate to non-important issues. | Arrange any simulation, if possible for learning better and a wow moment of realization, inspiration, insight, recognition, or comprehension. |

| **Retrospective/Continuous Improvement p-d-c-a cycle** | **Decide the proper technology support required?** |
|---|---|
| Conduct a quick (5 mins) retrospective at the end of every meeting to find good part and what needs to be improved, for more effectiveness in future. | Video conferencing/ zoom facilities or Team with bigger screen for large group. |
| Also conduct event Event- based retrospective at the completion of every major milestones. | White boards/ or if possible SMART type boards. |
| **Onboarding Consideration** | Network connectiviy for providing immediate access to the right information needed to perform the work. |
| Decide judiciously who should be in different Big Rooms discussions. No relevant person must be left out, at the same time none of the unnecessary person should be present. | Project management and Documentation sharing Software, required. |
| Decide a most convienient time which can be repeated, for ensuring everyone's presence. | Printing and plotting equipment. |
| Jointly decide the how to manage people, in case someone starts creating problrms. | Projector for any displaying of information, not available on the wall. |
| **Visual Information Goals** | |
| Make sure your visuals are interesting: colourful, simple, informative and Non-judgmental. Some recommendations are given for these for a Case Study in Annexture 5. | |

### 5.3.1 Key points to be considered for developing Effective Big Room

We also recommend for develop agreeable team behaviour in terms of 'ways of working together'. This should include the meeting rules, guidelines for how people can work together in hybrid (physical +virtual) environment.

### 5.3.2 Recommended Steps for setting Effective BIG ROOM

Figure 5.4 below provides the summarized key steps for setting the Effective Big Room. However, we also suggest to keep Process Improvement & debottlenecking meetings like VSM meetings separately on the similar lines but with different relevant teams.

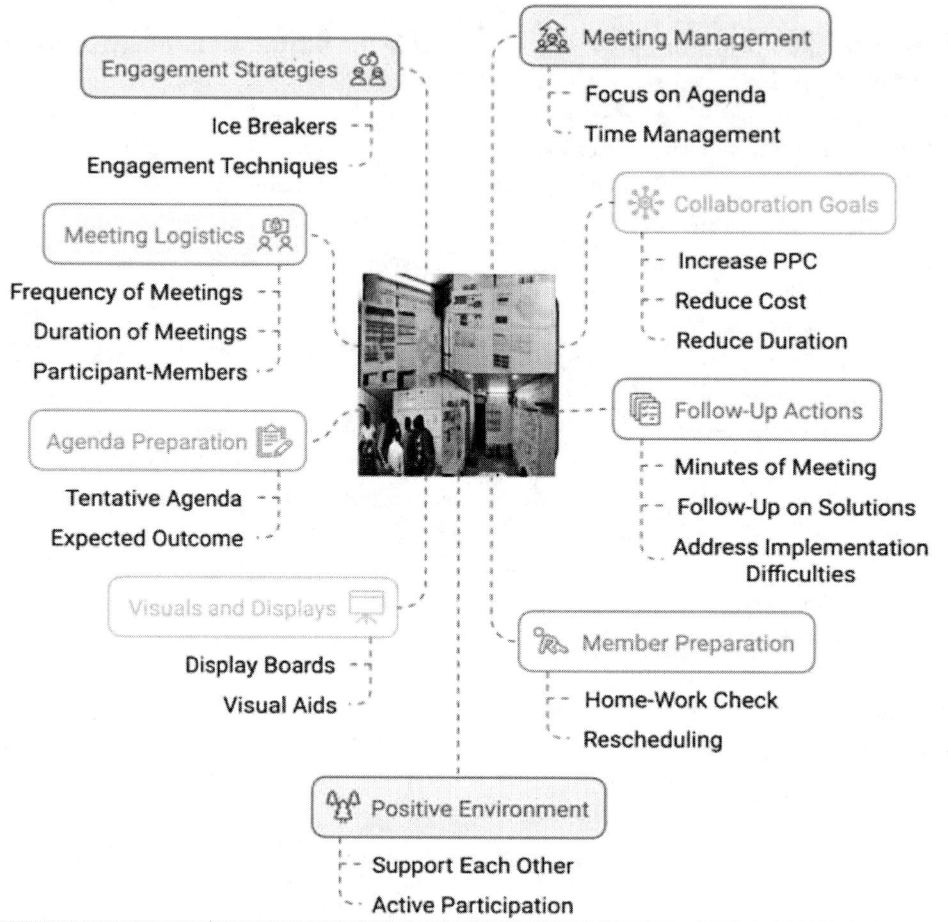


**Figure 5.4 Key steps for setting the Effective Big Room**

## 5.4 Key Measures of success:

Following parameters should be considered for assessing the health of BIG ROOM effectiveness:

- Visible chemistry and team spirit among the participants
- Learning & Coaching mindset displayed in the discussions
- Enthusiasm, Excitement and Active Participation of the participating members
- A good mix of seriousness and fun in the attitude of the people
- Transparency displayed (bringing issues in the open to resolve) Commitment to meetings (in terms of Punctuality, Advance Preparation, and positive participation in the discussions)
- Innovation & Creativity in finding solutions for improved processes for make-up the delays in the project, debottlenecking, waste reduction etc.

This is summarised below as a simple diagram shown in Figure 5.5.

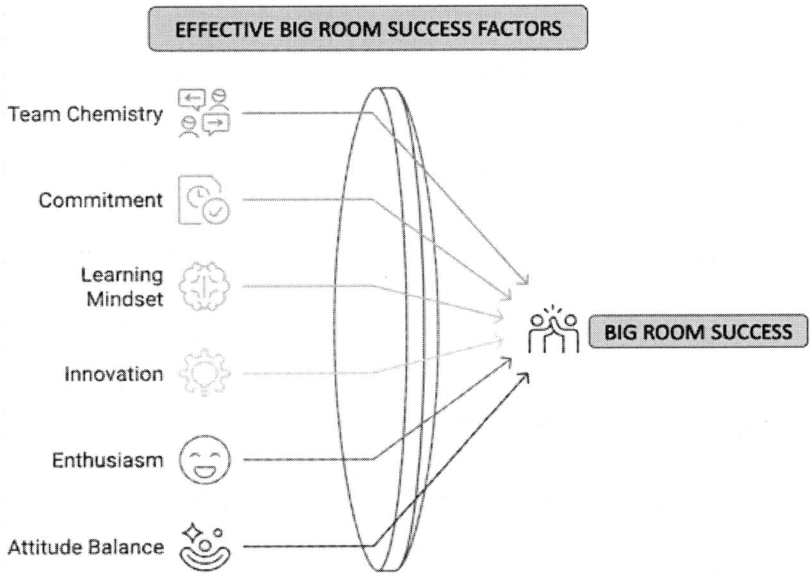


**Figure 5.5: Key Success factors for Effective Big Room**

Measurement can be done using a Web-Type Quality Matrix, with parameters captured as feedback after meetings of 60 minutes or more. Since most parameters relate to soft skills, HR teams of key stakeholders (Client, Contractors, Designers, Architects) should be involved to foster them through structured coaching and mentoring.

## 5.5 Some examples of Bad Practices of the Big Room

Members distracted by mobiles or side conversations, creating silos instead of one team, and showing disinterest. Others arrive unprepared without required data, or engage in unnecessary multitasking during virtual participation. Figure 5.6 below shows some of these behaviours. For better understanding most of these concepts, a real-life case study is provided in Annexure 5.

**Fig 5.6: Pics showing some undesirable Big Room behaviour**

## 5.6 ANNEXURE I: Case-Study

**Detailed Designing, Development and Operations of BIG ROOM for CPS in a Pharmaceutical Company:**

A pharmaceutical company was establishing a new state-of-the-art Research and Development (R&D) and manufacturing plant in Telangana. The facility aimed to enhance the company's capabilities in developing innovative pharmaceutical products and meeting growing market demands. The plant was equipped with advanced technology and infrastructure, enabling efficient drug discovery, development, and manufacturing processes.

The location in Telangana offers a favourable ecosystem for pharmaceutical research and development, with access to skilled talent, academic institutions, and research centres. The Lean Construction Planning and execution system was used for this project to ensure timely handover and provide great satisfaction to the pharmaceutical company's senior leadership team.

For CPS implementation, and conducting VSM and Waste Reduction Exercises, a large conference room was converted into a BIG ROOM and used regularly for more than two years. Most of the principles listed in the theory above were followed.The purpose of sharing this case study is to demonstrate the visual display aspect of the BIG ROOM. Toward the end, some key issues observed and recommended solutions are also included.

The next few pages present the visual display setup, along with the process to be followed for using relevant charts/templates. Where necessary, specific issues have been highlighted and solutions provided, Figures 5.7 and 5.8 may be referred to.

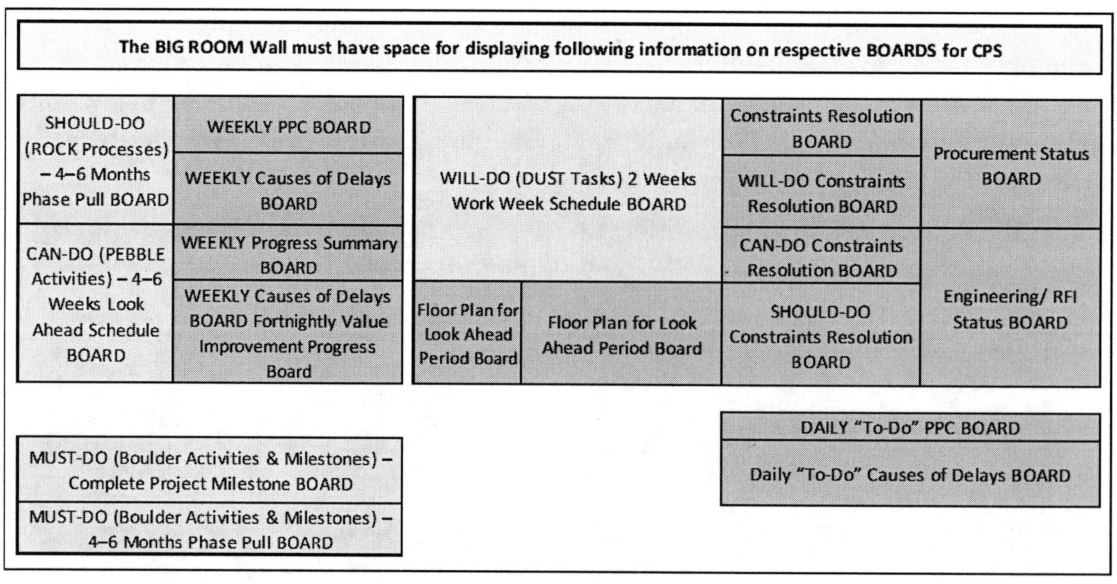


**Figure 5.7: Key information to be displayed in Big Room**

**Figure 5.8 Some displays of Big Room (Courtesy: Ferring Pharma, Hyderabad)**

**Some key issues observed, especially during the early stage of implementation:**

1. All required people did not attend the scheduled discussions, which often resulted in missing information during complete review and re-planning.
2. Many participants came unprepared for meetings and did not bring the necessary data. This frequently led to extensions of meetings up to 120 minutes, although they were planned for 60 minutes. Even then, some data would only be provided the next day.
3. Mobile phones were used despite repeated cautioning. A small penalty was later introduced for reading or responding to messages during meetings, but participants continued using them and did not mind paying the penalty.
4. Although delays, bottleneck processes, and lost time were identified, there was little effort made to meet separately and innovatively find collective solutions or improved processes. The general focus remained on blaming others.
5. At times, Contractors sent different representatives to different meetings who were not fully aware of the prior discussions. This caused disconnects among stakeholders.
6. There was low participation from Engineering and Procurement groups, primarily due to their locations outside Telangana State and poor communication infrastructure.

The information provided in the Figures 5.9 to 5.19 may be referred to for clarity.

**MUST-DO Lean MPS for Whole Project**

**MASTER BASELINED Lean MTS Schedule (Keeping 10% BUFFER Time at Project End ):27 Mont[illegible]** **Are Mo[illegible]on**

| | | | | | MAINPROJECT with AMIT | | | | | | | | | |
|---|---|---|---|---|---|---|---|---|---|---|---|---|---|---|
| **Major Deliverables** | **Timeline** | **Duration** | Q3-21 | Q4-21 | Q1-22 | Q2-22 | Q3-22 | Q4-22 | Q1-23 | Q2-23 | Q3-23 | Q4-23 | Q1-24 | Q2-24 |
| | | | | | Jan -mar | Apr-jun | Jul-sep | Oct-Dec | Jan -mar | Apr-jun | Jul-sep | Oct-Dec | Jan -mar | Apr-jun |
| Feasibility Study | | | | | | | | | | | | | | |
| Expected Basic Design | | | | | | | | | | | | | | |
| **Approval Reconstruction CAR (CRA-2)** | | | | | | | | | | | | | | |
| Preconstructionm Statutory Approvals | Sep-21 – Jan-22 | 5 | | | | | | | | | | | | |
| Detailed Engineering | Sep-21 – Jul-22 | 11 | | | | | | | | | | | | |
| Finalization of civil Contractor | Nov-21 – Feb-22 | 4 | | | | | | | | | | | | |
| Finalization of Key Equipment Options | Oct-21 – Apr-22 | 7 | | | | | | | | | | | | |
| **Approval Main CAR (CAR-3)** | | | | | | | | | | | | | | |
| Civil Completion: New Building | Jan-22 – Feb-23 | 14 | | | | | | | | | | | | BUFFER at the END |
| Civil Completion: Tankfarm / ETP | Feb-22 – Aug-22 | 7 | | | | | | | | | | | | |
| Major Procurement – Long Leads | Jan-22 – Mar-23 | 15 | | | | | | | | | | | | |
| Procurement Other Items | Feb-22 – Dec-22 | 11 | | | | | | | | | | | | |
| Mechanical Installation | Oct-22 – Jun-23 | 9 | | | | | | | | | | | | |
| HVAC Installation | Jun-22 – Feb-23 | 9 | | | | | | | | | | | | |
| Electrical & Instrumentation Completion | Oct-22 – May-23 | 8 | | | | | | | | | | | | |
| Finishing Activities | Oct-22 – May-23 | 8 | | | | | | | | | | | | |
| Post Construction Approvals | Feb-23 – Jun-23 | 5 | | | | | | | | | | | | |
| Commissioning | Apr-23 – Nov-23 | 8 | | | | | | | | | | | | |
| Qualification & Validation | Jul-23 – Jun-24 | 12 | | | | | | | | | | | | |
| Tech Transfer | Mar-23 – Dec-24 | 22 | | | | | | | | | | | | |
| Submission of Batch / PV Batch | Jan-25 – Aug-25 | 8 | | | | | | | | | | | | |
| EU Approval | Jul-25 – Sep-26 | 15 | | | | | | | | | | | | |

**Figure 5.9 Sample for MUST-DO Activities for Lean MPS for whole Project (High Level)**

**MUST-DO Lean MPS for Whole Project**

**PHASE PULL PLAN (6 Months) for detailing**

| | Activities | Duration (Month) | Feb-22 | Mar-22 | Apr-22 | May-22 | Jun-22 | Jul-22 | Aug-22 | Sep-22 | Oct-22 | Nov-22 | Dec-22 | Jan-23 | Feb-23 | Mar-23 | Apr-23 | May-23 | Jun-23 | Jul-23 | Aug-23 | Sep-23 | Oct-23 | Nov-23 | Dec-23 | Jan-24 | Feb-24 | Mar-24 |
|---|---|---|---|---|---|---|---|---|---|---|---|---|---|---|---|---|---|---|---|---|---|---|---|---|---|---|---|---|
| **Boulder Milestone** | **Sub Station & Metering Room – Power Live** | | | | | | | | | | | | | | | | | | | | | | | | | | | |
| Boulder Activity | Water tight building – Sub station | 5 | | | | | | | | | | | | | | | | | | | | | | | | | | |
| Boulder Activity | Key equipment delivery (Procurement + 1 Delivery-6) | 6 | | | | | | | | | | | | | | | | | | | | | | | | | | |
| Boulder Activity | Mechanical completion: Sub station equipment installation & Civil completion | 2 | | | | | | | | | | | | | | | | | | | | | | | | | | |
| Boulder Activity | Water tight building & Metering Room | 4 | | | | | | | | | | | | | | | | | | | | | | | | | | |
| Boulder Activity | Mechanical completion: Metering room installation & finished | 1 | | | | | | | | | | | | | | | | | | | | | | | | | | |
| Boulder Activity | External Power connection | 5 | | | | | | | | | | | | | | | | | | | | | | | | | | |
| Boulder Activity | Commissioning & Power Live | 1 | | | | | | | | | | | | | | | | | | | | | | | | | | |
| **Boulder Milestone** | **UG Water Tank System** | | | | | | | | | | | | | | | | | | | | | | | | | | | |
| Boulder Activity | Water tight building -Construction with final oivil finishing | | | | | | | | | | | | | | | | | | | | | | | | | | | |
| Boulder Activity | Key Prcurement delivery (Procurement + 1 +Del=3m) | | | | | | | | | | | | | | | | | | | | | | | | | | | |
| Boulder Activity | Mechanical completion=installation to pipeing | | | | | | | | | | | | | | | | | | | | | | | | | | | |
| Boulder Activity | Commissioning | | | | | | | | | | | | | | | | | | | | | | | | | | | |
| **Boulder Milestone** | **Potable water System Area** | | | | | | | | | | | | | | | | | | | | | | | | | | | |
| Boulder Activity | Key Prcurement delivery (Procurement + 1 +Del=3m) | | | | | | | | | | | | | | | | | | | | | | | | | | | |
| Boulder Activity | Mechanical completion=installation to pipeing | | | | | | | | | | | | | | | | | | | | | | | | | | | |
| Boulder Activity | Commissioning | | | | | | | | | | | | | | | | | | | | | | | | | | | |
| Boulder Activity | Qualification (OQ+PQ) | | | | | | | | | | | | | | | | | | | | | | | | | | | |

1. Understand 'Big Picture' & base-lined complete Project MPS Schedule & Key Challenges.
2. Refine base-lined Boulder Milestones as per Lean WBS guidelines.
3. Discuss Lean MTS with Design & Procurement Teams.
4. Resolve all issues (Long Lead Time & Complex Drawings).
5. Validate the allocated Resources (Manpower, Budget, Equipments, Materials, etc.).
6. Outcome: Required Lean MPS for the whole Project.

**Figure 5.10 Sample for MUST-DO Activities for Lean MPS for whole Project (detailed)**

SHOULD--DO Phase PULL

| Figure | | | PHASE PULL PLAN (6 Months) for detailing | | | | | | | |
|---|---|---|---|---|---|---|---|---|---|---|
| | Activities | Duration (Month) | Feb-22 | Mar-22 | Apr-22 | May-22 | Jun-22 | Jul-22 | Aug-22 | Sep-22 |
| **Boulder Milestone** | **Sub Station & Metering Room – Power Live** | | | | | | | | | |
| Boulder Activity | Water tight building – Sub station | 5 | | | | | | | | |
| Boulder Activity | Key equipment delivery (Procurement + 1 Delivery-6) | 6 | | | | | | | | |
| Boulder Activity | Mechanical completion: Sub station equipment installation & Civil completion | 2 | | | | | | | | |
| Boulder Activity | Water tight building & Metering Room | 4 | | | | | | | | |
| Boulder Activity | Mechanical completion: Metering room installation & finished | 1 | | | | | | | | |
| Boulder Activity | External Power connection | 5 | | | | | | | | |
| Boulder Activity | Commissioning & Power Live | 1 | | | | | | | | |
| **Boulder Milestone** | **UG Water Tank System** | | | | | | | | | |
| Boulder Activity | Water tight building -Construction with final oivll finishing | | | | | | | | | |
| Boulder Activity | Key Prcurement delivery (Procurement + 1 +Del=3m) | | | | | | | | | |
| Boulder Activity | Mechanical completion=installation to pipeing | | | | | | | | | |
| Boulder Activity | Commissioning | | | | | | | | | |
| **Boulder Milestone** | **Potable water System Area** | | | | | | | | | |
| Boulder Activity | Key Prcurement delivery (Procurement + 1 +Del=3m) | | | | | | | | | |
| Boulder Activity | Mechanical completion=installation to pipeing | | | | | | | | | |
| Boulder Activity | Commissioning | | | | | | | | | |
| Boulder Activity | Qualification (OQ+PQ) | | | | | | | | | |

1. Divide Boulders Activities for phase into Rock Activities.
2. Do collaborative iterative planning in 3 steps using Post-its.
3. Iteration 1: Reverse, backward planning.
4. Iteration 2: Resolving handling of conflicts & inconsistencies.
5. Iteration 3: Finalize the forward schedule for all Rock Processes.
6. Study full schedules to remove mismatch (dependency conflicts).
7. Perform constraints analysis; resolve constraints.
8. Commit the hand-over dates to next internal customer.
9. Outcome: Agreed SHOULD-DO Plan for next 4–6 months + list of inter-cluster Constraints and activities to be improved.

**Figure 5.11 Sample for SHOULD-DO Activities for Lean Phase PULL -1/2**

**SHOULD--DO Phase PULL**

| Figure | | | PHASE PULL PLAN (6 Months) for detailing | | | | | | | |
|---|---|---|---|---|---|---|---|---|---|---|
| | Activities | Duration (Month) | Feb-22 | Mar-22 | Apr-22 | May-22 | Jun-22 | Jul-22 | Aug-22 | Sep-22 |
| Boulder Milestone | **Sub Station & Metering Room – Power Live** | | | | | | | | | |
| Boulder Activity | Water tight building – Sub station | 5 | | | | | | | | |
| Boulder Activity | Key equipment delivery (Procurement + 1 Delivery-6) | 6 | | | | | | | | |
| Boulder Activity | Mechanical completion: Sub station equipment installation & Civil completion | 2 | | | | | | | | |
| Boulder Activity | Water tight building & Metering Room | 4 | | | | | | | | |
| Boulder Activity | Mechanical completion: Metering room installation & finished | 1 | | | | | | | | |
| Boulder Activity | External Power connection | 5 | | | | | | | | |
| Boulder Activity | Commissioning & Power Live | 1 | | | | | | | | |
| Boulder Milestone | **UG Water Tank System** | | | | | | | | | |
| Boulder Activity | Water tight building -Construction with final oivll finishing | | | | | | | | | |
| Boulder Activity | Key Prcurement delivery (Procurement + 1 +Del=3m) | | | | | | | | | |
| Boulder Activity | Mechanical completion=installation to pipeing | | | | | | | | | |
| Boulder Activity | Commissioning | | | | | | | | | |
| Boulder Milestone | **Potable water System Area** | | | | | | | | | |
| Boulder Activity | Key Prcurement delivery (Procurement + 1 +Del=3m) | | | | | | | | | |
| Boulder Activity | Mechanical completion=installation to pipeing | | | | | | | | | |
| Boulder Activity | Commissioning | | | | | | | | | |
| Boulder Activity | Qualification (OQ+PQ) | | | | | | | | | |

1. Divide Boulders Activities for phase into Rock Activities.
2. Do collaborative iterative planning in 3 steps using Post-its.
3. Iteration 1: Reverse, backward planning.
4. Iteration 2: Resolving handling of conflicts & inconsistencies.
5. Iteration 3: Finalize the forward schedule for all Rock Processes.
6. Study full schedules to remove mismatch (dependency conflicts).
7. Perform constraints analysis; resolve constraints.
8. Commit the hand-over dates to next internal customer.
9. Outcome: Agreed SHOULD-DO Plan for next 4–6 months + list of inter-cluster Constraints and activities to be improved.

**Figure 5.12 Sample for SHOULD-DO Activities for Lean Phase PULL -2/2**

| CAN-DO Look Ahead Plan | | | | | SixWeek Look Ahead Plan | | | | | | | | | | | | | | | | | | | | | | | | | | |
|---|---|---|---|---|---|---|---|---|---|---|---|---|---|---|---|---|---|---|---|---|---|---|---|---|---|---|---|---|---|---|---|
| | | | | | | | Week 93 | | | | | | Week 94 | | | | | | W95 | W96 | W97 | W98 | Constraints | | | | | | | | |
| | | | | | | | M | T | W | TH | F | S | M | T | W | TH | F | S | | | | | M | T | W | TH | F | S | | | |
| # | Location | Trade | Stones (As per Full Plan ) | Activity /Grid | Start | End | 02-Nov | 03-Nov | 04-Nov | 05-Nov | 06-Nov | 07-Nov | 09-Jan | 10-Jan | 11-Jan | 12-Jan | 13-Jan | 14-Jan | 09-Jan | 10-Jan | 11-Jan | 12-Jan | Dwg. | Matr | People | Machine | Clarif/ | Prev. Wrk | Safety/Wo | Weath./H | |
| 1 | Lab 2F | Civil | Plastering | South side wall | | | | | | | | | | | | | | | | | | | | | | | | | | | |
| 2 | Lab 2F | Civil | Plastering | East Wall | | | | | | | | | | | | | | | | | | | | | | | | | | | |
| 3 | Lab 2F | Civil | Plastering | Column Plastering | | | | | | | | | | | | | | | | | | | | | | | | | | | Clarification on Plastering height |
| 4 | Lab 2F | Civil | Plastering | Toilet Wall Plastering | | | | | | | | | | | | | | | | | | | | | | | | | | | |
| 5 | Lab 2F | Civil | Plastering | Lift wall Plastering | | | | | | | | | | | | | | | | | | | | | | | | | | | |
| 6 | Lab 2F | Mechanical | Pipe Back | U supportErection- Grid C-1/6-9 | | | | | | | | | | | | | | | | | | | | | | | | | | | |
| 7 | Lab 2F | Mechanical | Pipe Back | U supportErection- Grid C-1/9-11 | | | | | | | | | | | | | | | | | | | | | | | | | | | |
| 8 | Lab 2F | Mechanical | Pipe Back | U supportErection- Grid C-1/9-11 | | | | | | | | | | | | | | | | | | | | | | | | | | | 1. Base Plate Locations to be confirmed<br>2. Hilti Bolt to be delivered at site |
| 9 | Lab 2F | Mechanical | Pipe Back | U supportErection- Grid A-C/9-11 | | | | | | | | | | | | | | | | | | | | | | | | | | | |
| 10 | Lab 2F | Mechanical | Pipe Back | U support Fabrication - Cutting and Welding | | | | | | | | | | | | | | | | | | | | | | | | | | | |
| 11 | Lab 2F | Mechanical | Pipe Back | U support Fabrication - Painting | | | | | | | | | | | | | | | | | | | | | | | | | | | |
| 12 | Lab 2F | Mechanical | Pipe Back | U support Fabrication - Base | | | | | | | | | | | | | | | | | | | | | | | | | | | |
| 13 | Lab 2F | Mechanical | Pipe Back | | | | | | | | | | | | | | | | | | | | | | | | | | | | |
| 14 | Lab 2F | Mechanical | Pipe Back | | | | | | | | | | | | | | | | | | | | | | | | | | | | |
| 15 | Lab 2F | Mechanical | Pipe Back | | | | | | | | | | | | | | | | | | | | | | | | | | | | |
| 16 | Lab 2F | Mechanical | Pipe Back | | | | | | | | | | | | | | | | | | | | | | | | | | | | |
| 17 | Lab 2F | Mechanical | Pipe Back | | | | | | | | | | | | | | | | | | | | | | | | | | | | |
| 18 | Lab 2F | Mechanical | Pipe Back | | | | | | | | | | | | | | | | | | | | | | | | | | | | |
| 19 | Lab 2F | Mechanical | Pipe Back | | | | | | | | | | | | | | | | | | | | | | | | | | | | |

- validate Rock Process once again
- Elaborate Rock Processes into Peble Activities
- Define Sequence , duration , dependence ,key resources requirement for all pebble Activities.
- Resolve anyinonsistencies , in the Total CAN-DO plan and adds Look Ahead Constraints.
- Resolving Conflicts , Finaliize Schedule and agree on handling over dates between trades.
- Make Plans for VSM Process improvements etc. for debottlenecking.
- Outcome : Agreed CAN-DO look Ahead Plan +Constraints + Debottlenecking.

**Figure 5.13 Sample for CAN-DO Activities for Look Ahead Plan**

| WILL– DO Work WEEK Plan | | | | WEEKLY WORK PLAN-WEEK 45 (2-11 TO 8-11) | | | | | | | | | | | | | | | | | | | |
|---|---|---|---|---|---|---|---|---|---|---|---|---|---|---|---|---|---|---|---|---|---|---|---|
| | | | | Activity/Grid | | | Week 45 | | | | | | | Constraints | | | | | | | | | |
| | | | | | | | M | T | W | TH | F | S | SU | | | | | | | | | | |
| Sl no | Location | Trade | Stone Milstone(Based On Latest Schedule) | | Start | End | 02-Nov | 03-Nov | 04-Nov | 05-Nov | 06-Nov | 07-Nov | 08-Nov | Dwg | Matr. | People | Machine | Clarif/ | Prev. Wrk | Safety/Works | Weath/Holida | Remarks | Reasons For Non Completion |
| 1 | Lab 2f | Civil | Plastering | South Side Wall | 02-Nov | 03-Nov | | | | | | | | | | | | | | | | | |
| 2 | Lab 2f | Civil | Plastering | East Wall | 04-Nov | 05-Nov | | | | | | | | | | | | | | | | | |
| 3 | Lab 2f | Civil | Plastering | Column Plastering | 06-Nov | 07-Nov | | | | | | | | | | | | X | | | | | |
| 4 | Lab 2f | Civil | Plastering | South Side Wall | 02-Nov | 03-Nov | | | | | | | | | | | | | | | | | |
| 5 | Lab 2f | Civil | Plastering | East Wall | 04-Nov | 05-Nov | | | | | | | | | | | | | | | | | |
| 6 | Lab 2f | Civil | Plastering | Column Plastering | | | | | | | | | | | | | | | | | | | |
| | Lab 2f | Mechanical | Pipe Rack | u Support Erection-OE/9-11 | | | | | | | | | | | | | | | | | | | |
| | Lab 2f | Mechanical | Pipe Rack | u Support Erection-OE/9-12 | | | | | | | | | | | | | | | | | | | |
| | Lab 2f | Mechanical | Pipe Rack | u Support Erection-OE/9-13 | | | | | | | | | | | | | | | | | | | |
| | Lab 2f | Mechanical | Pipe Rack | U Support Fabrication Cutting Welding | | | | | | | | | | | | | | | | | | | |
| | Lab 2f | Mechanical | Pipe Rack | U Support Fabrication Painting | | | | | | | | | | | | | | | | | | | |
| | Lab 2f | Mechanical | Process Piping | Process Pipe Hydro Test | | | | | | | | | | | | | | | | | | | |
| | Lab 2f | Mechanical | Process Piping | Process Pipe insulation Grid C-E/6-9 | | | | | | | | | | | | | | | | | | | |
| | Lab 2f | Mechanical | Process Piping | Process Pipe insulation Grid C-E/9-11 | | | | | | | | | | | | | | | | | | | |
| | Lab 2f | Mechanical | Process Piping | Process Pipe insulation Grid A-E/9-11 | | | | | | | | | | | | | | | | | | | |

- Elaborate Rock Processes to Dust to Tasks For next 2 weeks, in line with ' Look Ahead' Plan, with no or manageable constraints.
- Detail duration,sequence,resources required for Dust Tasks.
- Ensure works are 'made ready' for week,by resolving constraints.
- Discuss inter trade dependencies,activity sequence,resource •requirement. Resolve the
- Outcome: Agreed WILL-DO Weekly plan (for 2 weeks)+ Resolve constraints.Week's Constraints.

**Figure 5.14 Sample for WILL-DO Activities for Work Week Plan**

| PPC BOARD | WEEKLY PERCENTAGE PLAN COMPLETE | | | | | GLOBAL ENGINEERING (MMM/YYYY) |
|---|---|---|---|---|---|---|
| | Weekly PPC | | | | | |
| | Cluster/Sub-Cluster | wk 1 | wk 2 | wk 3 | wk 4 | |
| | Civil | 81% | 79% | 75% | 64% | |
| | Fire Fighting | 100% | 100% | 100% | 60% | |
| | Electrical | 95% | 100% | 95% | 100% | 1. Compute the weekly PPC for Cluster/Sub-Cluster, based on Foremen daily data. |
| | HVAC | 78% | 100% | 100% | 100% | 2. Review Weekly Constraints and their resolution. |
| | Mecanical | 100% | 100% | 100% | 89% | 3. Study PPC trends & missed Dust Tasks and Pebble Activities. |
| | BMS/ELV | 100% | 100% | 94% | 97% | 4. Learn from these Interruptions and Constraints, prioritize and resolve. |
| | ETP | 0% | 100% | 100% | 100% | 5. Plan (separate meeting) to "make up" for the time lost due to incomplete Dust Tasks. |
| | GDLC | 100% | 67% | 100% | 100% | 6. Prepare inputs for action points for the next Make-Ready meeting. |
| | Clean Room | 100% | 100% | 100% | 100% | |
| | CQV | | | | | |
| | Complete Project | 89% | 93% | 94% | 84% | |

**Figure 5.15 Sample for PPC Board, for completed activities**

**Figure 5.16 Daily Stand-up Meeting by Foremen**

| PPC BOARD | WEEKLY PERCENTAGE PLAN COMPLETE | | | | GLOBAL ENGINEERING (MMM/YYYY) |
|---|---|---|---|---|---|
| Weekly PPC | | | | | |
| Cluster/Sub-Cluste | wk 1 | wk 2 | wk 3 | wk 4 | |
| Civil | 81% | 79% | 75% | 64% | |
| Fire Fighting | 100% | 100% | 100% | 60% | |
| Electrical | 95% | 100% | 95% | 100% | |
| HVAC | 78% | 100% | 100% | 100% | |
| Mecanical | 100% | 100% | 100% | 89% | |
| BMS/ELV | 100% | 100% | 94% | 97% | |
| ETP | 0% | 100% | 100% | 100% | |
| GDLC | 100% | 67% | 100% | 100% | |
| Clean Room | 100% | 100% | 100% | 100% | |
| CQV | | | | | |
| Complete Project | 89% | 93% | 94% | 84% | |

Compute the weekly PPC for Cluster/Sub-Cluster, based on Foremen daily data.
Review Weekly Constraints and their resolution.
Study PPC trends & missed Dust Tasks and Pebble Activities.
Learn from these Interruptions and Constraints, prioritize and resolve.
Plan (separate meeting) to "make up" for the time lost due to incomplete Dust Tasks.
Prepare inputs for action points for the next Make-Ready meeting.

**Figure 5.17 Causes of Variance for DUST Tasks**

The following new weekly report, displayed on the Value Improvement Board, was introduced for the PMO Consultant to address Issue #4, listed at the beginning of the Annexure. Although it achieved only partial success, it was recommended that this report be included from the outset as part of the contractual requirements for the Contractor/Sub-Contractors.

**Value Improvement BOARD**

| VALUE IMPROVEMENT EFFORTS & OUTCOME for the WEEK | | | | | | | | |
|---|---|---|---|---|---|---|---|---|
| Sr # | Area | ACTIVITY/ TASKS | # of Man-Hours Spent | TYPE of IMPROVEMENT | BENEFIT of Duration/ Costs | Implementation Action By | Target Date | IMPACT of PROJECT SCHEDULE & PPC % or COST |
| 1. Sub-Contractor/ Trade (like HVAC) | | | | | | | | |
| | | | | | | | | |
| | | | | | | | | |
| | | | | | | | | |
| | | | | | | | | |
| 2. Sub-Contractor/ Trade (like Electrical) | | | | | | | | |
| | | | | | | | | |
| | | | | | | | | |
| | | | | | | | | |
| | | | | | | | | |
| 3. Sub-Contractor/ Trade (like Civil ) | | | | | | | | |
| | | | | | | | | |
| | | | | | | | | |
| | | | | | | | | |
| | | | | | | | | |
| 4. Sub-Contractor/ Trade (like Mechanical) | | | | | | | | |
| | | | | | | | | |
| | | | | | | | | |
| | | | | | | | | |
| | | | | | | | | |

**Figure 5.18 Template for Value improvement board**

| Constraint List | | | | | | | | | | |
|---|---|---|---|---|---|---|---|---|---|---|
| Constraints | | | | | | | Activity | | | |
| # | Location | Constraint Type | Action By (Org) | Action By (Person) | Commitment Date | Constraints | Activity | Start | Actitvity By |
| 1 | Lab 2 F | Design | XYZ - Design | Ramesh | 04-Nov | Clarification on Plastering height | Column Plastering | 06-Nov | Civil |
| 2 | Lab 2 F | Prev Work | Symba | Ravi | 07-Nov | Plastering to be completed in lab area | U Support Erection - Base Plate Erection | 09-Nov | Mechanical |
| 3 | Lab 2 F | Design | XYZ - Design | Ramesh | 07-Nov | Base Plate Locations to be confirmed | U Support Erection - Base Plate Erection | 09-Nov | Mechanical |
| 4 | Lab 2 F | Material | Client-procurement | Samar | 07-Nov | Hilti Bold to be delivered at site | U Support Erection - Base Plate Erection | 09-Nov | Mechanical |
| 5 | | | | | | | | | |
| | | | | | | | | | |
| | | | | | | | | | |
| | | | | | | | | | |
| | | | | | | | | | |
| | | | | | | | | | |
| | | | | | | | | | |
| | | | | | | | | | |

**Figure 5.19 Template for Weekly constraints removal status**

**The other suggestions for resolving the implementation issues are as follows:**

1. Periodic discussions among HR departments — involving the Client, EPC Main Contractors, and major Sub-Contractors such as those for Civil Structure and MEP—should be held. These discussions, combined with involvement from their respective teams and joint sessions with others, must reinforce both the importance and the contractual obligation to adhere to the Client's 'way of working'. It was also recommended that this be included as a clause in the contract. This measure is expected to address Issues #1, 2, 3, 5, and 6.
2. Reward & Recognition systems, along with mutually agreed-upon Penalty mechanisms, should be introduced to encourage adherence to project norms. These systems are particularly important for Main and Design Contractors who significantly contribute to project value by embracing Lean Construction practices. It is suggested that these be embedded in the contract from the outset, when acceptance is more likely than at later stages.
3. A monthly 45-minute review should be conducted by the Senior Management Team (Steering Committee members) with the Core Lean Project Team. This review must include evaluating the effectiveness of Lean implementation, identifying challenges, and giving strategic direction. Additionally, the team should dedicate 15 minutes per month with Lean Experts to discuss implementation outcomes and any needed support. This is recommended to be integrated into the monthly Steering Committee meetings.

On the whole, it was felt by the Client that although not as effective as initially envisioned, the initiative did contribute meaningfully to the implementation of CPS and other Lean Construction methodologies. Without it, success would have been far more difficult to achieve.

# CHAPTER 6

# VALUE STREAM MAPPING: DRIVING PROJECT ACCELERATION

# 6 VALUE STREAM MAPPING (VSM)

A Value Stream refers to the sequence of activities required to deliver 'value' to the customer in a construction project. It encompasses all the processes, tasks, and resources involved in transforming raw materials, information, and effort into a final product or service. It includes end-to-end activities — both value-adding and non-value-adding — that are typically carried out to create a specific product or service, from the beginning of the project to the handover stage.

## 6.1 Need and benefits for Value Stream Mapping (VSM):

Value Stream Mapping (VSM) is needed to visualize workflows clearly, identify non-value-adding activities, and streamline processes. It helps teams understand how materials and information flow across production or service systems.

VSM enables organizations to reduce waste, enhance efficiency, and improve communication. By mapping current and future states, it promotes strategic thinking and guides targeted interventions that boost overall productivity. For infrastructure projects, VSM helps align multiple stakeholders, reduce delays, and enhance resource allocation. It fosters transparency in coordination, making complex timelines and dependencies easier to manage collaboratively.

### 6.1.1 What is Value Stream Mapping (VSM):

Before understanding Value Stream Mapping, it is important to understand the meaning of Value Stream.

This concept applies not only to the value delivered to the external customer, who pays for the product or service, but also to the internal customer. The internal customer is the next trade in the sequence — the party to whom the previous trade hands over the work-front for subsequent value-adding activities or operations.

As illustrations, some typical key Value Streams that are repeated several times during the construction of a real estate project include:

- Floor-to-floor slab cycle — casting columns, beams, and slabs on one floor, then repeating for each subsequent floor in the tower
- Installation of windows and doors in one apartment — repeated for each apartment in the tower
- Completing floor tiling in one room — repeated for all rooms across the building
- Installing false ceilings — either floor-wise in commercial buildings or apartment-wise in residential towers.

Each of these illustrates the repetitive nature of Value Streams and the importance of understanding them before implementing Value Stream Mapping techniques.

**Sometimes the Value Stream can also be non-repetitive, such as:**

- **Excavation of Earth** to prepare for the raft foundation of a multi-storeyed building
- **Completion of Raft Foundation** for a multi-storied tower construction — this task occurs only once but typically takes a very long duration to complete.

In the case of Linear Infrastructure Projects, following are some of the examples of the important Value Streams:

- Completing piling work for one span of a bridge or flyover — repeated several times for the full bridge length.
- Launching the required number of girders for one span of the bridge — repeated across all spans needed to complete the bridge.
- Completing one kilometre of asphalt coating — repeated 100 times for a 100 KM road.
- Completing 500 meters of concrete slabbing — repeated 300 times to cover a 150 KM long road.
- Installing 500 meters of broad-gauge railway line — repeated 600 times for a 300 KM railway stretch.
- Conducting land survey for one kilometre of road — repeated 300 times to cover 150 KM, using a Total Station Machine.

In linear infrastructure projects, some key Value Streams may also be non-repetitive in nature. For example, completing end-to-end excavation for a 150 KM State Road — a task influenced by the varying geography and soil conditions along the route.

You can also observe that the duration for these Value Streams may span hours, days, or even weeks, depending on how each Value Stream is defined. Upon deeper analysis, you'll find that many activities are associated with physical material movement or material flow, while others relate purely to information transfer or information flow.

These activities consist of several essential value-adding processes, such as concrete pouring. However, many tasks may also fall into non-essential, non-value-adding categories—like waiting for a crane or waiting for the crane driver to begin work. Such non-value activities — both in number and in duration — unnecessarily extend the cycle time, leading to project delays.

Any improvement in key Value Streams can enable the project to accelerate toward completion and reduce the overall timeline. Value Stream Mapping (VSM) helps achieve exactly that — by reducing the duration or cycle time of the bottleneck Value Streams and pushing the project forward.

VSM is equally applicable to business processes such as procurement and design workflows, helping to debottleneck them and reduce the duration of those Value Streams as needed.

It is important to highlight that Value Stream Mapping (VSM) covers only the macro-level sequence of steps in the complete value chain of a process. This mapping is done to understand — at a high level — how the work is designed, produced, procured, and delivered to the customer. Typically, this end-to-end value stream involves multiple trade organizations, departments, sections, or individuals, working together to complete the process. VSM helps in **hunting down the eight wastes** mentioned earlier by examining each step of the value chain and challenging the status quo.

Once identified, efforts are made to eliminate as many of these wastes as possible—without requiring significant capital investment — to improve **process efficiency** and reduce the **cycle time** of bottleneck or critical path processes.

- The key benefits of this approach include:
- Reduction in **Construction Process Cycle Time**
- Reduction in **Business Process Duration**
- **Cost Reduction**
- Enhanced **Capacity and Throughput**

Ultimately, VSM serves as a powerful tool for driving both operational and strategic improvements across construction and business workflows.

## 6.2 What is Need for Value Stream Mapping:

In a construction project that involves several Clusters/Sub-Clusters of teams, Value Stream Mapping (VSM) can be used to challenge and evaluate the time, effort, and duration contributions of every member within those cluster teams. This means critically assessing each team member — treated as an 'internal customer' — in terms of their overall value-addition to the Value Stream that serves the users of the project outcomes.

Through this approach, VSM can highlight multiple non-value-adding efforts within the process that may be eliminated or reduced. It enables the identification of redundancies and inefficiencies by questioning each step and challenging existing practices. VSM is particularly effective in helping to reduce the end-to-end process cycle time and significantly improve output within the given project duration.

As an illustration, let's assume a Value Stream has 'n' Activities to perform, as shown in figure below. P1, P2, ...., Pn are those activities. Each Activity will have a 'duration' like P1's duration is LT1 (lead Time 1), P2 has LT2 and so on. The total duration/ Lead Time of the complete Value Stream is summation of all LT1 to LTn. If we go deeper into each duration time of the activity, it can be noticed that, each time, like LT1, consists of 4 parts. These are:

i. **Basic Operations Time** (which really adds the 'value' to Value Stream), like applying plaster to Brick wall in wall plastering process.

ii. **Set-up Time**: like preparing Cement mixture for the plastering in a separate container

iii. **Logistics Time**: Bringing the Cement mixture near the wall, or bring cement to the mixing vessel etc.

iv. **Waiting time:** Mason waiting for cement mixture, mason waiting for clarification on plastering thickness, as the RCC Column and walls are not plum. Uniform thickness of 13 mm specified in the drawing would give a poor aesthetic finish to the wall.

Figure 6.1 below depicts this break-up of different time duration classification.

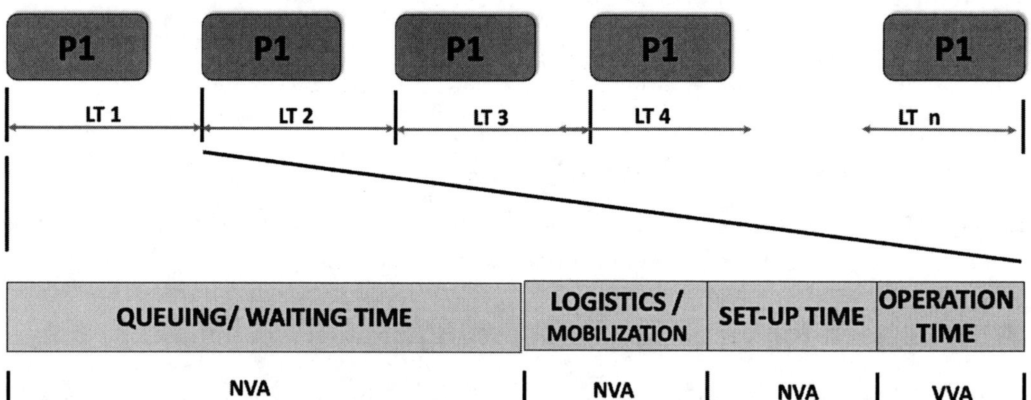


**NVA (Management Related):** **Non Value Adding Activities : Eliminate/ Merge using 5W1H and 5 WHYs**
**VA. (Engineering Related)** **: Value Adding Activities**
**LT 1 : Lead Time duration of Process Activity 1 ; Likewise others**

**Figure 6.1: A general Value Stream, consisting of 'n' process-steps.**

Research and past experience show that 'Waiting Time' has the largest magnitude among all four classes of time. This waiting represents a major waste, as delays caused by it do not add any value to the customer. Hence, if all waiting times across the *n* activities can be removed or reduced, the overall lead time can be shortened. This can be achieved by merging some activities, performing certain tasks in parallel, engaging more skilled personnel, or exploring low-cost automation for selected activities. Many construction and business processes have achieved reductions of 80% to 90% in cycle times through the wise application of such strategies. Therefore, Value Stream Mapping (VSM) is a highly effective method for de-bottlenecking the Boulder and Rock Processes, mentioned in Chapter 3 of the CPS framework.

VSM also enables enhancement of overall production capacity and the rate of output — particularly crucial when dealing with bottlenecks, expensive equipment, or scarce high-skilled resources. Ultimately, this leads to cost reduction, quality improvement, and faster project delivery.

### 6.2.1 Benefits of critical analysis of VSM

Based on our cumulative experiences, we strongly recommend to use VSM effectively for the Construction projects, to get following benefits mentioned in the diagram below:

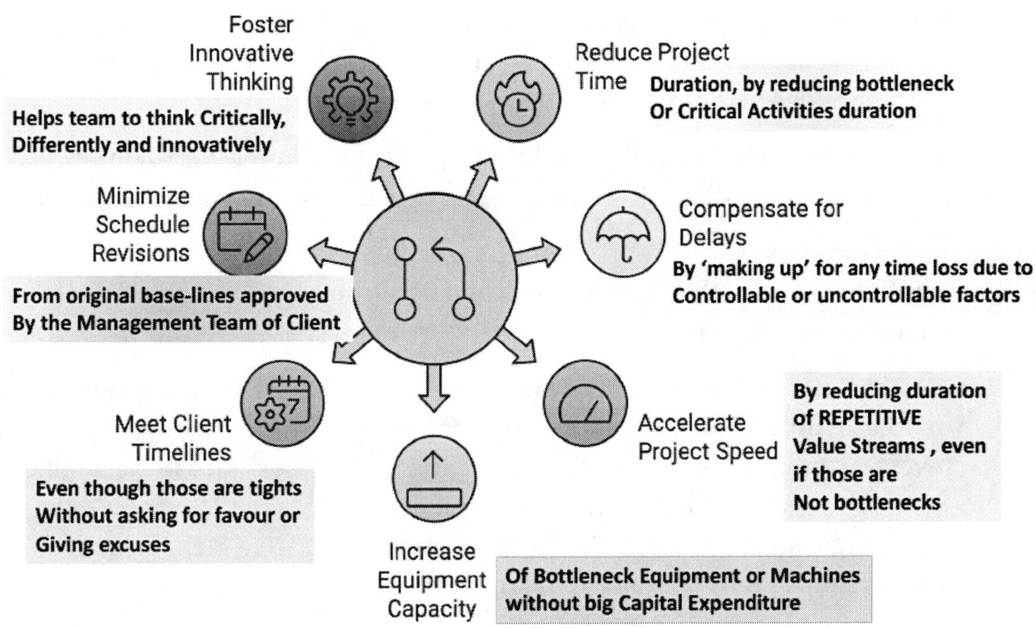


**Figure 6.2: Benefits of critical analysis of VSM**

## 6.3 Process for VSM for Accelerating Construction Processes

The process for Value Stream Mapping (VSM) to accelerate construction or business processes begins by identifying bottleneck areas; particularly Boulder/Rock processes within CPS, that require speed enhancement. The scope of the VSM study is then defined proactively or based on emerging needs. Next, the value stream is established to gain insight into delays and waste. This is followed by a critical analysis of processes with significant inefficiencies to uncover improvement opportunities. VSM is used strategically to recover lost time caused by controllable or uncontrollable risks. The details of these processes are described in the paragraphs below.

### 6.3.1 Identify the Bottleneck Processes (from Boulder/ Rock processes in CPS), which need to be accelerated

During CPS Lean MRP Planning (refer to Chapter 3) for the complete project, or Pull Phase Planning for the next 5–6 months, it is essential to identify those Boulder or Rock Milestones that appear difficult to achieve within the scheduled timeline. These milestones need to be compressed in duration to meet the promised delivery schedule. Use the BIG ROOM for this purpose by inviting all relevant Cluster and Sub-Cluster teams, bringing together their collective experience, domain knowledge, and problem-solving abilities to overcome constraints collaboratively.

**Follow the steps outlined below to address these challenges effectively.**

- Decide the CROSS-Functional and CROSS-Clusters Teams to be invited for this discussion. Ensure each relevant Department/ trade/ supplier is involved in this process, and is part of the Team.
- Have deliberations on 'all' possible frequently repeating large duration Value Streams or the bottleneck Value Streams which need the VSM Analysis, in order to accelerate the project.
- Prioritize the Top 2-5 Value Streams and agree on specific Teams for each of the Value Streams identified.
- Also finalize the Cross-Cluster Teams for each of the VSM study chosen, and agree on the TARGET Reduction of Duration, without traditionally and routinely investing in high Capital money automation processes. Please. remember that VSM does not start looking for Engineering solution first by higher-end technological solution, but tries to first find an innovative Management Solution, which requires only human capital through innovation.
- Prepare an agreed schedule for each chosen team for conducting VSM and achieving the target duration for those.

### 6.3.2 Define the Scope of VSM Study. (pro-actively or need-based)

It is always better to act proactively to reduce the time and cost of selected Value Streams. However, when there are schedule slippages — due to various reasons including uncontrollable factors like prolonged heavy rains — it is advisable to apply Value Stream Mapping (VSM) at those points to help recover lost time. For each VSM stream selected, designate one Team Leader responsible for driving the exercise. The VSM methodology should be explained clearly to all relevant Team Members, ideally with illustrative examples to aid understanding.

**Follow the steps listed below to initiate and execute the VSM process effectively:**

- VSM Team Leaders should call the Cross-Cluster Teams for discussions for the Value Stream Chosen, in the BIG Room.
- Review, customize and finalize the current VSM template/ structure, for the 'macro-level' Value Stream chosen. Always remember why were those chosen and 'target' reduction of time duration including the budget constraints, if any.
- Explain the big picture, the purpose once again, and the templates to be used to all the VSM team members. If required, provide a couple of examples from the past.
- Explain the key tools and methodology to be used. Explain how each template is to be used, with past examples.
- Challenge the contribution of 'each internal customer' to the final value delivered to the external customer. Think critically to eliminate non-essential contributions, or at least minimize the number of steps and/or the duration of each step.

### 6.3.3 Establish & get INSIGHT of the Value Stream, for Wastes and Delays identification perspective.

Draw all the steps on the wall of the BIG ROOM, as per the agreed template. Ensure that all high-level value-adding activities are included in the process. It is better to use 'Post-its' for this, as any corrections or changes required at any stage can easily be accommodated. Avoid going too deep into the tasks. It is better to keep the number of activities on the wall to 15–20, and not more than that. This will also ensure that the team does not go too deep into sub-processes or tasks, as is often done in Methods Study.

Ensure that all inter-cluster and inter-sub-cluster handovers of the work-front to the next trade are depicted in the visual process. In fact, wherever there are such handovers, delays are bound to occur, and a considerable amount of waiting time will likely be involved due to the queue of pending work lying in front of the next trade.

Follow the steps listed below in Figure 6.3, as per the VSM one-page template for the "Current State."

After drawing the Current State, follow the process steps of Figure 6.3 and suggest and define the new *faster and better* 'Future' process. Pilot the future process and resolve all the issues of implementations; standardize & implement the new improved process. Also Identify Activities, which have large durations/high costs which need to be relooked.

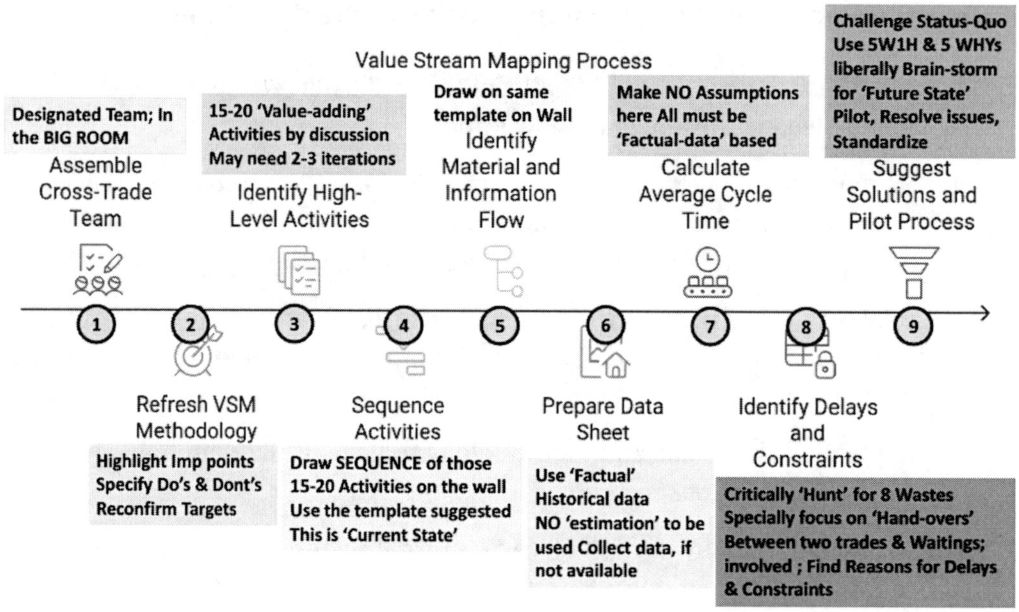


**Figure 6.3 Key Steps in Conducting Value Stream Mapping (VSM) Process**

### 6.3.4 Analyse and challenge the Process, specially where there are larger wastes and delays to find improvement opportunities

Sharpen your critical and fundamental questioning capabilities and use techniques like 5W1H m 5 WHYs etc. extensively. Challenge the current state and all the identified big opportunities for reducing duration and or the costs. Brain storm to find better ways of doing activities, by changing people, equipment, technologies, low-cost automation, more parallel processing etc. Be open to innovative ideas for improvements. In brief, this is shown here in Fig 6.4. Two case studies in Annexure will make these points more understandable.

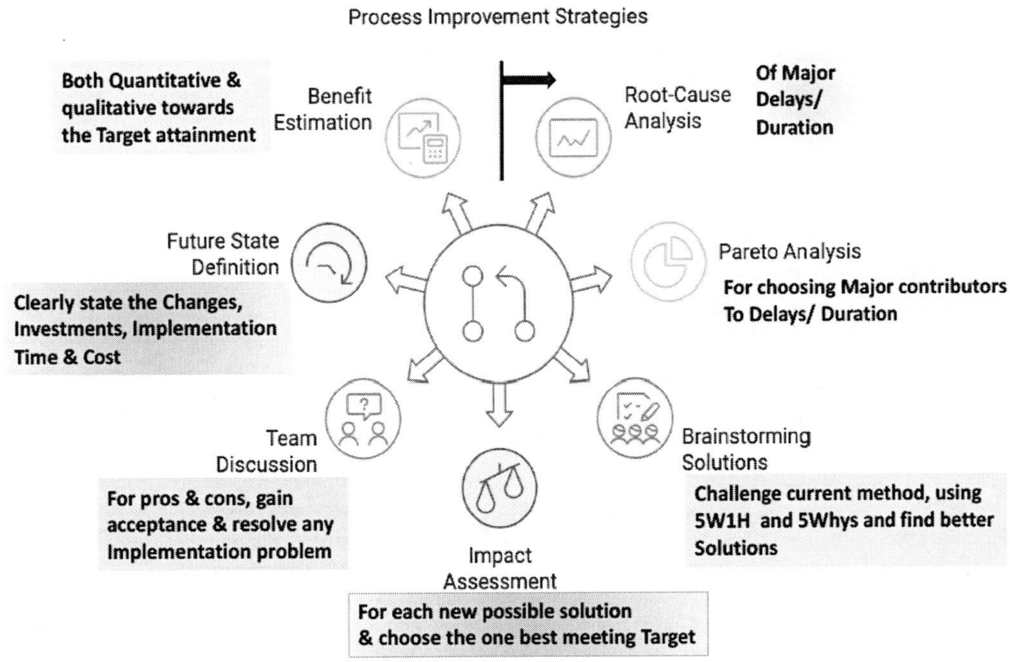


**Figure 6.4: Summary of finding big opportunities for improvements.**

### 6.3.5 Get Management's approval to assign a Person, for the accepted change to implement.

Prepare the business case for justification for the changes for any small additional investment etc. State clearly the Costs involved and the Benefits (both Tangible and Intangible) Use all your customer-facing skills to get the idea approved. In brief,

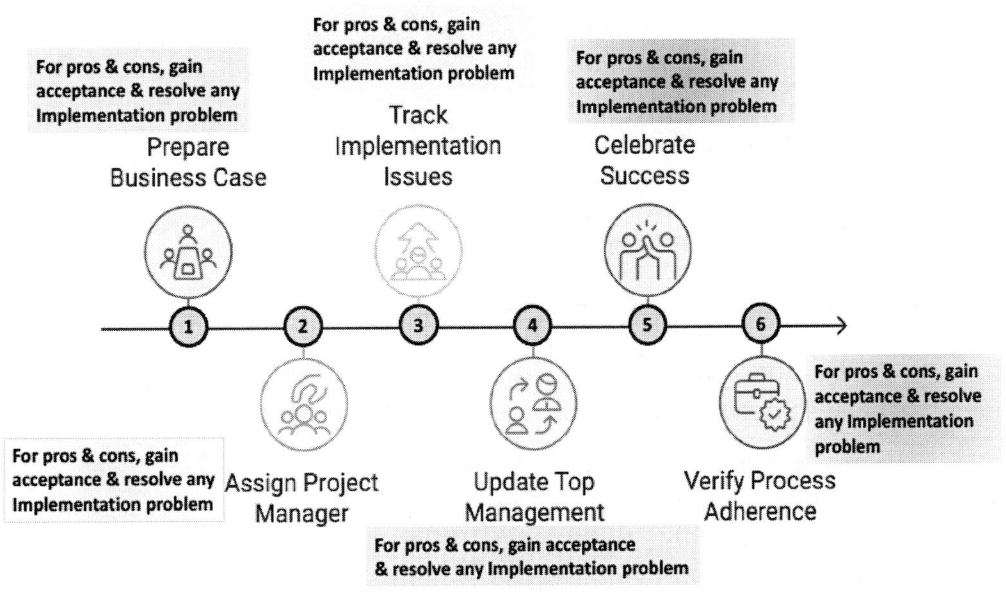


**Figure 6.5: Summary of Steps for getting management's approval**

### 6.3.6 Using VSM to 'make-up' any time loss in past, due to controllable or uncontrollable Risks occurred

As mentioned earlier that there may be time lost due to uncontrollable reasons like, Nature's unexpected behaviour (like more than planned rains, some political disturbances etc.). In such cases also VSM can be used to 'make-up' for at least some of the time-lost, and provide a better adherence to the original schedule.

## 6.4 Key Measures of Success:

Following are the indications, which will assure you that your team's initiatives have been successful.

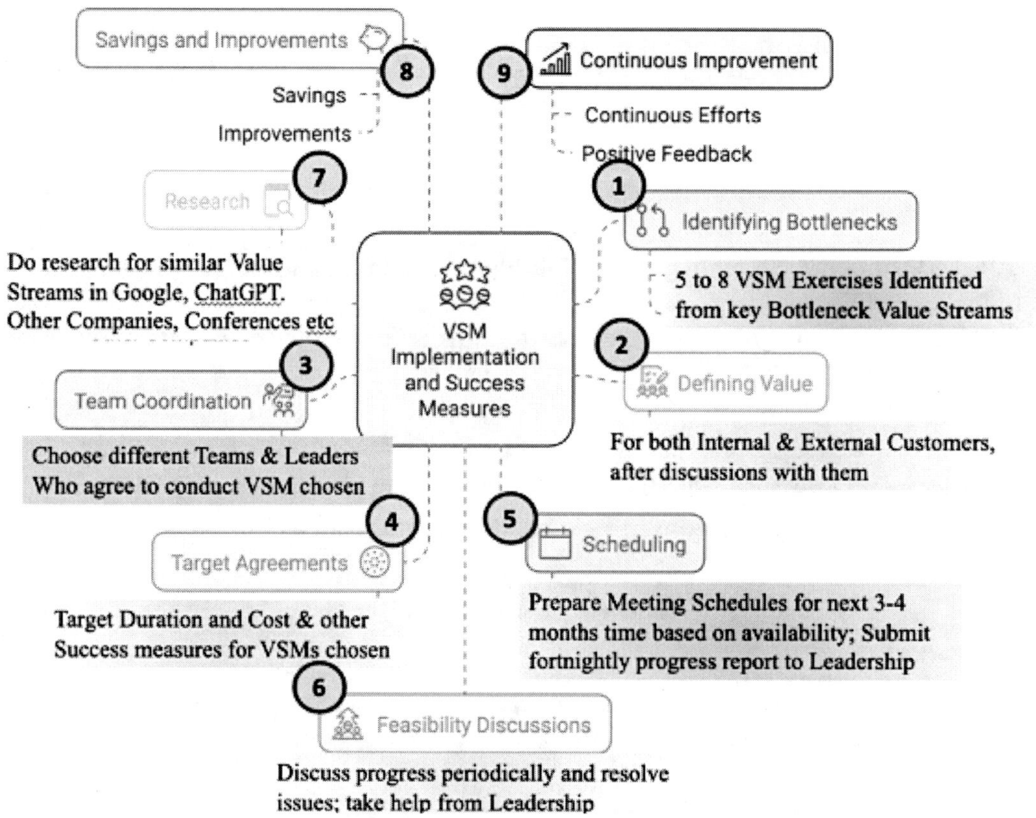


**Figure 6.6: Key Measures of VSM success:**

One qualitative measure of success should also be that the Company and site people are talking positive about the efforts invested in improving things."

Two case-studies below will make you appreciate some of the concepts discussed above, in much better fashion. We recommend to read these two case studies in the Annexure Section below.

## 6.5 Annexure I: Case Study on VSM – A River Bridge

**The Project:**

A 2.8-kilometre-long important river Bridge, joining two important cities namely, Bhubaneswar and Cuttack in Odisha in India, was being built by Odisha Works Department. It had 75 Spans of about 35 meters each, designed for two lanes traffic. Like many Government Construction Projects, it was a delayed project, even after twice revising the completion dates in past. In Sept 2017, it was noticed that it was supposed to be completed within 5 months, but due to various delays and current construction processes, it will take minimum 8 months more to complete.

The new team of Lean Construction (internal Engineers and external Lean Mentor), took-up this challenge to avoid the delay and successfully experimented to complete the balance 8 months project work in just 5 months, without extra budget. This enabled it to be actually inaugurated on 1st April, 2018, an important day for Odisha State being its State Formation-day. This was 3-months ahead of originally publicly announced inauguration day of 1st of July 2018.

The case study explains briefly how this apparently impossible mission was achieved successfully using Value Stream Mapping (VSM) innovatively. It also mentions the various issues emerged and how were those resolved in bureaucratic set-up to accelerate the completion of the project by almost 60%.

**The Objectives of VSM in this project:**

i. To accelerate the project progress rate for the remaining part of the project by 60-70% within same budget. To identify the most critical Bottleneck Value Stream(s), which is impacting project delay the most. Launching of four Girders on remaining spans was identified as the most critical bottleneck Value Stream. It was also because, the Girder Launching was an automated using a very special and expensive equipment, named Girder Launcher. And you can't increase the number to double, on the same Bridge, due to physical constraints.

ii. To demonstrate the utility of Lean Tools like Value Stream Mapping along with leveraging Project's collaborative work culture through motivating multi-disciplinary team members from client & contractors' teams.

iii. To 'hunt' maximum number of wastes, that is non-value adding activities, by the multi-disciplinary team and try to eliminate those in the whole end-to-end repetitive Value Stream of Girder Launching process followed for every Span.

## Outline of the Methodology followed:

Key stakeholders from State Works Department & Main Contractor's sides were identified form a core VSM team. It was deliberated among all, about what can really motivate each one of them. These motivational needs were integrated with the way the project management was done, involving internal team of Bureaucrats and external team of contractors.

Value Stream Mapping was used innovatively for identifying and eliminating all major wastes and accelerating the flow of project till completion. This will be clearer, as we explain more details in next paragraphs.

Current practice is shown in Figure 6.7 and 'Current State' Value Stream for Girder launching is shown Fig 6.10 below, having a Span-wise (four Girders) launching cycle time was of working 20 days per Span of the bridge.

All the non-value adding activities and the wastes, like waiting and rework etc., were identified. Team did the brainstorming for different, better and faster ways. 5W1H was used vigorously. Some of the key suggestions which were accepted, were:

- Design a strong steel structure to mount on the girders to make the four girders very sturdy, so that the Girder Launcher can travel over it safely, even if the cross-girder is not ready. This ensured that the Girder did not have to wait for 11 days of making RCC Cross-Girder> Cross Girders were required for safe movement of Girder Launcher to the next span, after completing one Span of the Bridge. The duration for RCC cross-girder construction and curing took 11 days, as shown in the Fig 6.12. Please refer the diagrams and the figure and see the steel structure within red ellipse mark.
- With the Steel Structure ensuring the sturdiness of the four Girders, the Girder could travel over it to transfer Girders for the next span, even without Cross-Girder being completely ready. Thus, the installation of four Girders was made as overlapping activities to complete the whole duration almost 70% faster. This is also clear from Figure 6.8 and 6.9.
- Keep the 'set-up' of 4 activities ready in advance, so that these do not result in waiting for other tasks following it. Increase the labour team for 5 of the activities at very low investment, to accelerate the completion of those activities
- Automate (using low-cost, like using electric hoist, instead of mechanical hoist for lifting heavy material) 4 activities to speed up work pace. Work on the 5 activities in parallel, instead of doing those in sequence, to reduce the cycle time.
- Create the 'Future State' VSM, and once it is agreed by the team, pilot it to learn and improve it. This is shown in Figure 6.11 below. Rough Scheduling of Span-wise Girder launching before and after changes is shown in Figure 6.12.
- Implement it with full-fledged manner to get the benefits. The Pareto Chart below shows the improvement in cycle time reduction.

Figure 6.7: Girder Launcher at work

Figure 6.8 Sturdy Cross Frames used to make Main Girders safe for enabling movement of the Girder Launcher for next span.

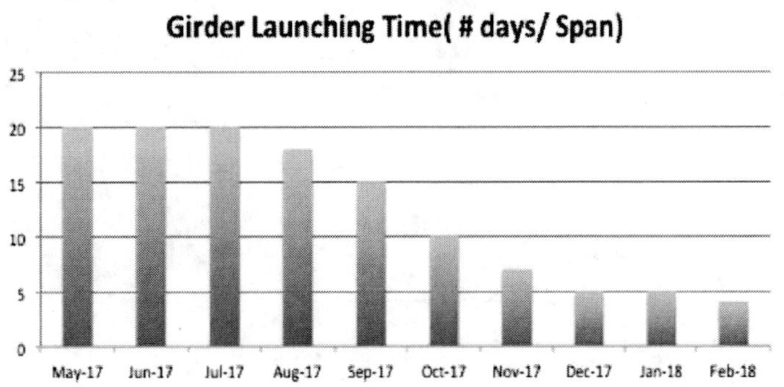


Figure 6.9: Improvement of Span Cycle type over months

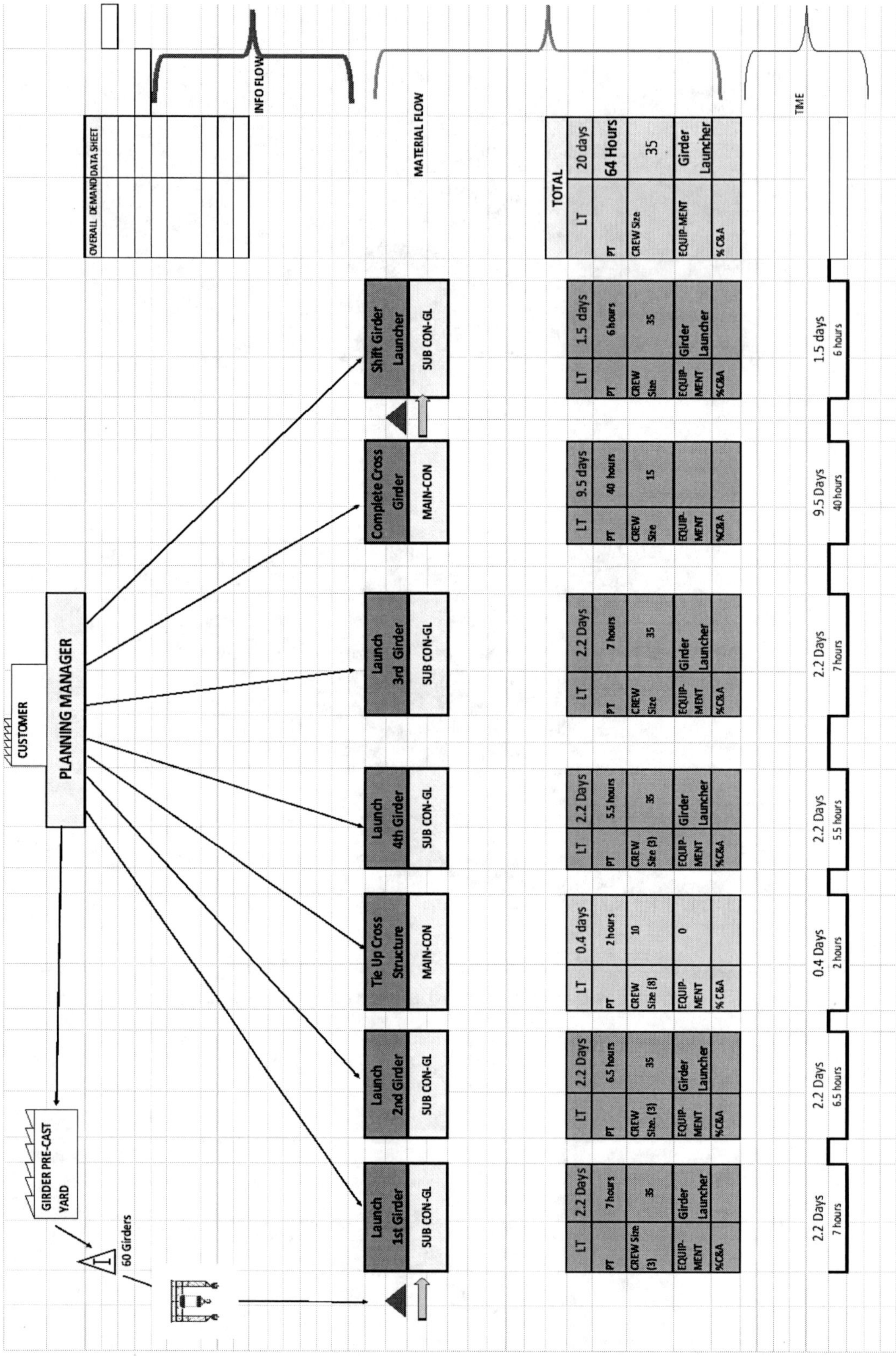


**Figure 6.10: Current State of Girder Launching Value Stream**

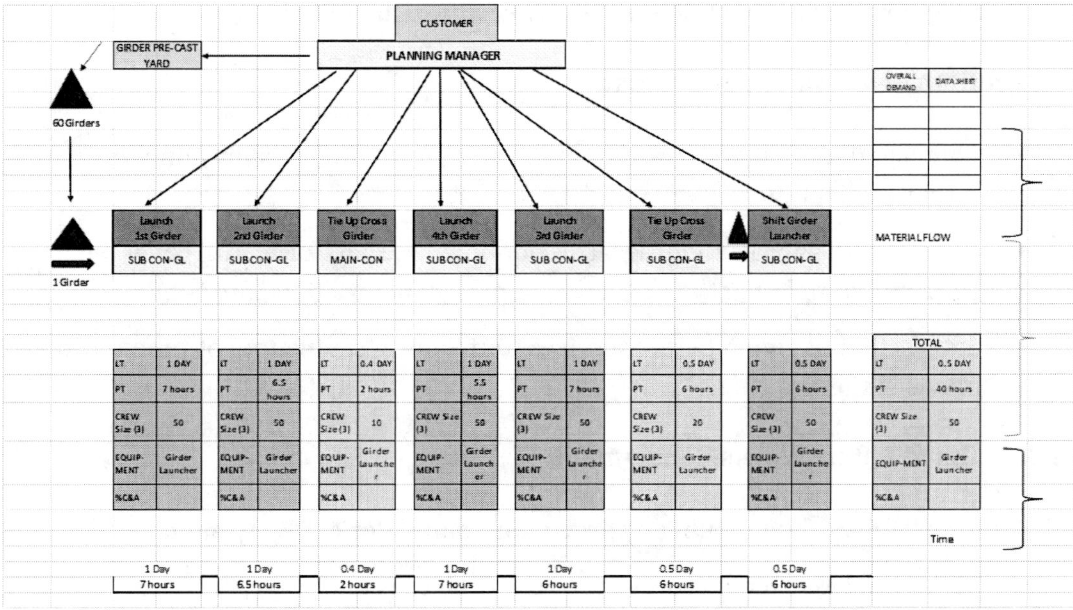


**Figure 6.11: Future State of Girder Launching Value Stream**

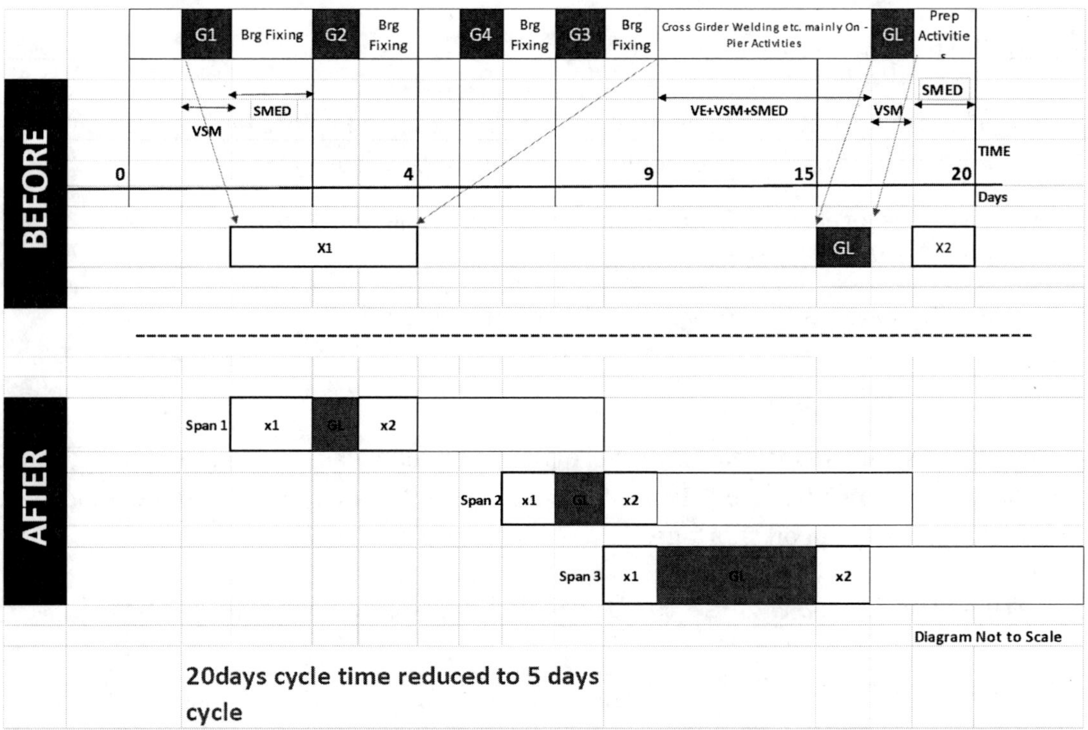


**Figure 6.12: Rough Scheduling of Span-wise Girder launching before and after changes**

**The overall benefits of this VSM exercise are summarized as below:**

**1. Benefit to Government of Odisha:**

- Completed the Project 3 months in advance, from the officially announced date of formal inauguration.
- Good Public appreciation because of above and lots of media coverage, for the first time completing the Government project before time, for public.
- Satisfaction of giving better services to people & saving lots of time for logistics between the two cities.

**2. Benefits to Public Works department:**

- Chief Minister's and public appreciation to the Works Department.
- Big and positive media coverage to the Project Team.
- Excellent Project team's image in the Works department
- Sponsorship of 3 team leaders to two international Conferences, to present this case.

**3. Benefits to Contractors & Consultants:**

- Increased profit, due to several savings to his fixed Project Budget, like Site Cost for 3 months and Early Completion Bonus for the project.
- Listed as preferred Contractor for future projects of Works Department and hence their business growth.
- Public applauds for Contractor, which was quite rare.

**4. Benefits to End-Users:**

Petrol & Time savings, since earlier travel of 36 km by road was reduced to 3 km by bridge that too 3 months in advance. This was estimated to be about 1 million Litres, which was very high in the days of oil shortages.

## 6.6 Annexure II: VSM Case Study: Raft Foundation (for 14 Floors Tower)

**<u>The Project:</u>**

This case refers to a commercial-cum-residential complex in the UAE. There were four buildings of 14 floors each, one retail mall, another commercial building with many shops for lease, one hotel, and one residential building having apartments. The total value of the project was equivalent to ₹900 crores, with a floor area of more than 1.2 million square feet. The project duration for construction and handover was agreed at 24 months, with a severe delay penalty clause. A large number of workmen had to be imported from India to optimize project costs.

Somehow, the start of the project was delayed — primarily due to indecision from the client's side — although the contract with a specified completion date within 24 months had already been signed. Because of this, there was a 2-month delay in starting the raft foundation work for the 14-storeyed towers.

However, the client insisted on adhering to the original contractual completion date, thereby allowing only 22 months for the main contractor to complete and hand over the project. This became especially difficult and carried high risk due to the heavy penalty clause included in the contract.

This was a significant problem, but at that stage, the main contractor did not want to lose the contract and decided to attempt accelerating project progress by leveraging Lean Construction concepts.

This scenario led to the following problems and challenges. How some of these were overcome is the focus of this brief case study.

### Problems & Challenges Observed

A good deal of time was spent on the site for initial observations and discussions with key internal and external stakeholders — including the main contractors, some of the key sub-contractors, the main smart steel supplier, and even some of the foremen involved at the project site. This was done to understand the current problems in detail for the project at that stage.

### The following were the key issues identified:

- There is a heavy penalty involved for delaying the handover of the project — to the tune of approximately ₹18 crores. Therefore, the project timeline must be reduced without incurring significant additional investment.
- Another very big constraint was the mobility of the RMC vehicles to the project site, which was located in the middle of a densely inhabited residential area. Due to their restricted movement, bringing a huge quantity of RMC to the project site was planned only on Fridays. In Dubai, heavy vehicle movement was permitted from 10 p.m. on Thursday, throughout Friday, and up to 5 a.m. on Saturday. Since the total requirement for each foundation was approximately 9,000 cubic meters, only four pours were planned for the raft foundation, to accommodate this restriction. Figure 6.13 gives the drawing of the four pours, and Figure 6.15 shows the schedule for the four pours, with a current planned duration of approximately 18 weeks (121 days, to be exact).

- There was statutory inspection and approval of the Rebar Lay by the Dubai Municipality representative, before Cement pouring was allowed. Once he did not approve the Steel Rebar Layout on Thursday, and the whole concreting plan for one pour was delayed by one week. This was possible some other time as well.
- As per the original plan of the main contractors, the column, beam, and slab work should have started by now. Accordingly, they had imported a large number of carpenters from India, as the visa and other requirements for bringing people on work permits took a considerable amount of time. This led to a large number of carpenters being idle for most of the time, as the ongoing work involved only steel rebar, which did not require many carpenters. This effectively meant that the steel rebar work progressed far slower than required, while a significant number of carpenters remained idle.
- Four days of Gemba walk at site, also revealed that the Fitters start the real work after about three hours of time punching. This was mainly because the required Steel Bar for that day were not available in time, despite a large amount of inventory of Smart Steel in the Yard. The Supplier was very regular in supplying 60 Tons of Steel every alternate day, as per the contract. But they were not supplying the ones that were required, because they also did not know what is required for the next day, or next 2 days.

Next section explains the focus on VSM study, and the recommendations which came out of this. Finally, we explain the benefits of the whole study.

### The Focus of VSM and the recommendations arrived for the project:

The VSM Study Team was formed, involving members from Design, the Main Contractor, the Civil Structure Contractor, the Procurement Representative, the Project Planner, and the Lean Expert. Some illustrative figures are included for better understanding, although not all are from the actual project due to company policy. Figure 6.13 shows the drawing of the four pours, and Figure 6.14 depicts the concreting of the raft foundation for one pour. The following are the key points and the final agreed recommendations that emerged from the critical examination and innovative proposals.

- It was realized for the current state of the project, the real VALUE for the Main Contractor is not the Construction COST but is Construction TIME, because of the severe penalty for delays. Therefore, if the time duration can be reduced even at some extra cost, it will be fine.
- We realized that the smart steel required was being delivered to the site on alternate days, without concern for Friday restrictions. It was then found that heavy-duty vehicles could actually move in the city at night, from 10 p.m. to early morning at 6 a.m. With this knowledge, it was agreed to increase the number of pours from four to seven and to create smaller pours, so that regular weekdays could also be utilized for pouring. Figure 6.16 shows the drawing of the seven

pours plan. However, this led to increased man-hours and material consumption due to the extra construction joint length of 335 meters resulting from the seven pours. It also required special additional care to ensure the quality of the joints. These details are provided in Figure 6.17. Nevertheless, this strategy led to 33 days of early completion by effectively leveraging regular cement concreting days.

- On the analysis because of Dubai Municipality's (DB) representative rejection, it was found that the standards for amount of 'overlap' of steel bars in the Foundation Cage for the pour were different. What was being followed by Designers and Quality Team was different than what was being followed by DB. There were three different standards. That was brought to one common standard, after discussions with all relevant authorities, including DM.
- The carpenters were made to understand the project difficulties and the delay. They eventually agreed to learn and perform the wire-binding work for the cage, which is actually a fitter's job. It was clarified to them that they were not being converted into fitters, but would only need to learn and perform one specific fitter's task — laying and wire-binding. A local trainer was engaged for 10 days to train the carpenters properly and also to supervise and guide them for another 10 days. In return, and within the company's policy, they worked for 10 hours only, although they were present for 12 hours. This allowed them to earn extra income by taking on multi-tasking responsibilities during a period when there was no work assigned to them. A small metallic rod tool with lugs welded at the required spacing between the rebars was also provided to them — to ease their work and prevent mistakes.
- Bar Bending Schedule, popularly known as BBS, surprisingly had no Schedule (meaning time-table) attached to it. It is simply a Bill of Materials. In a joint meeting with Supplier, the Planner was requested to provide the Date-wise Dispatch rolling schedules for the required Smart Bars in next two trips. The supplier will prepare the Trailers' logistics as per that only to make 60 tons. This will minimize time lost in locating and identifying steel for the site. In fact, the required material was transferred to the site the previous working shift, to eliminate long waiting time for that.

All these modification and elimination of non-value adding activities achieved big gains. These benefits are listed in Figure 6.17

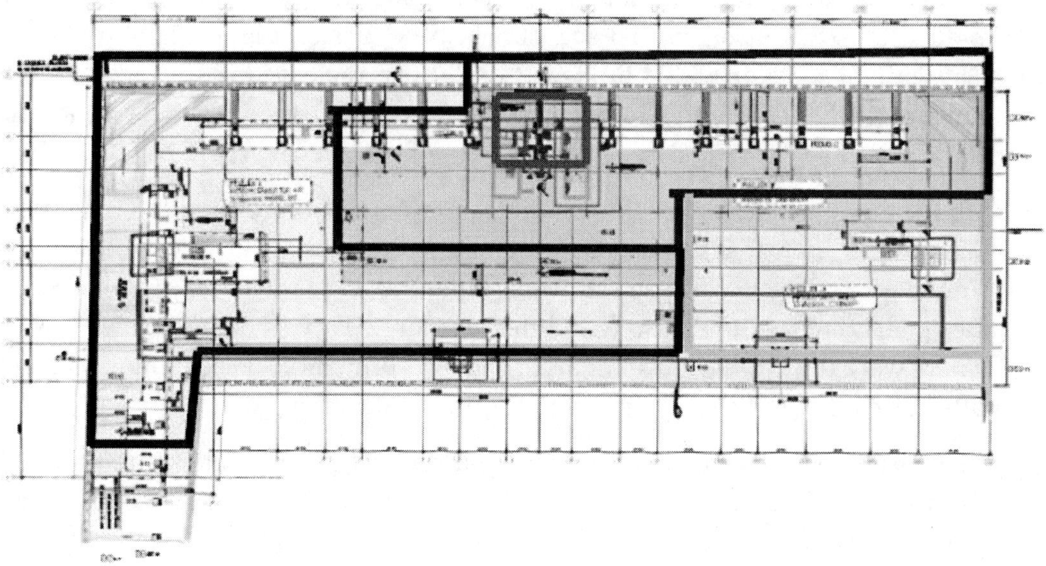

Fig 6.13 The Four POUR Plan for the Raft Foundation (Current way)

Fig 6.14 Concreting for one of the pours (Illustrative, not from real project-site)

| Project XYZ | | | Can Do? (Enter | | Week 1 | Week 2 | Week 3 | Week 4 | Week 1 | Week 2 | Week 3 | Week 4 | Week 1 | Week 2 | Week 3 | Week 4 | Week 1 | Week 2 | Week 3 | Week 4 | Week 1 | Week 2 | Week 3 | Week 4 | Week 1 | Week 2 |
|---|---|---|---|---|---|---|---|---|---|---|---|---|---|---|---|---|---|---|---|---|---|---|---|---|---|---|
| 15-11-2014 | | | | Duration | sun | sun | sun | sun | sun | sun | sun | sun | sun | sun | sun | sun | sun | sun | sun | sun | sun | sun | sun | sun | sun | sun |
| Activities | Start Date | Finish Date | | | 21-09-2014 | 28-09-2014 | 05-10-2014 | 12-10-2014 | 19-10-2014 | 26-10-2014 | 02-11-2014 | 09-11-2014 | 16-11-2014 | 23-11-2014 | 30-11-2014 | 07-12-2014 | 14-12-2014 | 21-12-2014 | 28-12-2014 | 04.01-2014 | 11-01-2014 | 18-01-2014 | 25-01-2014 | 01-02-2014 | 08-02-2014 | 15-02-2014 |
| Blinding (Bli) | 27-09-2014 | 30-Sep-14 | | 3 | | | | | | | | | | | | | | | | | | | | | | |
| Water proofing (W/P) | 01-10-2014 | 03-Oct-14 | | 2 | | | | | | | | | | | | | | | | | | | | | | |
| Rebar for raft (Rr) | 04-10-2014 | 01-Nov-14 | | 28 | | | | | | | | | | | | | | | | | | | | | | |
| Raft Concrete (Rc) | 15-10-2014 | 06-Nov-14 | | 22 | | | | | | | | | | | | | | | | | | | | | | |
| | | | | | | | | | | | | | | | | | | | | | | | | | | |
| Blinding (Bli) | 20-10-2014 | 25-Oct-14 | | 5 | | | | | | | | | | | | | | | | | | | | | | |
| Water proofing (W/P) | 23-10-2014 | 29-Oct-14 | | 6 | | | | | | | | | | | | | | | | | | | | | | |
| Rebar for raft (Rr) | 30-10-2014 | 20-Dec-14 | | 51 | | | | | | | | | | | | | | | | | | | | | | |
| Raft Concrete (Rc) | 02-12-2014 | 25-Dec-14 | | 23 | | | | | | | | | | | | | | | | | | | | | | |
| | | | | | | | | | | | | | | | | | | | | | | | | | | |
| Blinding (Bli) | 06-12-2014 | 10-Dec-14 | | 4 | | | | | | | | | | | | | | | | | | | | | | |
| Water proofing (W/P) | 09-12-2014 | 12-Dec-14 | | 3 | | | | | | | | | | | | | | | | | | | | | | |
| Rebar for raft (Rr) | 13-12-2014 | 10-Jan-15 | | 28 | | | | | | | | | | | | | | | | | | | | | | |
| Raft Concrete (Rc) | 31-12-2014 | 15-Jan-15 | | 15 | | | | | | | | | | | | | | | | | | | | | | |
| | | | | | | | | | | | | | | | | | | | | | | | | | | |
| Blinding (Bli) | 30-12-2014 | 03-Jan-15 | | 4 | | | | | | | | | | | | | | | | | | | | | | |
| Water proofing (W/P) | 01-01-2015 | 05-Jan-15 | | 4 | | | | | | | | | | | | | | | | | | | | | | |
| Rebar for raft (Rr) | 06-01-2015 | 20-Jan-15 | | 14 | | | | | | | | | | | | | | | | | | | | | | |
| Raft Concrete (Rc) | 10-01-2015 | 22-Jan-15 | | 12 | | | | | | | | | | | | | | | | | | | | | | |
| | | | | | | | | | | | | | | | | | | | | | | | | | | |

**Figure 6.15: The schedule for key activities of the four Pours**

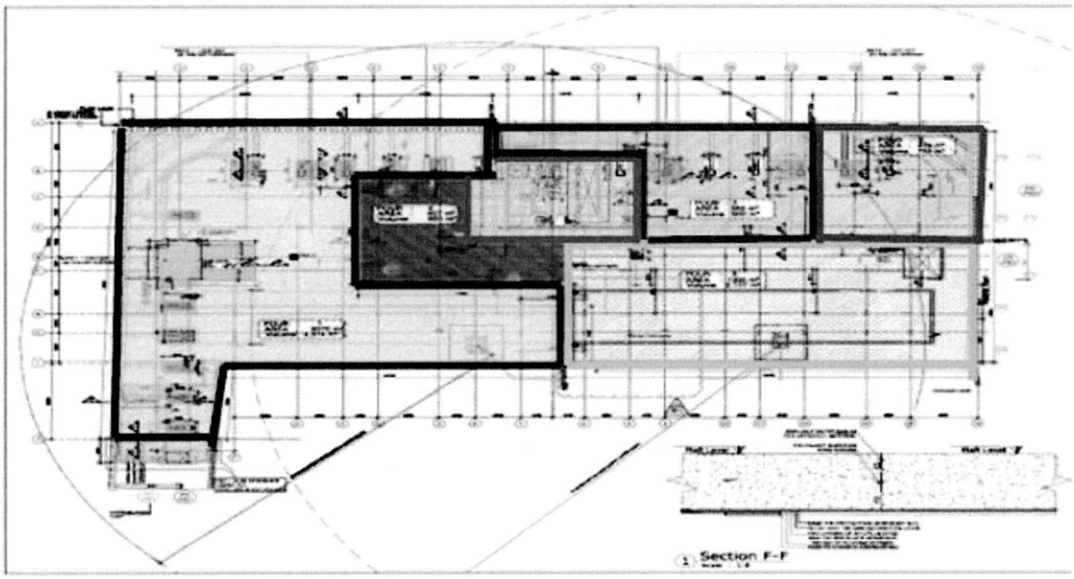

**Figure 6.16 The Seven POUR Plan for the Raft Foundation (Proposed way)**

| POUR PLAN COMPARISON | | | | | | | | | | | | | |
|---|---|---|---|---|---|---|---|---|---|---|---|---|---|
| | | | Steel Manpower engaged | | | | | | Shuttering Manpower engaged | | | | |
| Pour # | Rebar Duration ( Days) | Rebar Quantity (Tons) | Fitters | Helpers | Trained arpenters for Steel | Concreting Duration (Days) | Concreting Quantity (CuM) | Shutterin g Quantity (Sq. M) | Carpenters | Helpers | Trained Fiter for Shutterin g | Steel Productivity (For Tying Wires) | Shuttring Productivity |
| 4 | 121 | 903 | 1230 | 61 | 351 | 72 | 8603 | 1926 | 150.4 | 26 | 527 | 0.55 | 0.81 |
| 7 | 88 | 908 | 753 | 56 | 307 | 69 | 8603 | 2323 | 131.4 | 24 | 323 | 2.73 | 4.86 |
| | 33 | | | | | | | 397 | | | | | |

**Figure 6.17 The comparison of the current and proposed Status after VSM**

## The overall benefits of this VSM exercise are summarized as below:

### 1. Key Benefits achieved for Main Contractor:

- Project Team realised that in this case the Construction TIME was more VALUABLE here than the Construction COST, as delay penalty was very high.
- Raft Foundation Duration Time reduced from originally planned 121 days to 88 Days. This reduction of 33 Days for each of Four Raft Foundation that is about 27%. This also significantly increased the chances of completing the project in 22 months.
- This reduction of raft Foundation duration saved potentially more than 18 Crores of Indian Rupees for the Main Contractor, because of avoiding potential future delay of the project by 33 days.
- Workmen Group Size was optimized and reduced the manpower required by 20%. This increased the workers 'utilization (for Carpenters, and Fitters and their Helpers) significantly. This also increased Steel Rebar Productivity by more than 50%.
- Removed the confusion about the Quality Standards among different agencies, and brought uniform Quality Codes for everyone.
- Achieved better collaboration and relationship with the Smart Steel Suppliers, and avoided unnecessary handlings of heavy Steel Bars at the Project Site Yard.
- Redefined the traditional meaning of BBS, and started included Time Schedule, along with the basic Bill of Materials in that terminology, when they share the BBS Document with the Suppliers.
- Reduced the level of stress for everyone working at the Project-site

## 2. Key Benefits to Workmen:

- Workers received on-the-job training, which was quite rare those days. In fact, even now, there is low focus on Workers' training.
- Workers Increased their Productivities
- Workers also increased their earning, without working overtime.
- Workers accepted to work on 'Task-Based' Multi-Skills as a practice in future too, if some small incentives are provided to them. This is a good practice in Construction.

# CHAPTER 7

# PROACTIVE DEFECTS PREDICTION & PREVENTION

# 7. Proactive Defects Prediction & Prevention

In modern infrastructure and construction domains, proactive defect prediction and prevention are vital for mitigating risks before they manifest. It fosters team accountability, sharpens decision-making, and supports reducing impacts linked to poor quality. It adds value to quality management, not as not a cost burden, rather as a strategic enabler for delivering durable assets with optimized lifecycle performance.

## 7.1 Need and Benefits of Quality Management- for Improved Quality at Lesser Cost:

The need for quality management stems from its ability to systematically control variability, ensure compliance with design intent, and reduce rework costs. By integrating real-time analytics and risk-based monitoring, teams can foresee potential failure zones and address them early in the lifecycle. The benefits of effective quality management include improved project outcomes, reduced cost overruns, and enhanced resource efficiency. It aligns with Lean principles by minimizing non-value-adding activities and driving continuous improvement.

The Sections below illustrate these facts further.

### 7.1.1 Need for Better Quality Management

**The snags or defects in any construction are expensive and problematic because of the reasons like:**

- Lack of skilled & qualified manpower to 'critically' inspect the defects or snags. People are not formally trained to identify the defects *with a critical eye*. Hence, often defects are missed to be captured by the people conducting snagging.
- Capturing, rectifying, tracking, and finally accepting defects is a tedious and highly manual process, despite many stakeholders using Excel sheets. The actual status of a defect — *whether it is rectified, partially rectified, difficult to rectify, or not rectifiable at all* — often becomes a point of dispute among different stakeholders during coordination and review meetings.
- The rectified quality of any defect is generally *not as good as the original first-time-right execution*. Therefore, differences often arise among stakeholders regarding the final quality of the building at the time of handover
- Rectification of defects *incurs significant costs*, resulting in shrinkage of profit margins for contractors. There is no prevailing culture or practice of predicting the quantum and type of defects/snags in advance, which could enable their prevention. Most traditional CAPA (Corrective and Preventive Actions) approaches fail to address the needs of the current real estate structure — both in terms of quality assurance and the required speed of execution.

- Rectification also adds further to the *delay in project completion* and causes *inconvenience to customers* once the facility is occupied and operational. In addition, such delays may often incur penalties, leading to further reduction in profit margins. This is especially significant after the introduction of RERA statutory regulations by the Indian Government.
- Lastly, one of the most accepted reasons for bad quality of products is also the poor workmen skills and poor supervision.

Because of all these reasons, there is strong need to move from the philosophy of reactive approach of "**Finding & Fixing**" the defects, to proactive approach of "**Predicting & Preventing**" the defects.

### 7.1.2 Benefits for better prediction & prevention-based Quality Management

The key benefits lie in first predicting, and then preventing, critical defects during construction. To achieve these benefits, it is essential to understand the concept of Defects Lifecycle Management as it applies to construction.

This is explained in detail below in Figure 7.1.

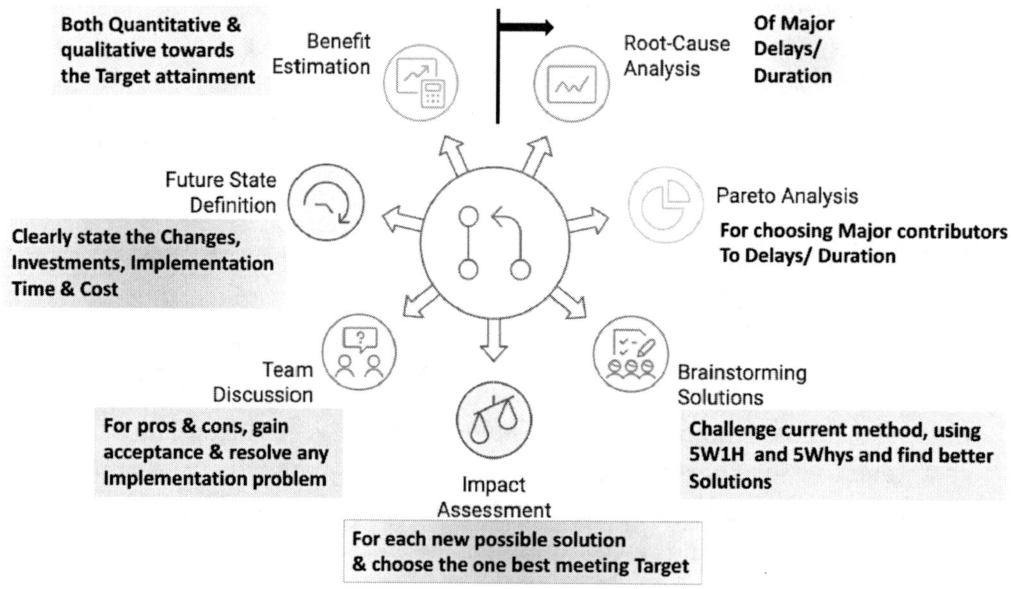


**Figure 7.1: Benefits of Predicting & Preventing Defects**

### 7.1.3 Introducing Defects (Lifecycle) Management for Construction:

The five stages of the Defects Lifecycle Management are shown below, in Figure 7.2 .

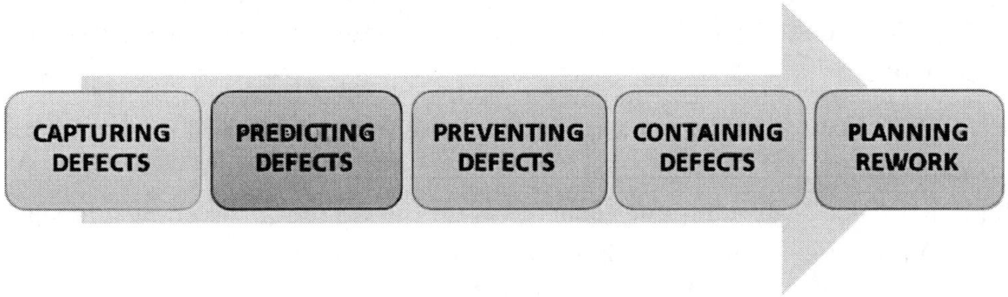


**Figure 7.2 Five Stages of Defects Lifecycle Management.**

**Capturing Defects**

In the first step, defects are identified and captured by experts visiting the project site at various stages of completion — from foundation to handover. These defects, both interior and exterior, must be fixed before handing the property over to the owners. About 80–90% of defects are similar across most building projects, yet many are missed, leading to wasted effort later. After collecting this data, defects can also be proactively identified using historical data from similar projects — such as issues in work packages like plastering, false ceiling finish, or items like switch sockets and dado tiling. This preparation helps avoid recurring problems. The following steps are recommended to be followed for this:

**Predicting Defects**

Predicting defects is about *forecasting potential issues* before starting the project, or during the construction process, for the remaining work at any stage. If you can forecast the *number and type of defects* in a given project — even before its start — you can plan to *prevent the most critical ones, thereby saving project costs* associated with poor quality and delays.

**Therefore, this step asks the questions like:**

1. Can we predict the total number of defects in a complete project, even before it starts? That is can we say, a 25 storied Residential premium tower with 150 premium apartments is expected to have 10000 defects at various stages of completion?
2. Can we predict the total number of defects in the total blockwork required for this building, even before the building work starts?

3. Can we predict the total number of defects for the building in total tiling work? Likewise, for each work-item?
4. Can we predict the total number of defects in interior plastering of the building in the next two months of work scheduled, so that those can be prevented and better controlled?

All this is possible but unfortunately no one does this, and pays a big price for defects capturing and then rectifying. This forecasting can be based on Historical Data of the similar project or by sampling techniques on the current project. Besides predicting the type and number of defects, this step can also be leveraged for forecasting the total rework efforts, total cost of rework and even total project delays because of that rework. More details on the process of predicting are provided in next Section.

**Preventing Defects**

After knowing the type and number of defects, the focus must be to prevent the defects which are critical, expensive to repair or very occur frequently. This would result in handing-over the buildings to owners, faster and of better quality, also leading to cost saving for the contractors. Section 7.2 provides 14 effective methods to prevent defects.

**Containing Defects**

This step is primarily to improve the accuracy of prediction. The predicted numbers and actual number of defects are compared, the reason for this variance is established (like poor skilled workforce, very complex design etc.). If required the 'Defect Density' parameters are updated for future, for more accurate predictions.

**Planning Rework**

While rework is a reality, unfortunately the workforce and time required for it are not formally accounted for in the project planning process. It is advisable to *integrate expected rework into the production plan of CPS (Collaborative Production System).*

Based on Defect Lifecycle Management, the steps are suggested in the next Section to proactively prevent defects.

## 7.2 Process for Defects Management for Construction

The process for defect management for projects during construction involves the following steps, and helps in reducing delays, reworks while managing the defects in a structured manner.

### 7.2.1 Proactively Capture Defects with A Gemba Walk for Defects.

To reduce project delays, rework costs, and improve quality, *prioritize the most frequently occurring defects* across various work packages and work items. You can utilize historical data from similar past projects, or collect sampling data from the current project. Additionally, identify and prioritize those defects that have been reported as *causes of variance in the project schedule* during *the look-ahead planning period of the next 4–6 weeks.*

**Follow the steps listed below in Figure 7.3 for proactive capturing defects.**

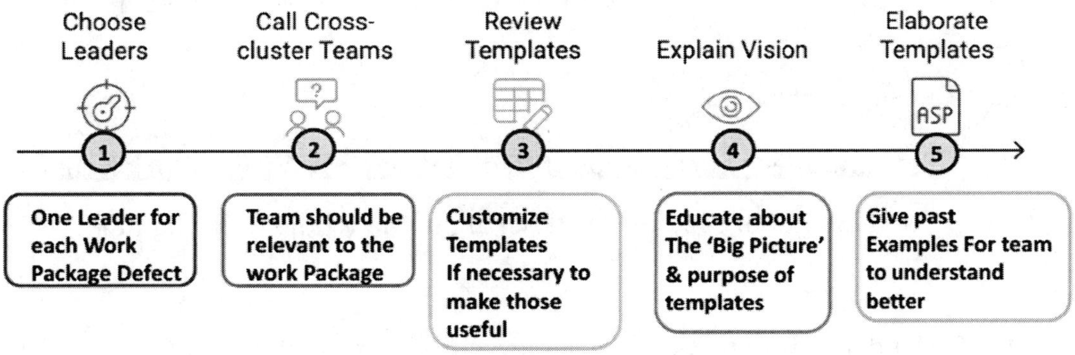


**Figure 7.3: Process of Gemba Walk to capture Defects**

## 7.2.2 Predict The Potential Number, Types Of Defects and Efforts for Rework for The Rest of The Project.

The Prediction can be based on either historical data of defects in the similar projects, or based on sampling done on the ongoing project at different level of completion. The steps involved are shown below in Figure 7.4.

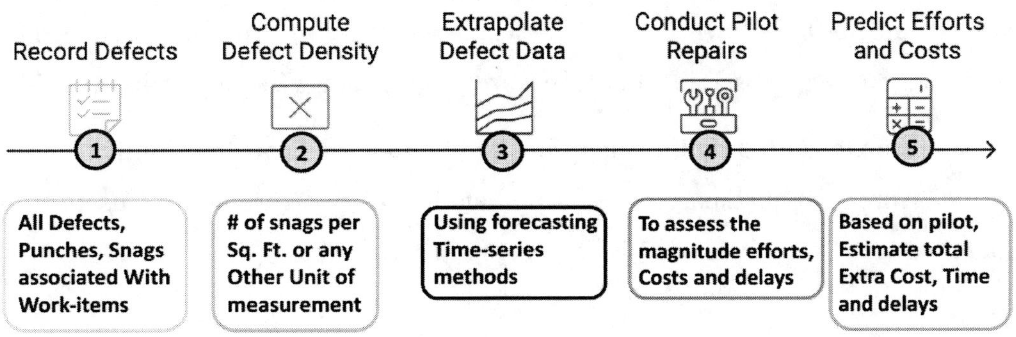


**Figure 7.4: Process of Predicting Defects**

To predict the number of defects, a key parameter used is 'Defect Density' (also referred to as 'Snag Density'). This metric is conceptually similar to 'Bug Density' in IT software development. Defect Density is defined as the number of snags identified per unit of relevant measurement. In building construction, this unit can be the total constructed area in square feet. Alternatively, it can be expressed in terms of the quantity of specific work items associated with the type of snag observed. Below are mentioned two examples of these definitions:

- Total Defect Density = $\dfrac{\text{Total number of defects identified in the whole building}}{\text{Total constructed floor area of the building in Sqm/ Sqft}}$

- Defect Density of Electrical Wiring =

$$\frac{\text{Total number of defects identified in Electrical Wiring Work Item}}{\text{Total Length of the Electrical wiring in the whole Building in Mtr}}$$

In fact, it depends upon you, to what level of accuracy you want to measure and predict the type of defects and number of defects. It may look more accurate to you to talk in terms of Item-wise snags, but then the cost of collection of the data at that level will be high but the accuracy of the prediction will be low, because of availability of quantum and quality of data. You have to trade-off between the accuracy and the cost involved.

Figure 7.5 gives an idea about how, you can drill down the Building Structure to more detailed level, for understanding and predicting the specific type of repetitive snags at those levels. This figure just shows an illustrative number of four levels, and does not show the complete Drill-Down Tree for the whole building. But you can get an idea from this. Table 7.1 gives the possible unit of measures for the Snags Density at different levels.

One simple rule to define Snag Density for Buildings construction is to use the Unit of Measurement for snags, same as what unit of measurement is used for the commercial Piece-Rate of various Work-Items in the Contract.

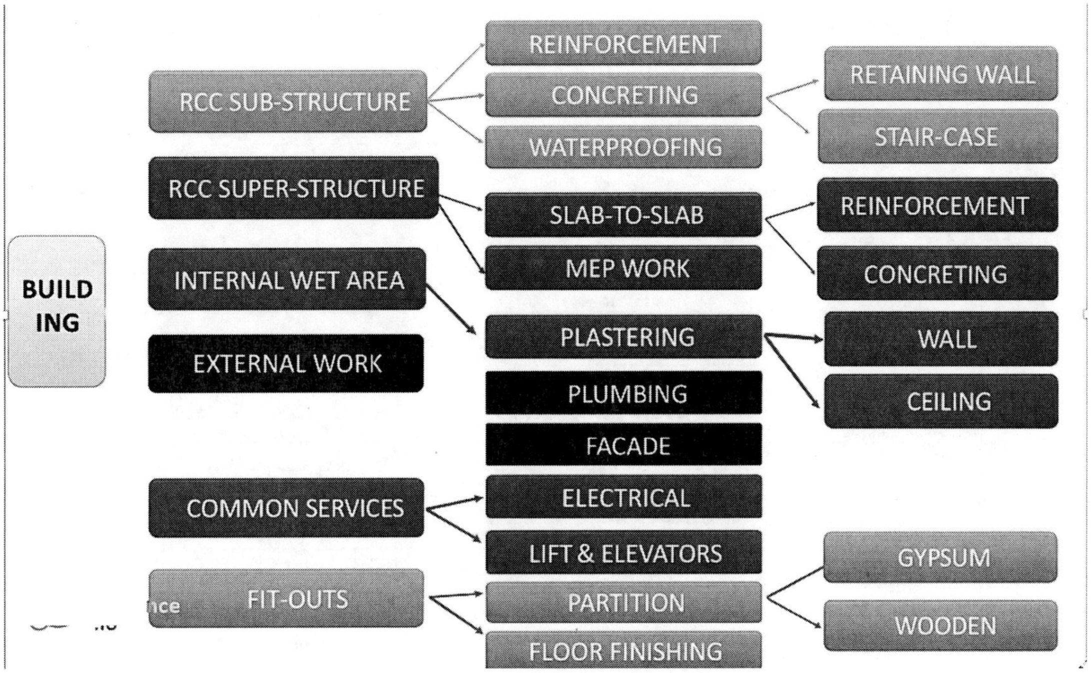


**Figure 7.5: Four levels of details to consider Defect Density**

You can refer Tables 7.1 to 7.5 in this chapter, for some more understanding of the type of defects for various Work-Packages and Work Item.

Once the defect density definition is understood, predicting the number and type of defects is derived from the traditional forecasting methods. In construction, the prediction can be based on two different ways:

- Prediction based of Historical data
- Prediction based on Sampling

**Prediction based on Historical Data:** This can be used if the construction company has maintained the snagging data of previous several projects. The type of prediction would be based on the level of defect-density data available. We also explained this in the diagram and the table above.

**Table 7.1 Defining different levels of Building Details for Defect Density (Illustrative only)**

| SAMPLE DEFINITIONS OF DEFECT DENSITY in CONSTRUCTION | | | |
|---|---|---|---|
| **LEVEL Number** | **Item Description in brief** | **Possible Unit of Measure** | **Possible Definition of Defect Density** |
| 1 | Whole Building | Total Constructed Floor Area , in Square Feet (SFT) | Snags per Square Feet |
| | | 500 Apartments | Snags per Apartment |
| | | | |
| 2 | RCC Sub Structure | Total Amount of Concreting (Cubic Meter: CM) | RCC Snags/ Cubic Meter |
| | RCC Super Structure | Total number of Floors / Slabs | Snags per Slab |
| | Internal Wet Area | Total Constructed Floor Area , in Square Feet (SFT) | Snags per Square Feet |
| | External Building Work | Total Constructed Floor Area , in Square Feet (SFT) | Snags per Square Feet |
| | Common Services | Total Constructed Floor Area , in Square Feet (SFT) | Snags per Square Feet |
| | Fit-outs | Total Constructed Floor Area , in Square Feet (SFT) | Snags per Square Feet |
| | | | |
| 3 | Re-inforcement | Total amount of Steel Used in Metric Tons | Snags per Ton. Of Steel |
| | Concreting | Total Amount of Concreting (Cubic Meter: CM) | RCC Snags/ Cubic Meter |
| | Water proffing | Total Area of Water proofing Square Meters (SM) | Snags per Suare Meters |
| | | | |
| | Slab-to-Slab Soper-structure | Total number of Floors / Slabs | Snags per Slab |
| | Internal Plastering | Total amount of Plastering (Square Meters SM) | Snags per Sq. Meters |
| | Fasade work | Total number of Floors | Snags per Floor |
| | Fasade work | Total amount of Fasade work (Square Meters SM) | Snags per Sq. Meters |
| | Lifts | Total Number of Lifts | Snags per Lift |
| | Floor Finish | Total Constructed Floor Area , in Square Feet (SFT) | Snags per Square Feet |
| | Partition Wall | Total number of Partitions | Snags per Partition |
| | Partition Wall | Total area covered by Partition Walls (Square Meters SM) | Snags per Sq. Meters |
| 4 | Stair Case | Total number of Steps in the stairs | Snags per step |
| | Retaining Wall | Total area covered by Retaining Walls (Square Meters SM) | Snags per Sq. Meters |
| | | | |
| 5 | Windows | Number or area (Square Meters) | Snags per number or per Sq. M. |
| | Doors | Number or area (Square Meters) | Snags per number or per Sq. M. |
| | Laying of Pipes | Length in meters | Snags per meter |
| | Electrical Fittings | | |

**For such forecasting, we always assume that:**

1. The historical data used for snag-density for quality of finishes, is from the same type of buildings and its design complexities (for example, the snag density for wall finish may not be same for a five-star hotel building and a MIG apartment building)
2. The type and level of skills of the workmen used in the current building project is like the one in the previous projects.

**In general, if:**

1. $dd_1, dd_2, dd_3, \ldots\ldots dd_n$ are the **defect densities** of work-items number 1, 2, 3,……, n
2. $cr_1, cr_2, cr_3, \ldots\ldots cr_n$ are the **cost of repairs** for a unit item of work-items number 1, 2, 3,……, n
3. $WI_1, WI_2, WI_3, \ldots\ldots WI_n$ are the total number of units of **work-items** number 1, 2, 3,……, n, required in the whole Building project

Then, total Number of Defects (ND)in the whole project,

$$ND = \sum dd_i \times WI_i \quad (i= 1 \text{ to } n) \quad \ldots\ldots\ldots\ldots\ldots\ldots(1)$$

And, Total Cost of Rework (TCR) for the whole project

$$TCR = \sum cr_i \times dd_i \times WI_i \quad (i= 1 \text{ to } n) \ldots\ldots\ldots\ldots\ldots\ldots(2)$$

These details are used in the case study discussed in Annexure 7.2 in great details. Annexure 7.1 gives an example about how to predict the number of snags for a new building using historical data.

### 7.2.3 Prioritize the correcting of the defects, analyse the causes to prevent forthcoming defects

The next step is to prioritize the defects based on criticality criteria, specially from Business Impact and Rework Cost point of view. Follow the steps listed below in Figure 7.6.

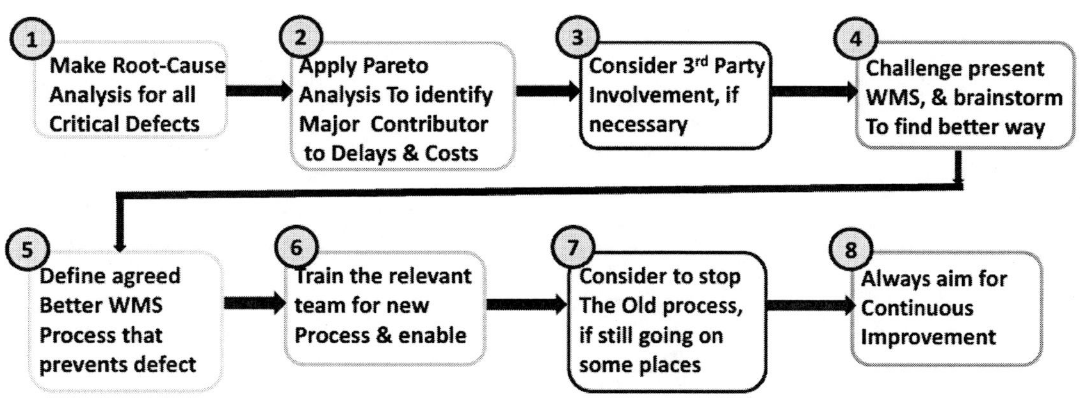


**Figure 7.6: Process of prioritising Defects for prevent**

### 7.2.4 14 Ways to prevent Defects to happen and avoiding rework

This section outlines fourteen methods to prevent or minimize potential snags in construction. Some common methods can be enhanced with minor adjustments, while other newer methods are adopted from sectors like Manufacturing and IT. The goal is to make all methods more effective.

## 7.3 Effectiveness of Methods

Brief descriptions of these fourteen methods are explained below.

### 7.3.1 Smart Master Defects Album

We are surprised to find that the construction industry repeatedly makes the same mistakes in the same project or in similar projects, leading to the same list of snags and defects. Lessons from past projects are often ignored, causing each new project to start from scratch.

Around 150 common work-items, such as blockwork, tiling, and window fixing, cover 90% of building projects, while infrastructure projects like railways or roads have different items. Defects are usually similar, so educating Engineers, Supervisors, and Workers about common defects is key. A learning manual can include defect criticality, rework efforts, and stages of origin-like design or procurement. Figures 7.7 and 7.8 provide examples of defects in building finishes and road or bridge construction.

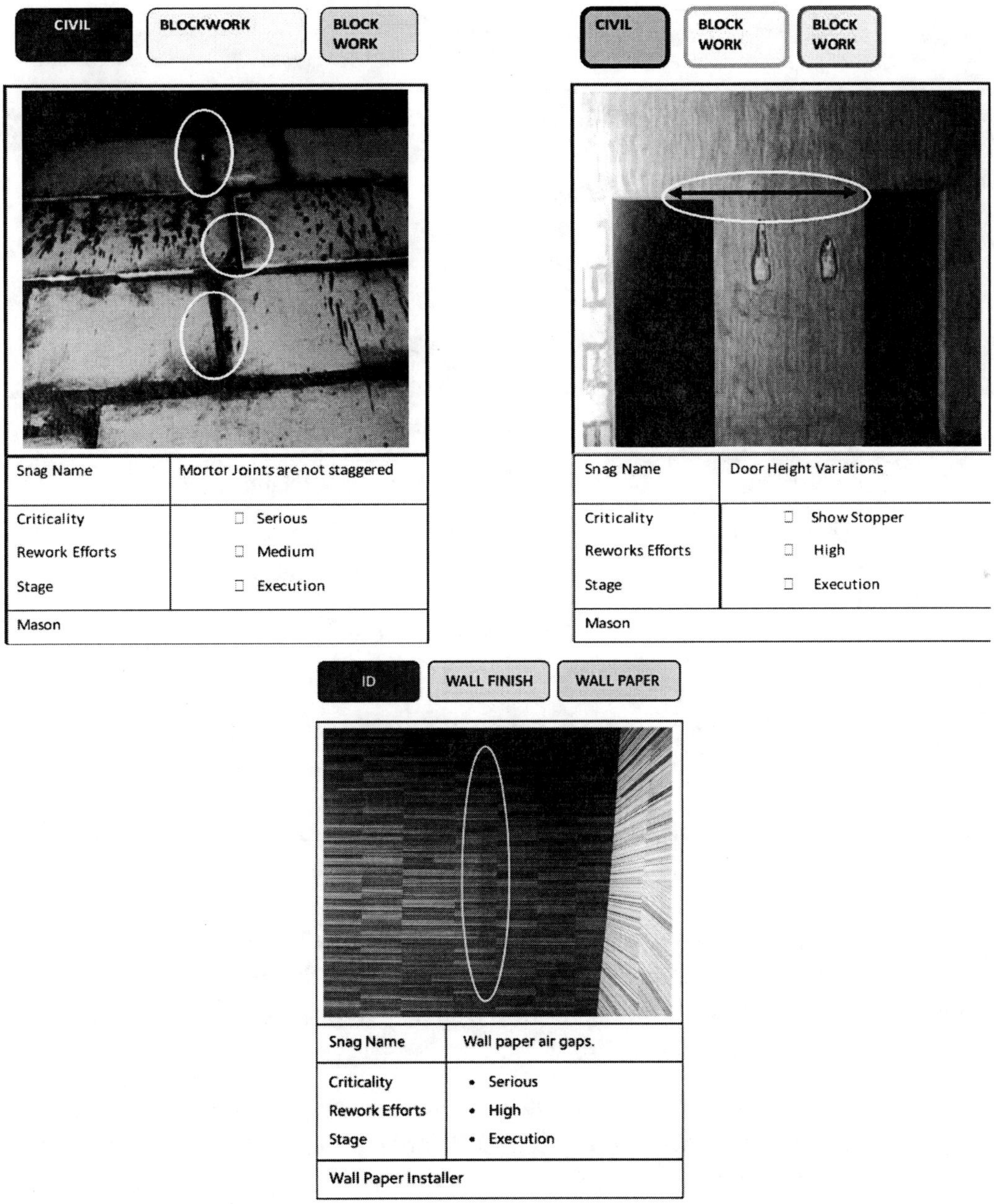


**Figure 7.7: Sample page from Smart Snag Album for real estate projects**

The Snags Album can be made into a digital format like a PDF, shared via cloud servers or pen drives. Interns or engineering trainees can compile past project reports, offering a cost-effective solution and valuable learning for them. With some investment, the album could include a search feature for easy access on any device, shifting the focus toward snag prevention rather than just fixing them.

CIVIL | PAVEMENT | PAVEMENT

| Snag Name | Crocodile Crack |
|---|---|
| Criticality<br>Rework Efforts<br>Stage | • Serious<br>• High<br>• Execution (incl. BC, WMM |
| Pavement Contractor | |

CIVIL | PAVEMENT | PAVEMENT

| Snag Name | Joint Crack |
|---|---|
| Criticality<br>Rework Efforts<br>Stage | • Serious<br>• High<br>• Execution |
| Pavement Contractor | |

**Figure 7.8: Sample page from Smart Snag Album for infra projects**

### 7.3.2 Smart pre-construction, during & post construction Checklists

You must have seen, Checklist method is widely used to prevent defects and snags, emphasized by main contractors as part of their Quality Management System to win contracts. It ensures key tasks are done correctly before, during, and after operations, reducing reliance on memory or skills. A sample checklist for plastering by a professional Indian organization is shown in Table 7.2.

Though useful, the system is manual and often seen as cumbersome. Hard copies are carried to worksites, which aren't ideal for paperwork, leading to memory-based checks. The process becomes more about compliance than using the knowledge embedded in checklists, with the focus on reducing NCRs rather than improving quality.

Checklists often lack important details like mortar mix ratio, curing time, or maximum waviness, leaving quality control to workers' judgment. Defects usually become apparent only when further work like plumbing or electrical begins, and version control is often inadequate.

Checklists are rarely updated, leading to quality issues. High-quality finishes are expected, but outdated checklists cause problems. For example, poor lighting during plastering caused undulations to go unnoticed, later visible when wallpaper was applied, resulting in costly rework and disputes over expenses. In a case, we were involved, it becomes a big issue when the interior decorator put the wallpaper on plastered wall, and the final finish of wall-paper was rejected by the Customer. A lot of waste (or rework and rectification) was involved. There were issue and argument about who would bear that extra cost. The civil contractor who did the plastering some two months back, was not with the project by then. We observed many similar cases in India, Middle East, and Sri Lanka.

We recommend the following suggestions to improve the Checklist method of defect prevention by making them Smart:

- **a.** Organizations should maintain updated Master Checklists based on experiences from their various other projects at different locations.
- **b.** Checklists should include all essential information to ensure output quality, with a focus on the appropriate instruments and tools for inspection required in the process.
- **c.** Organizations should lease affordable cloud space (e.g., AWS) to store master Checklists, accessible on mobile devices, ensuring the latest version is available to everyone. Refer to fig 7.5 for a sample.
- **d.** Organizations should consider using a mobile app to provide dynamic checklists at the worksite, allowing real-time updates with minimal data entry.

**Table 7.2 Sample Checklist Building Interior Plastering (for Illustration only)**

| | Checklist for Plastering | | | |
|---|---|---|---|---|
| | Project | | Doc. No. | |
| | Client | | Project Location | |
| SITE | | LOCATION | | DATE |

| | | YES | NO | NA |
|---|---|---|---|---|
| | CHECK POINTS BEFORE PLASTERING | | | |
| 1 | Check Drg. No./ Date & Revision for all other service like electrical, plumbing etc., | ☐ | ☐ | ☐ |
| 2 | Safety access platform available? | ☐ | ☐ | ☐ |
| 3 | Check for hacking/ Button marking / Mortar thickness | ☐ | ☐ | ☐ |
| 4 | Is Provision of chicken mesh for conduits, column brick work joints & other done? | ☐ | ☐ | ☐ |
| 5 | Is Surface prepared - Ensure moss/ efflorescence is removed & Surface is Dampened | ☐ | ☐ | ☐ |
| 6 | Is the Material required properly stacked - Cement, sand at various locations | ☐ | ☐ | ☐ |
| 7 | Is Prewetting of surface done? | ☐ | ☐ | ☐ |
| 8 | Is Quality of Sand is maintained using relevant screening silt? | ☐ | ☐ | ☐ |
| 9 | Check the Beam, B/W joints packing with Nito Bond AR | ☐ | ☐ | ☐ |
| 10 | check for Door Frame Projections | ☐ | ☐ | ☐ |
| | CHECK POINTS DURING PLASTERING | | | |
| 1 | Is the Platform for mortar mixing on tray is ready and free from unnecessary items | ☐ | ☐ | ☐ |
| 2 | Are the thickness/ level & line & right angle correct? | ☐ | ☐ | ☐ |
| 3 | Is Proportion of mix correct? | ☐ | ☐ | ☐ |
| 4 | Is the water proofing compound added? | ☐ | ☐ | ☐ |
| 5 | Is the surface roughing for the first coat Proper ? | ☐ | ☐ | ☐ |
| 6 | Is arrangement for collection of mortar spills made? | ☐ | ☐ | ☐ |
| 7 | Is dead mortar removal is planned properly? | ☐ | ☐ | ☐ |
| 8 | Ensure there is no waviness in the surface? | ☐ | ☐ | ☐ |
| 9 | Check for grooves/ drip moulds | ☐ | ☐ | ☐ |
| 10 | Will you ensure application of cement slurry on concrete surface/Space for tile area, cutting for skirting area? | ☐ | ☐ | ☐ |
| 11 | Check for loft thickness and top surface in dead level W/Robe size | ☐ | ☐ | ☐ |
| | CHECK POINTS AFTER PLASTERING | | | |
| 1 | Curing water distribution for various locations | ☐ | ☐ | ☐ |
| 2 | Check for hollowness | ☐ | ☐ | ☐ |
| 3 | Check for cracks | ☐ | ☐ | ☐ |
| 4 | Check of plastering date for curing | ☐ | ☐ | ☐ |

Remarks of Quality Assurance Engineer:

| Signature | | Date | Remarks/ Comments |
|---|---|---|---|
| Checked By: | | | |
| Approved By: | | | |

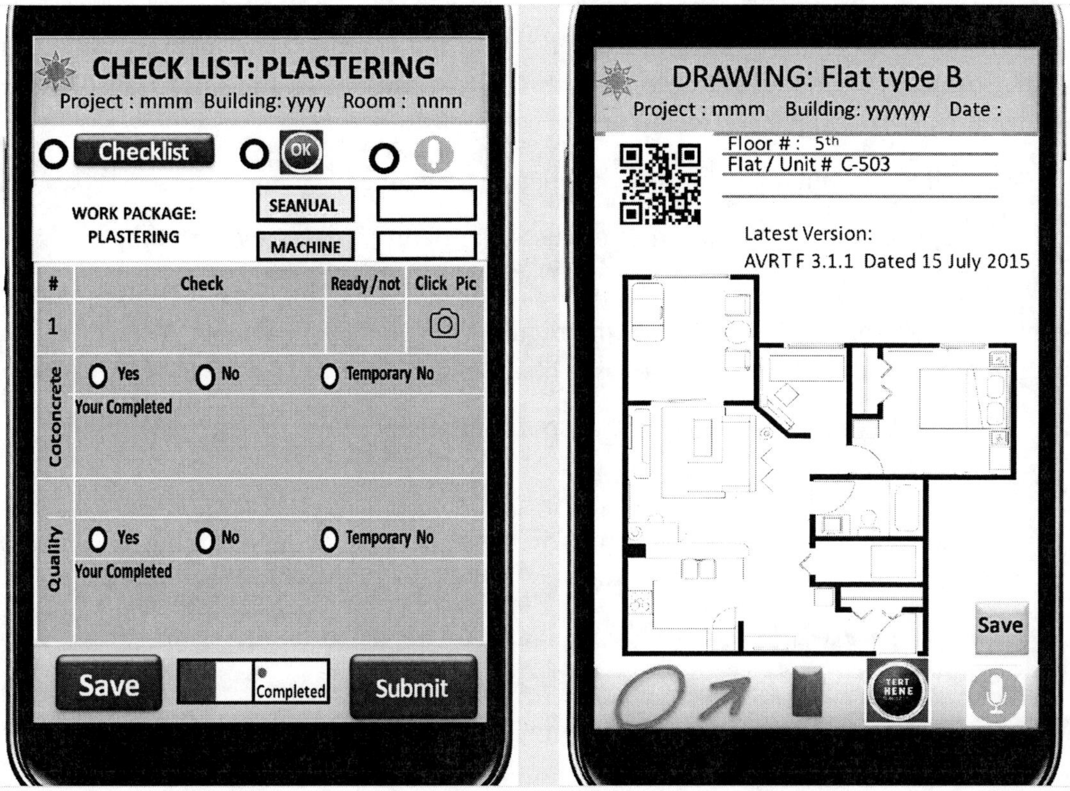


**Figure 7.9: Sample Screen of SMART CHECKLISTS**

**e.** If required, you can attach Pictures of work-items with comments for clarity, and voice recordings can replace written text for convenience. This can be done using an App like shown above. Relevant drawings can be accessed and zoomed for better understanding. Status reports, such as completed checklists, those in progress, and those pending for over 15 days, can also be generated.

### 7.3.3 Smart Work Methods Standards (WMS)

WMS is another widely used tool by contractors to define the best construction processes, helping prevent defects and snags. It's actually a key selling point for quality management systems. WMS lists construction activities and their optimal sequence for main work packages, based on best practices from past projects, ensuring consistent quality

Like Checklists, WMS aims to prevent reliance on memory by ensuring everyone follows proven processes. However, in practice, it's often not used effectively. Table 7.3 shows a WMS from a building project, while Annexure 7.1 provides one from a bridge project. Despite its usefulness, many issues hinder WMS effectiveness, as outlined below.

We are also offering some suggestions to organizations to ponder and to invest and make this good system also more effective.

WMS often provides incomplete or vague information. For instance, tiling in building projects often faces problems with tile tolerances and the final wall or floor dimensions. A better approach would be:

- Lay tiles without adhesive as per approved drawings.
- Get approval from the designer or customer before final tiling with mortar.

New-age customers prioritize sustainability, which in construction involves conserving natural resources and maintaining ecological balance. For tiling, a "Sustainability" section could include tasks like:

- Reducing tile breakage during transport and installation
- Minimizing mortar waste
- Reusing broken tiles and mortar for beautification or landscaping within the project
- Ensuring no waste materials are sent to external dumping sites, preserving ecological balance

Like checklists, most WMS documents are not regularly updated, despite opportunities for improvement through technological innovations.

**Following suggestions are recommended to take care some of the problems with WMS and make these WMS Smart.**

a. Organizations should regularly update the Master WMS based on lessons from various completed projects, whether residential, commercial, or infrastructure like railways and roads.

b. The WMS should include information on new technology, tools, and machinery that improve quality, productivity, and sustainability

c. Master WMSs should be stored on a low-cost cloud server, accessible from mobile devices like phones and tablets, ensuring the latest version is available to all users.

d. Organizations should invest in short instructional videos of WMS processes to be shown during Tool-Box meetings, ensuring workers follow optimal practices before tasks.

e. A mobile app should allow crowdsourcing ideas from colleagues at other sites. Workers can share problems with pictures, seeking expert advice for better solutions, and at least two crucial WMSs should be improved in each project through collaboration.

As a part of continuous improvement, at least couple of crucial WMS must be improved in every project to get the benefit of several brains on the project. Following diagram has summarised these points:

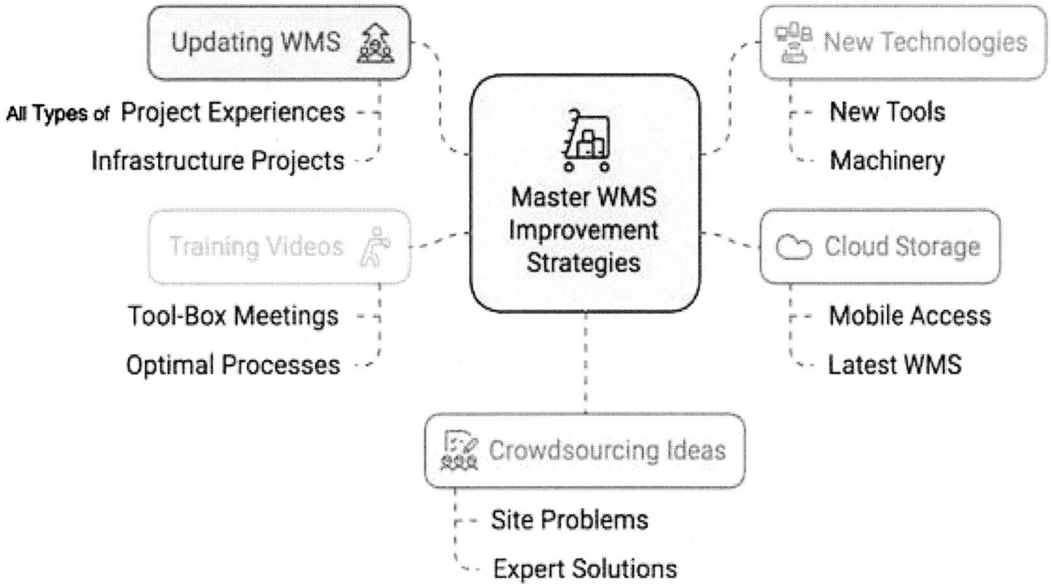


**Figure 7.10 Process of improving WMS continuously**

**Table 7.3 Work Method Statement for Installation of Floor Tiles (For Illustration only)**

| Ref Code | **Description of Task Activity:** Installation of Floor Tiles |
|---|---|
| | **Project Name & reference:** |
| | **Responsibilities:** Project Manager, Construction Manager, Site Supervisors |
| **Task #** | **Task / Activities Description** |
| 1 | Ensure proper completion of the screed and water-resistant surface preparation (for wet area), as per approved method statements and shop drawings. |
| 2 | Inspect Tiles (maximum size variation allowed ±3 mm; No colour variation allowed). Remove rejected ones from the area. |
| 3 | Extend the tile work into recesses and under and behind equipment; form a complete covering as per the approved shop drawings. |
| 4 | Mark position of applicable movement joints and bed in with tile glue prior to tile lay. |
| 5 | Cut and drill the tiles using appropriate machine. Trim and abut the edges and file to get good finish. Fit the tiles closely to electrical outlets, piping, fixtures & other extrusions so that the plates, collars or covers overlap tile joints. |
| 6 | Apply adhesive on working areas as per manufacturer's suggestions. |
| 7 | Lay tiles as per the approved shop drawings. Align joints when adjoining tiles on floors, base walls and trim are of the same sizes. |
| 8 | Fix tiles into place by pressing the dry tile with a twisting/circular motion to ensure good contact. Obtain good surface level using spirit level. |
| 9 | Remove all excess materials and adhesives from tiles by a damp cloth before it sets. |
| 10 | Give minimum 24 hours for curing before grouting. |
| 11 | Remove and replace any defective tiles before grouting. |
| 12 | Ensure no movement of people during grouting to avoid discolouration to the grout. |
| 13 | The interface of the floor access panels and wall tile finish shall comply with the approved detail drawings. |
| 14 | Clean the tiles by dry and clean cloth after completion of tiling. |
| 15 | Protect the floor with 4mm corrugated vinyl, joints taped with duct tape, folded up 150mm at the walls, to include skirting protection till handover. |
| **Task #** | **Labour / Materials / Equipments / Tools** |
| 1 | Agreed and adequate number of workers must be allocated as per main contractor's requirement. |
| 2 | Ceramic Tiles as per different sizes and quality in BOQ in the approved drawings. |
| 3 | Other materials like: Tile Adhesives: Thin Bed Epoxy, Thick Adhesive Mortar, Ready to Use Pasty Tile Adhesive, Polymer Modified Tile Grout, Water Cleanable Epoxy Tile Grout Plant / Tools. |
| 4 | The Machines Required: Grinding Machine with Water Cutter, Glue mixing drill. |
| 5 | The Tools required: Basic hand tools (rubber hammer, spirit level, aluminium bar, right angle), Ruby cutter, Dumpy level. |
| 6 | The licenced Machine Operators for all machines. |

**Table 7.3 Work Method Statement.... (Contd.)**

| Task # | Safety / Health / Environment |
|---|---|
| 1 | Get the required Work Permits, before starting. |
| 2 | Install the required Signage, Sign Boards and barricades at working area. |
| 3 | Install physical prevent access any area which is drying. |
| 4 | Conduct Toolbox Talk covering working at heights, manual handling, site emergency procedures, housekeeping and working in or around plant to ensure everyone's safety. |
| 5 | Define High risk specific areas. |
| 6 | Ensure wearing agreed safety gears like safety helmet, boots, neon vest, gloves, safety glasses, dust mask and hearing protection, as appropriate. |
| 7 | Use water in the disc cutting & drilling machine to suppress dust. |
| 8 | Use elevated fall arrest equipment when working above 2 meters. |
| 9 | Report every accident and near miss immediately to the concerned safety officer, to take further appropriate action. Use accident reporting method. |
| 10 | Inspect critically the working area, before starting the work. Relevant site engineer / foremen / safety officer, to ensure all safety precautions are implemented and also identify any hazard that may still exist. |
| 11 | Keep a full time Safety supervisor (foremen/charge hand) on location to monitor safety compliance. |
| 12 | Conduct regular housekeeping to clean-up before every break and final clean before work end every day. Dispose the waste as per contract. |
| 13 | Provide adequate task lighting using non-breakable covered lights. |
| 14 | Ensure daily inspection of all machinery being used by competent person. |
| 15 | Remove the machinery which fails the inspection for maintenance. |
| 16 | Stop the work in case significant risk is identified which can harm the operatives, work area re-evaluated and made ready for safe working. |
| 17 | For better environment, perform all works in accordance with LEED requirements. |

### 7.3.4 Digital 8-10 minutes Training Videos for Toolbox Meetings

A major issue in the global construction industry is the *lack of trained workers*. Low skill levels lead to poor workmanship, causing defects and requiring later repairs.

Organizations often face a dilemma: training workers is costly, both in terms of training expenses and lost work time. 2-10 days of long training sessions are also ineffective as much of the information is quickly forgotten. We suggest a cost-effective solution using digital technology—"**Just-in-Time, Just-Enough-Training**." Short 8-10 minutes videos on common processes like plastering, painting, or tiling can be shown during morning Toolbox meetings. This ensures workers receive relevant training for tasks they'll perform that day, preventing many defects. A sample screenshot been shown in Figure 7.11.

Suppliers and tech companies often create useful short videos, which can also serve as training material. You may like to see one such video for single-person floor tiling process at https://www.youtube.com/watch?v=PeVsOR13XT4. This is a good video for learning. However, to be more effective from quality perspective, it is better to produce the 8-10 minutes videos of quite common work-items at least, and those should cover the following important points as well.

- Right Tools & Instruments to be used for the WMS
- The common mistakes workmen make, and impact of those mistakes on quality, cost, time
- Limitations of the processes
- Tips for the better quality
- Tips for the better productivity
- Tips for better safety
- DO's and DON'T's to be practiced for that process

(Courtesy: Polmaster Construction & Tile Ltd., Australia)

**Figure 7.11: Sample Screenshot for Training Video**

These daily videos — just 8 to 10 minutes long — containing the essential knowledge related to the work that the workmen are going to perform on the same day, could be especially useful in minimizing defects and snags. Further details on how to make these videos more engaging and interesting for effective adult learning are provided in Chapter 10.

### 7.3.5 Trades' Group Technology for Synchronised uniform rate (Takt Time) of project Construction

In building construction, a large number of specialty subcontractors are typically involved during the interior finishing stage, *sometimes increasing up to 90*, depending on the project's scale and complexity. Due to the interdependent nature of their tasks, communication and coordination among them are often inadequate, resulting in frequent snags and execution issues.

Tasks can be categorized as *sequential, reciprocal, and parallel.* Sequential and parallel tasks are well known and handled by project management tools like Primavera and MS-Project. We have added a new term **'Reciprocal tasks'**, which are difficult to incorporate for planning. For instance, if you have to get a concealed conduiting in a room, then there are series of many tasks which need to reciprocate between a civil subcontractor for plastering and electrical subcontractor for concealed wiring. These reciprocal tasks, which require coordination between subcontractors (e.g., chiselling wall, concealed wiring and plastering), are harder to plan and often left to judgment, as incorporating them into software would overload the task list. Some other examples of reciprocal work would be like Flooring, plumbing, waterproofing and electrical for a bath room.

Reciprocal tasks shuttle between key subcontractors, and the quality of one subcontractor's work impacts the others. Subcontractors tend to focus 'only' on their contractual scope, blaming others instead of addressing project issues.

Two examples of team-based approaches that focus on shared responsibility (group technology) are discussed below.

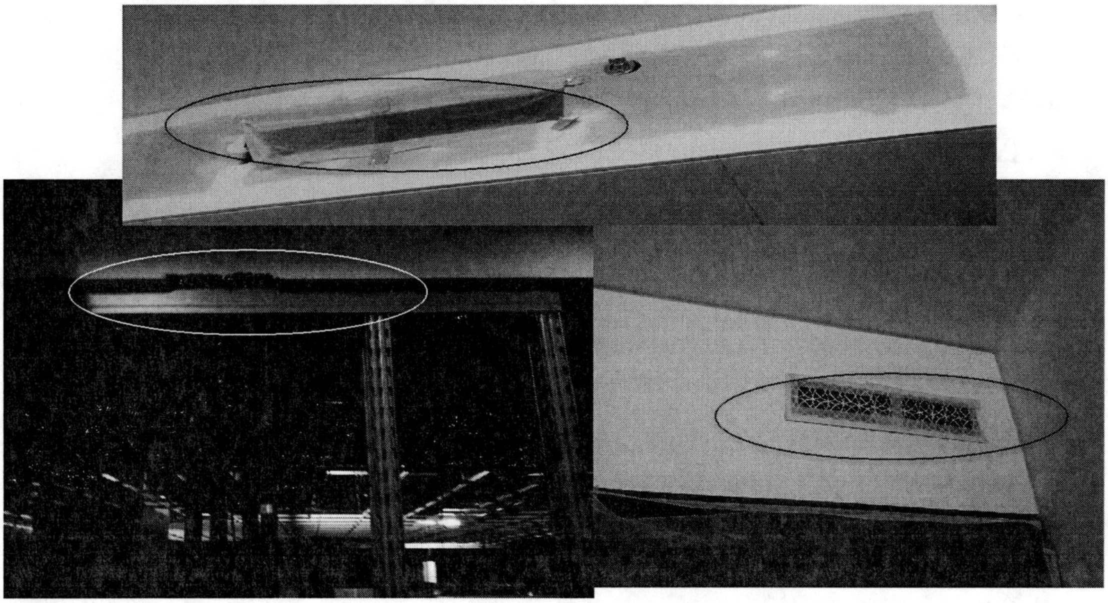

**Figure 7.12: Simple 'Group Technology' concept**

### Case-Let 1: AC Grill and Wall Paper issue in a Premium Hotel

In one case, the HVAC subcontractor was responsible for installing AC ducts and an external grill, while the interior designer handled the wallpaper. A common snag involved dirty fingerprints on the wallpaper, caused by HVAC workers fitting the grill after the wallpaper had already been installed. To resolve this, both subcontractors were made jointly responsible for the issue. They agreed that the interior team would also fit the grill, which only required a press-fit and no specialized skill — effectively reducing the problem.

Similarly, assigning joint responsibility to the civil finish and electrical subcontractors helped avoid later snags, such as the type shown in the window in Figure 7.12. This approach, inspired by Group Technology (GT) from the manufacturing sector, emphasizes multi-skilling and joint responsibility. A major Indian developer applied this concept, gaining significant benefits.

### Case-Let 2: Construction many towers of Apartments

The project team, quality assurance, and main contractor created a flow chart mapping skills, sequences, and dependencies in finishing activities. Using Group Technology techniques, they identified eight highly interdependent activity groups involving many reciprocal tasks.

To foster teamwork, the contract was structured differently. Payments were based on overall team performance rather than individual work packages. Completion milestones, like the "First Fix" for all wet areas on a floor, were set to ensure the subcontractors saw project completion as a joint effort, not just their individual tasks., This measurement could be in terms of in terms of number of units of work completed, where everyone work-package involved in that is completed. For this purpose, the work completion was defined like the examples below:

- Internal First Fix of wet area for all the apartments on one floor
- Internal Second Fix of wet area for all the apartments on one floor
- Internal Finishing of wet area for all the apartments on one floor

Subcontractors such as plumbing, electrical, flooring, and masonry worked together to complete these milestones. Payment rates were set for all subcontractors based on common goals, promoting synchronized work and collaboration. WMS was revised to align with the milestones, assigning tasks to specific subcontractors and improving visibility. Resource planning and adjustments were discussed among subcontractors, with a common quality checklist helping foster accountability and peer pressure.

This approach reduced coordination efforts and improved safety and quality performance by 70%, turning subcontractors into a cohesive team. Results included:

1. Greater consultation during quality checklist completion.
2. Fewer non-conformities in daily audits.
3. Valuable improvement suggestions from subcontractors.
4. Higher cleanliness standards.

These examples show that adopting Group Technology concepts can significantly prevent defects by enhancing teamwork, though it is challenging to implement.

### 7.3.6 Greater use of Jigs & Fixtures for repetitive work

Both building and linear construction projects involve many repetitive tasks. However, subcontractors often overlook the benefits of using jigs and fixtures for faster, higher-quality installations, which is a common practice in manufacturing. Two examples from construction illustrate this. In one project, designers specified a rectangular groove around three sides of main doors (refer to Figure 7.13). Workers used small, uneven wooden pieces (12-15 inches long) to create the groove during plastering, resulting in poor quality. A simple solution involved creating two aluminium U-channel tools (6 feet and 2.5 feet) with handles for grip, allowing the groove to be shaped in one go, improving both quality and speed.

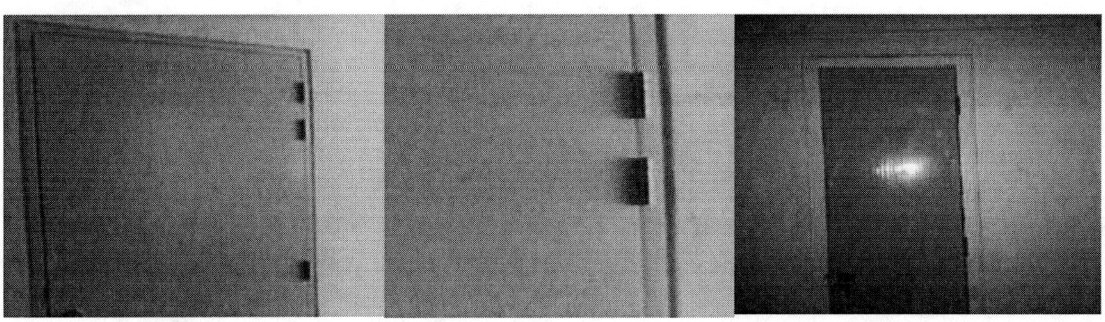

**Figure 7.13: Groove Defect around Doors**

Another example involved fixing approximately 2,000 uPVC windows (of three different sizes) in twin towers of 30 floors each. These windows were factory-manufactured, but when some were fitted on-site, it was observed that a large number did not fit properly.

Extensive rework, including breaking the civil work, had to be done to accommodate the windows. You may refer to Figure 7.14 for an illustration of the type of wall-breaking required.

We found this to be a fairly common problem in many building construction projects. It resulted in situations similar to the one shown in Figure 7.14, which depicts a similar issue with a lift door installation. Eventually, a simple aluminium template frame fixture, like the one shown in red in Figure 7.15, was fabricated and provided to the civil contractor for use in the unfinished part of the window installation. This solution virtually eliminated the quality and rework issues.

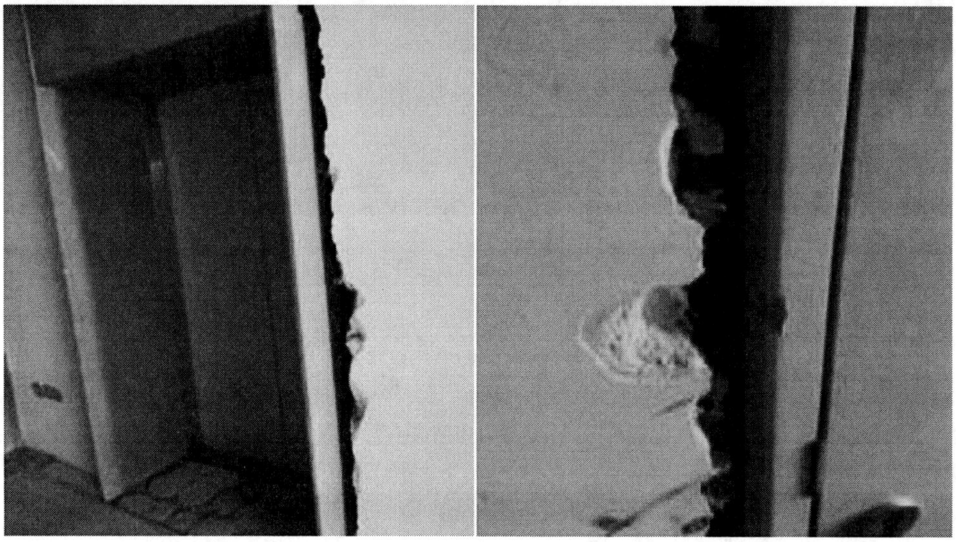

**Figure 7.14: Breaking Wall to accommodate Windows/ Doors**

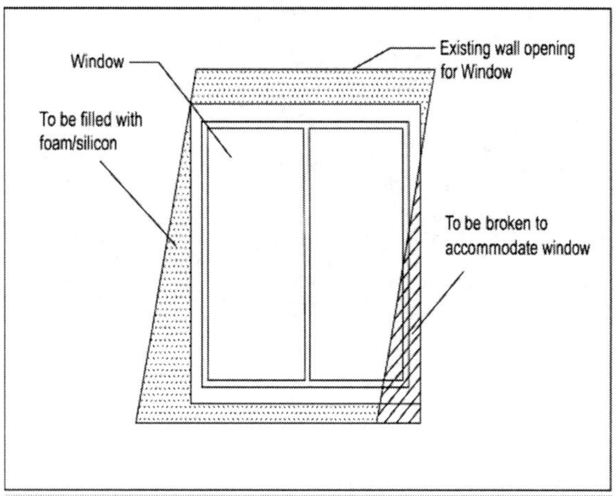


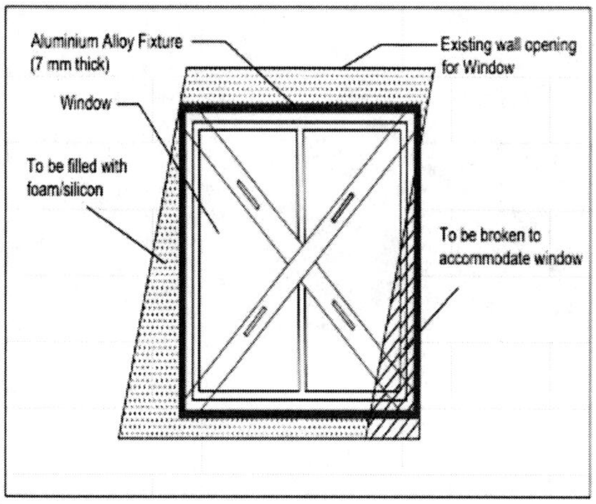


**Figure 7.15: Mismatch Window & Walls Figure 7.16 Window Frame Fixture**

In a river bridge project in India, the repetitive task of making RCC crash barriers was done entirely on-site, including steel bending, reinforcement, and concreting. This slow process resulted in defects like poor crash-side contour and honeycombing, making it cumbersome for workers and supervisors. A better solution would have been to create steel cages and fixtures off-site and then assemble them on-site. We don't have the actual pictures for this, but have tried to show as illustration in Figures 7.17 and 7.18 below.

We recommend investing time at the project's start to identify repetitive tasks and design appropriate tools and fixtures. Many of these can be pre-made off-site for improved quality, reduced costs, and faster completion.

Figure 7.17 Forming for RCC Crash Barrier

Figure 7.18 Factory-made RCC Crash Barrier at Site

(illustrative, not actual)

### 7.3.7 Highlighting Defect-prone area in Drawings

Architects and designers, based on the complexity of their designs and past experiences, are the best people to *predict potential workmanship defects*. While preparing detailed drawings, they can anticipate possible mistakes and highlight areas of concern. Specifying exact tolerances (e.g., linear or angular dimensions) instead of vague references to standards can help supervisors and workmen avoid mistakes. Most people don't frequently consult codes and standards, so clear drawing annotations can minimize errors.

Using document-sharing apps like Oracle's Aconex or Microsoft's SharePoint, drawings can be shared and annotated in real-time on mobile devices, ensuring everyone is aware of potential issues. A sample screen for such as a case is shown in the Figure 7.19 below (taken from Aconex).

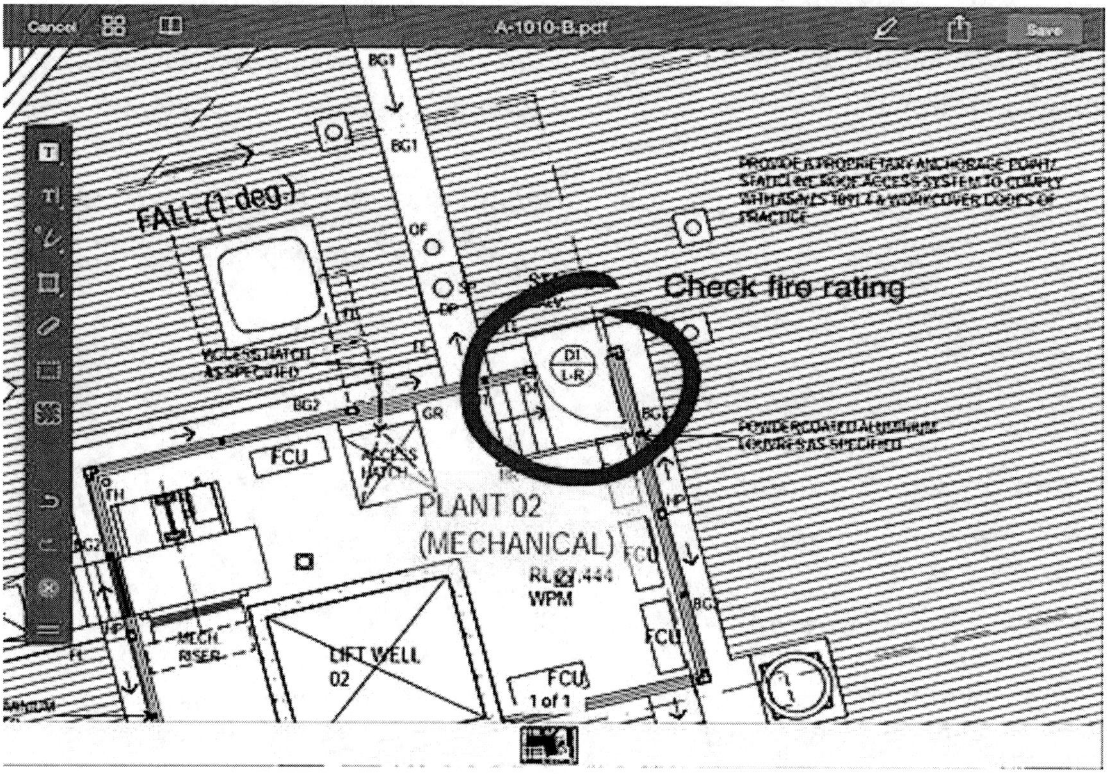


**Figure 7.19: Sample Highlighter on the Drawing**

### 7.3.8 Root Cause Analysis (RCA) & CAPA

Root Cause Analysis (RCA) is a widely used method in the construction sector to identify the underlying causes of defects or problems. A root cause is defined as the factor whose removal prevents the recurrence of the issue. However, in practice, much of the focus tends to be on causal factors, which, when addressed, may improve outcomes but do not fully prevent future occurrences.

Corrective and Preventive Actions (CAPA) are then implemented to resolve current defects and prevent similar ones in future projects.

A Fishbone Diagram (Figure 7.20) was created to analyse quality defects in doors at a commercial project. It categorized causes into six key areas:

- Design
- Material Quality
- Project Management
- Aesthetic Finishes
- Tools/Machines
- Worker Methods

This analysis helped prevent similar snags in future projects by enabling a structured review of contributing factors.

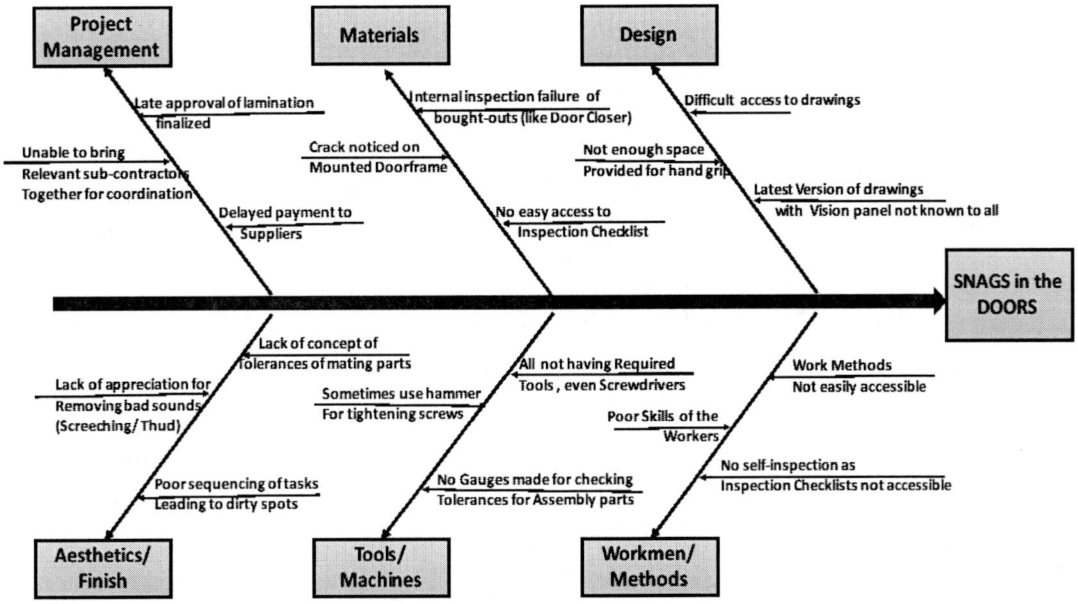


**Figure 7.20: Fish-Bone Causal Diagram for the Door Defects (Illustrative)**

Some causes were further analysed using the 5 WHYs method to uncover the root cause. A specific issue involved doors imported from Singapore, whose replacement was costly and time-consuming. The problem arose when doors, delivered late at night, were manually transported to the 12th floor due to unavailable lifts (Figure 7.20). During this process, the doors were damaged while passing through the stair-case, as unskilled labourers handled the task unsupervised.

After the analysis, it was decided that in future projects, a supervisor should be present during late deliveries of expensive materials to guide workers and prevent damage. Additional corner packing might also be required for protection in similar situations.

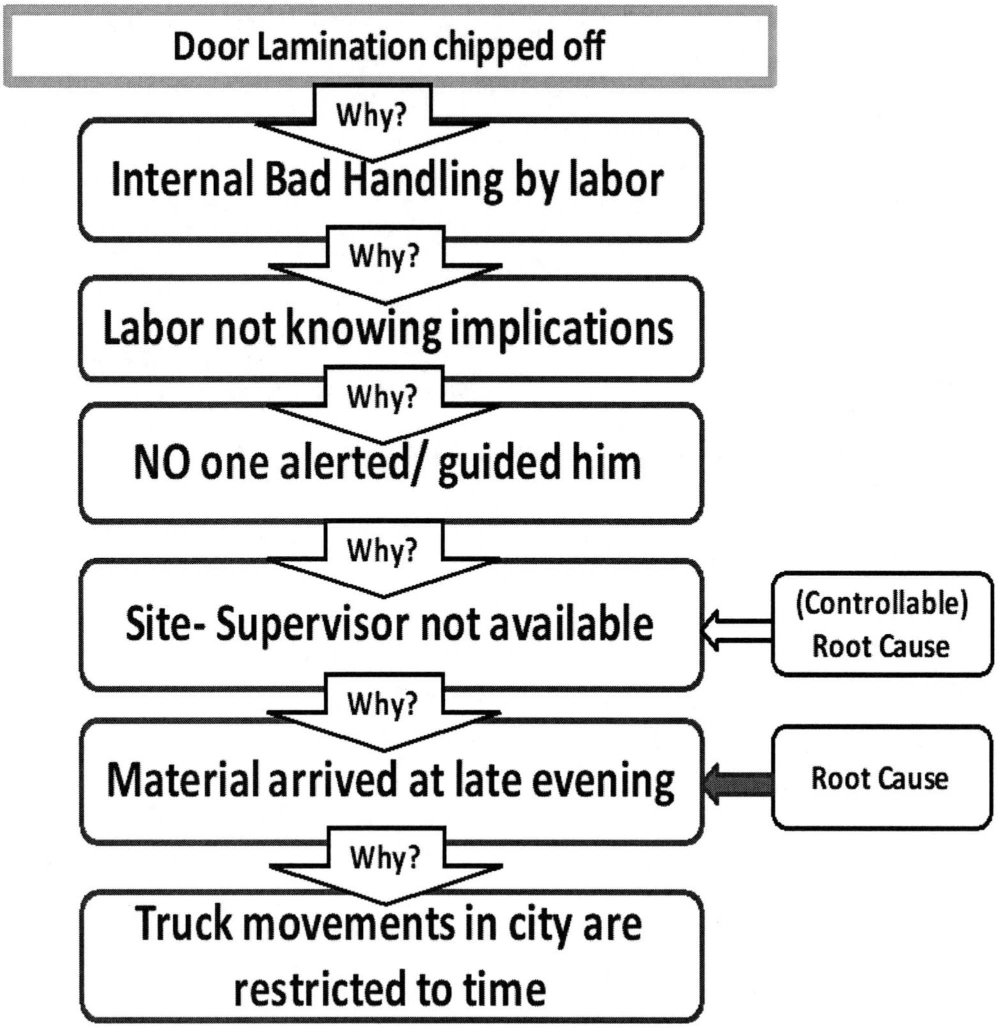


**Figure 7.21: 5 WHYs Root Cause Analysis (For illustration only)**

Many other checklists were also prepared for Project Management, Procurement, Design, and Site in-charge to avoid various causes in the next projects.

### 7.3.9 Use of alternate Technologies

Over the last two decades, new technologies have emerged that significantly influence the quality, cost, and timeline of construction projects. We have already discussed various technologies related to building and road construction, such as formwork and pavement systems. Additional technologies — including BIM (Building Information Modeling) and Virtual Reality, which support precise design — are detailed in Chapter 9.

Table 7.4 lists the essential technologies that should be rapidly adopted. Leveraging these innovations is crucial to meet the evolving quality expectations of New-Age customers.

**Table 7. 4 'Must-adopt' technologies for Quality of 21st Century**

| MUST ADOPT Technologies | | | |
|---|---|---|---|
| **Process** | **Sr #** | **Technology** | **Brief Description** |
| Architecture & Design | 1 | Virtual Reality | Preparing Finished building Mock-ups, customization and Walk-throughs to involve Customer in designing the Dream Projectand realizing it (e.g., SmartVizX.com) |
| | 2 | Mass Customization / Product & Interior Configurator | Customize the floorplan of the Flat )or at least Interior design of the flats), for each user, and workout quotation and execution based on each customer's need (e.g., LivSpace.com) |
| | 3 | Computer Aided Design | 2D Modeling for design and drawings |
| | 4 | BIM (Errorproofing) | 3D modeling of the various parts of the buildings, for finding alternative solutions to choose, finding design and possible executions defects (likepipe interferences) in advance and improving; Errorproofing & Cost reduction |
| | 5 | LED | For decorative designers lighting (e.g. Efftronics.com) |
| Form Work | 6 | Metal Alloys FORMS (like Myvan, Peri) | Very good surface finish eliminating the need for plastering the walls, and multi usage to save overall cost for forming. |
| Concrete Pour | 7 | Ready Mixed Concrete (RMC) | |
| Blockwork | 8 | Load Bearing RCC Walls | |

### 7.3.10 Mini-Mock-ups (for Defects-prone frequent & expensive Operations)

Creating mock-ups for selecting building materials like carpets, furniture, and lighting is a common practice in construction, allowing customers to make informed decisions about the final look and feel. Full mock-ups of apartments or hotel rooms help customers visualize the end product. However, issues with final finishes often arise because the skilled workers who create mock-ups differ from the less-skilled workers who complete the project. This can result in poorer quality finishes.

However, the same concept can also be adopted for use for construction of the difficult portion of the building, or where 3-4 different specialist sub-contractors are engaged, and whose work is interdependent. You can understand this better through a real-life example.

Figure 7.22 below shows the finished guest room of a premier hotel building project, where the skirting on the room was a complex design. The requirements for finish are also shown in Figure 7.22.

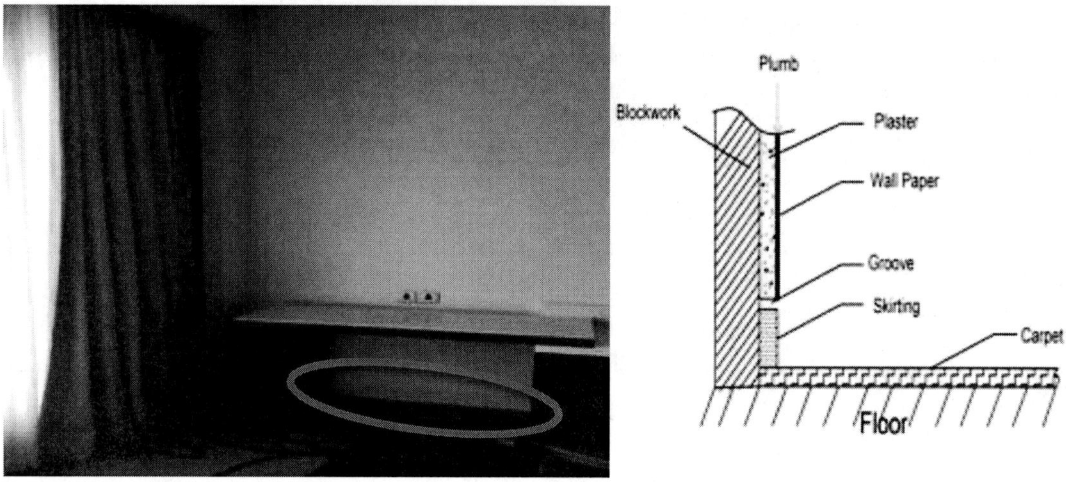


**Figure 7.22: Problem of Skirting not in plumb after wallpaper installation**

Achieving plumb between the vertical wall with wallpaper and the wooden skirting was challenging, especially with a 20-mm thick carpet on the floor, as shown in Fig 7.22. The work suffered due to the poor skills of the labor, leading to defects like undulations and uneven edges (Fig 7.23). There were debates about how much undulation would remain after the wallpaper was pasted and how to maintain a uniform groove width.

**Figure 7.23: Problem of Uneven edge of Skirting**

To address these issues, the Project Director, Interior Decorator, Civil Finishing subcontractors, and the Lean Consultant decided to create a mini-mock-up in one guest room (Figure 7.23). This mock-up, built by the same team working on the project, allowed them to observe the process and quality. After reviewing it, they suggested changes, including a wooden tool to ensure proper dimensions. This tool was then used to improve plastering in other rooms, ensuring a better groove finish and cleaner wallpaper edges.

This mock-up wasn't for selecting finishes but for identifying construction issues and determining the best methods to achieve quality using the available workforce. This approach, when used wisely, can prevent many defects.

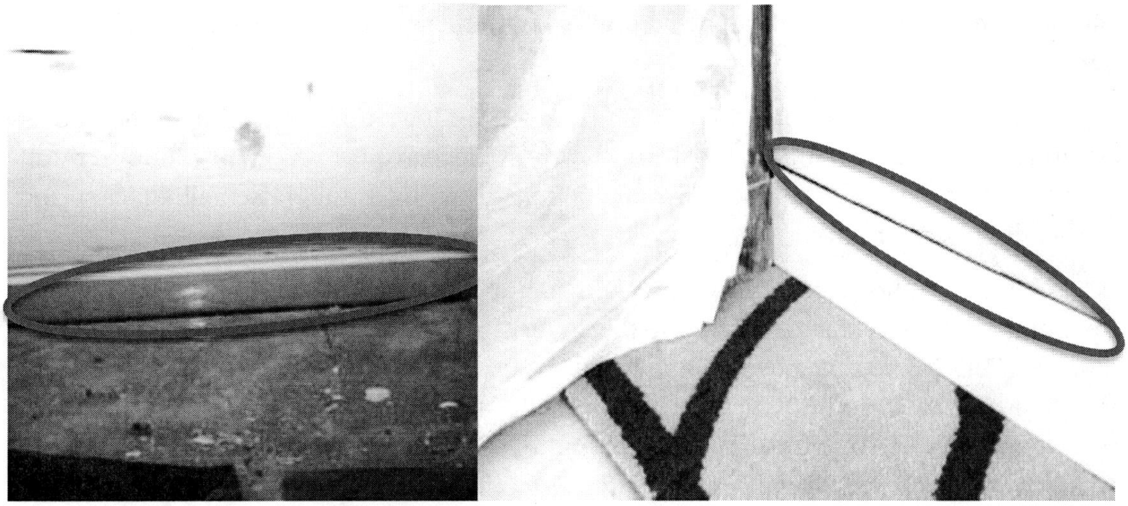

Figure 7.24: Mini-Mock-up for better constructability

### 7.3.11 First-Piece Inspection by Client

In manufacturing, it is common practice to produce a complete component involving various trades, followed by a quality inspection conducted by a team comprising the **Quality Consultant**, **Customer**, **Designer**, and **Contractor**. This critical review helps identify opportunities for process modifications.

In construction, this approach can be applied to **repetitive work items** such as **window installation** or **electrical work**. Observations from such inspections enable planning of **corrective actions**, including the introduction of **new tools**, **additional training**, or changes in the **sequence of operations**, helping ensure that defects do not recur.

For linear projects, one unit — such as a kilometre of road — can be completed and inspected for defects before progressing to the next segment. This process is similar to a mini mock-up, and can also be applied to repetitive tasks such as doors, windows, or tiling. In one project involving 360 lift doors, a mismatch between the frame and the structure necessitated repairs. One door was rectified, inspected, and used as a model for the remaining installations, effectively preventing further rework. Daily stand-up meetings among the relevant trades helped ensure that the new processes were consistently followed.

For Linear Projects, one meaningful unit length of the project can be completed in all respects. For example, for a road project, completion of one kilometre of one major work item says creating Bitumen Macadam (BM) could be done and then it must be critically inspected to find all possible defects. Based on that, CAPA actions can be defined, and then only rest of the BM can be completed with these inputs, to prevent many future potential defects.

This method is something similar to Mini-Mock up. While mini mock-up is generally meant for any difficult end-product, first piece inspection can be followed by all repetitive work items, like doors, windows, fans, skirting, tiling, false ceiling, covering wall with wallpaper etc.

In one of the projects, there was a mismatch between the lift entry wooden frame (manufactured in the factory) and the blockwork, as the plastering had been completed much earlier. Due to this mismatch, the civil structure had to be broken to accommodate the door. There were 360 such lift doors in the twin towers that needed repair and rectification. To address this, the best method of repair was discussed and agreed upon, and one door was fully rectified and offered for inspection to the customer. Once the repaired door was accepted by the quality team and the customer, the worker team was trained on the correct material and method for filling the gaps, as well as the appropriate tools and sequence of operations to follow. Only after this training, the same process was replicated for the remaining 359 doors, which significantly reduced potential rework. A 20-minute daily stand-up meeting was held at the site among the relevant trades to ensure that the new standard processes and correct tools were being consistently used to prevent further defects and rework.

Self-certification, common in manufacturing, assures products meet customer requirements. Unfortunately, it's rarely adopted in construction. Instead, customers often point out defects at handover, which reflects poorly on workmanship. Contractors should inspect and correct defects themselves before offering the product for customer inspection. This practice would also benefit linear projects, where on-site testing is difficult and costly. This will be particularly useful in Linear Infra Projects, which often do not have good facilities for testing at site, and testing every time at remote area adds to the cost and the delays. This would also improve the supply chain efficiencies of construction projects,

One Bangalore developer uses a mobile app to allow workers to report defects by taking photos and tagging locations via QR codes. A quality expert reviews these defects twice weekly, and workers earn financial incentives for identifying them. This approach results in fewer customer complaints and better-quality outcomes. A sample app screen is shown in Figure 7.25.

**Figure 7.25: Sample Screen of Snag Capture for prevention**

## 7.3.12 Multi-Disciplinary Problem-Solving Workshops

This is basically a multi-skills collaborative team work, comprising of the people from different contractors, subcontractors, and suppliers also often involving the designer to review any "critical" defects and find a method to correct the defect and also prevent that in the future. This involves the following simple steps.

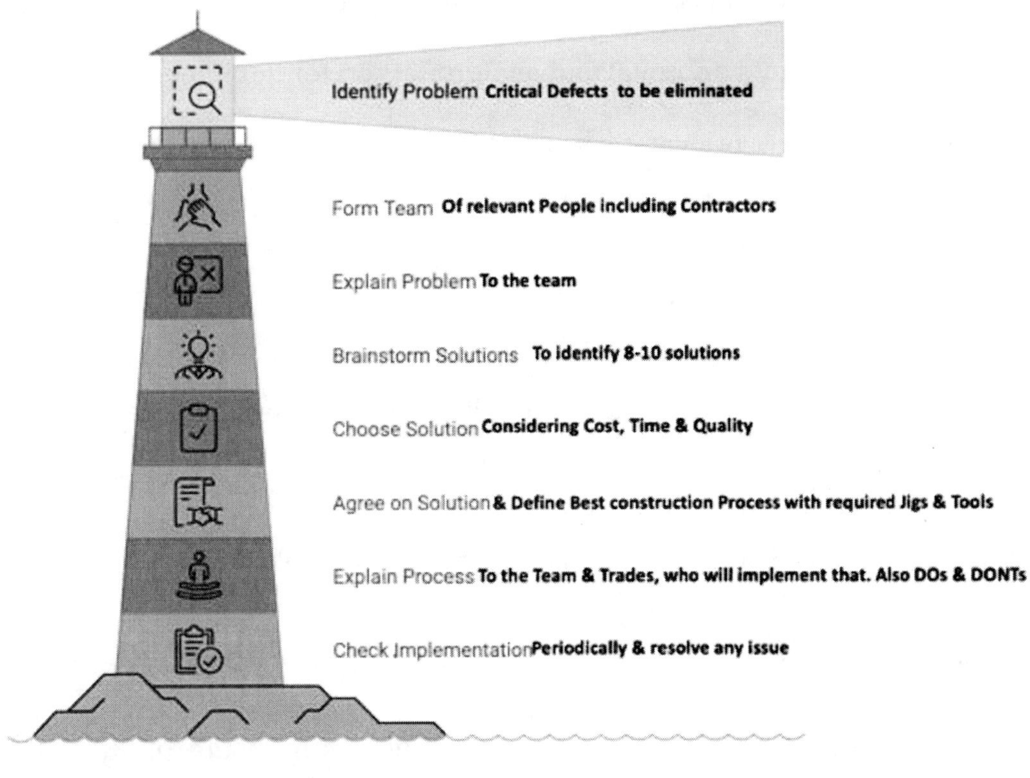

**Figure7.26: Process representation of Problem Solving Workshops**

Often Value Engineering also is carried out in the same way. Many of the problems related with snags were resolved using these workshops. For example, to resolve the problem of mismatch of factory-made windows shown in Figures 7.14 & 7.15, it was found that about 68 windows had the problem of misalignment at the time of the observation, and those were to be broken for fitting the uPVC Windows. Of course, for preventive action, 5 window frame template fixtures were made and given to the workers. But to minimize the cost and time loss in the 68 windows, a problem-solving workshop was conducted.

It is always better to collaborate for problem solving for the major defects and find the corrective and preventive methods. The role of the Project Director is very key to make this happen.

### 7.3.13 CABB: Customer Approval Based Build

This innovative approach, inspired by Test-Driven Development (TDD) from Agile Project Management in software development, is highly effective in addressing both the stated and unstated quality expectations of modern customers. Design and Architecture experts, along with Main Contractors and Quality Management Consultants, should seriously consider adopting this method to achieve exceptional product quality and finishes in building projects. A sample CABB (Component Acceptance Based Build) for the installation of 110 doors is provided in Table 7.5 below.

### 7.3.14 Get Management's approval and assign Person for implementation

For this purpose, make a presentation to the Decision-making Team with following information

- What are the top priority issues with the Quality Defects Chosen
- What are the solutions agreed for their prevention?
- What are the Quantified tangible benefits from them (Cost & Time savings)
- What are all possible intangible benefits, like looks & finishes and Brand.
- Prepare the Cost –Benefit report from the Current and Future practices, which will improve the Value to the Customers, and other stakeholders

After approval, ensure that those are implemented. Occasionally follow-up to ensure proper implementation.

## 7.4 Key Measures of success:

Some of the following measures can be taken as the measure of successful implementation, as shown in Figure 7.27:

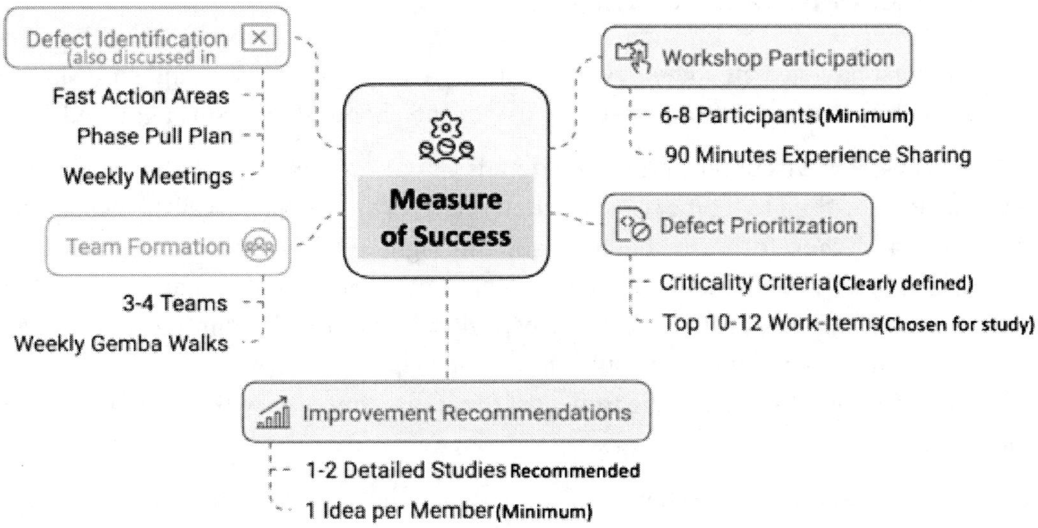


**Figure 7.27: Key Measure of Success.**

**Table 7.5: CABB for Doors**

| CABB for Door | | | | | | | | | | |
|---|---|---|---|---|---|---|---|---|---|---|
| Location: | | Date: DD-MM-YY | No: 1 | | Complete mock-up to be done for D4 & D6a | | | | | |
| Inspection By : | | Project: XXXXXX XXXX | | | | | | | | |
| **Description** | | **Specification/ Approval** | | | | | | | | |
| **Aesthetic / Visual** | | D1 | D2 | D3 | D4 | D5 | D6 | D6a | D7 | D8 |
| **Parameters to be checked at Site** | | Proprietary Aluminium Frame Glass Door | Fire Rated Door with GI Door Frame | Fire Rated Door with GI Door Frame | Proprietary Aluminium Frame Glass Door | Proprietary Aluminium Frame Laminate Door | Proprietary Aluminium Frame Laminate Door | Proprietary Aluminium Frame Laminate Door | Proprietary Aluminium Frame Glass Door | Proprietary Aluminium Frame Laminate Door |
| 1 | No Spots, Dust, Color variation in frame finish and laminate; glass to be clean for glazed doors and without scratches | | | | | | | | | |
| 2 | Door Alignment-tolerance +/- 3 mm | | | | | | | | | |
| 3 | No sound while operating | | | | | | | | | |
| 4 | Open/close smoothly- check door closer speed, locking completely | | | | | | | | | |
| **Functionality/ Accessories** | | | | | | | | | | |
| 1 | Door Closer | OZONE | OZONE | OZONE | OZONE | OZONE | OZONE | OZONE | OZONE | OZONE |
| 2 | Provision for Access Control | N/A | YES | YES | N/A | N/A | N/A | N/A | N/A | YES |
| 3 | Kick Plate | N/A | N/A | N/A | N/A | N/A | N/A | N/A | N/A | YES |
| 4 | SS Sheet | N/A | N/A | N/A | N/A | N/A | N/A | N/A | N/A | YES |
| 5 | Hinge and screws | N/A | 4 nos | 4 nos | N/A | 4 nos | 4 nos | 4 nos | N/A | 4 nos |
| 6 | Pivots | top & bottom | N/A | N/A | top & bottom | N/A | N/A | N/A | top & bottom | N/A |
| 7 | Handle | OZONE | OZONE | OZONE | OZONE | OZONE | OZONE | OZONE | OZONE | OZONE |
| 8 | Door Lock | OZONE | OZONE | OZONE | OZONE | OZONE | OZONE | OZONE | OZONE | OZONE |
| 9 | Door stopper | OZONE | OZONE | OZONE | OZONE | OZONE | OZONE | OZONE | OZONE | OZONE |
| 10 | Vision Panel Provided | N/A | YES | YES | N/A | YES | N/A | YES | N/A | YES |
| 11 | Film/mesh on vision panel | 3M design | Fire Rated Glass | Fire Rated Glass | 3M design | N/A | N/A | N/A | 3M design | N/A |
| 12 | Swing direction as per drawing | YES | YES | YES | YES | YES | YES | YES | YES | YES |
| 13 | Fire Test Certificate | N/A | YES | YES | N/A | N/A | N/A | N/A | N/A | N/A |
| | Door Type | D1 | D2 | D3 | D4 | D5 | D6 | D6a | D7 | D8 |
| | No of Door | 1 | 5 | 1 | 30 | 9 | 7 | 21 | 11 | 1 |
| | No of Door Checked | | | | | | | | | |

## 7.5 Recommendations & Conclusions

You must have learnt some new methods of defects prevention, and some known methods, which can be used more effectively, with the changes we have suggested.

We have also discussed in greater details in earlier chapters that, it is better to prevent the defects, than to let it occur and then rectify. You can never get the best quality by rectifying, and therefore you have to get the right quality first time, every time. Moreover, the cost of rectification is specially very high in the case of Linear Infra projects. Therefore, our recommendation is that for every type of defect you notice, identify and invest some time and money, in required, for the best 2-3 methods to prevent those defects. This will not only reduce the total cost of quality, but also will make the new Age stakeholders happier.

After discussing in details about the 'Product' dimension of Construction quality, in the next section, we will discuss the 'Process' dimension of the construction quality.

## 7.6 References


1. Subhash Rastogi & Akhilesh Laghate (2021). "Quality Management of *Construction in 21st Century; Leveraging LEAN Philosophy and Digital Technologies", published by Innovative LEAN Mentoring Academy, Hyderabad, India*
2. *Subhash Rastogi (2017), "Construction 4.0: The 4th generation Revolution", Proceedings of Indian Lean Construction Conference, ILCC 2017, July 28-29, Chennai. India.*
3. *FICCI, (2019) "Paving future roads for India", Infrastructure Advisory, http://ficci.in/spdocument/23155/FICCI-White-Paper-%20Paving-future-roads-for-India.pdf.*
4. *Subhash Rastogi & V. Srinivasa Rao (VSR), "SNAGGERR: A Lean-digital mobile app for capturing, predicting, preventing & closing the defects/ defects for Tall Buildings.", IJRET: International Journal of Research in Engineering and Technology, Volume: 05 Special Issue: 20, Nov-2016.*
5. *Subhash C Rastogi (2015). DTBB: Designers Test based Built: A new approach to Quality Assurance, Proceedings of national conference ILCE 2015, Indian Lean Construction Conference 2015, Feb 2015, Mumbai, India pp 224-230.*


## 7.7 ANNEXURE I

### Defects Prediction & Prevention Case-Study: Residential Twin-Towers

In this model, whenever any work-Item work is over by 5-10% of the total work, the snagging can be done for this work-item. That would provide the rough estimate of the defect density for that work-item, based on the amount of work completed. Based on the data collected so far, the prediction can be made for the rest of the project. This method is found more popular and useful in most of the buildings projects we have done, in India, UAE and Sri Lanka. The steps involved in this process are written below:

1. Define the accuracy level desired for the Snag Density
2. Select snagging time when about 5-10% the work is completed of the building or the work-package or work-item(s), as decided in the level of accuracy desired in point 1 above
3. Conduct snagging critically, and record all the snags associated with various work-Items under observation
4. Compute the relevant Snag Density parameters
5. Extrapolate the number of snags in that category for the remaining 90-95% of the project
6. If required, repeat the whole process at different stage of project completion for more accurate results and better benefits.

We are providing below, two simple examples to use both the methods.

**Example # 1: Use of historical data for the whole project, at a high level**

**Residential Twin Towers is planned in Southern part of India. It is having a total of 400 Apartments of 2BHK and 3 BHK.**

**Total number of 3BHK apartments= 250**

**Total number of 2BHK apartments= 150**

**Total Constructed Floor Area= 900,000 SqFt**

**Total Open Landscaped area = 100,000 SqFt**

**Figure 7.28: Building Project for Example # 1 (pictures and data are only illustrative)**

From the past 20 similar quality residential projects by the same builder, the overall high-level Snag density is estimated at:

- Defect Density for Constructed areas = 0.0048 / SqFt

- Defect Density for Landscaped areas = 0.001 / SqFt

**Therefore,**

- Expected grand total of snags in constructed area= 900000x0.0048 = 4320
- Expected grand total of snags in landscaped area= 100000x0.001 = 100
- Total number of snags expected in the project = 4320 + 100 = 4420

Hence, this project shown in Figure 7.28 is expected to have a total of 4420 snags at the time of handing over, based on the past experiences of the company. This figure can be used to plan and manage these either to prevent or to plan rectification.

**Example # 2: Use of sampling data for the whole project, at a high level**

For the same project listed in the example above, if we do not have the historical data about the similar projects. In that case apartments of about two or three floors can be completed first, and then the snagging can be done on that sample to find how things are done. This looks difficult to most builders, but we have found very useful in all the cases, because, completing few apartments in all respects, brings out many possible errors of the various drawings, many problems of sequencing of actual work, and provide opportunities to improve the construction processes, for preventing the defects to occur. The detailed case study discussed in this chapter earlier confirms this.

Assuming, on snagging the few completed 2BHK and 3 BHK apartments in this building, we estimate the following Defect densities:

- Defect Density for 3BHK Apartment = 11.2 /Apartment
- Defect Density for 2BHK Apartment = 9.1 /Apartment
- Defect Density for Landscaped area = 0.001 / SqFt

**Then,**

- Expected grand total of snags in all 3BHK Apartments= 250x 11.2 = 2800
- Expected grand total of snags in all 2BHK Apartments= 150x 9.1 = 1365
- Expected grand total of snags in landscaped area= 100000x0.001 = 100
- Total number of snags expected in the project = 2800+1365+100 = 4265

Hence, this project is expected to have a total of 4265 snags at the time of handing over, based on the past experiences of the company. This figure can be used to plan and manage these either to prevent or to plan rectification

These figures can further be used to go deeper into work-item wise snag density and for predicting the efforts required, the extra cost required and also the additional delay caused due to this poor quality.

## 7.8 ANNEXURE II: Defects Prediction & Prevention Case-Study: Twin-Towers Hotel-Cum-Apartments

Real Estate construction has many issues dealing with Real Estate Gated Community with several towers or other high-rise commercial Towers requiring global standards of finished Quality. Some of these include snagging done independently by several stakeholders (Contractors, Quality Teams, PMCs, Architects, Sales Teams and Clients). Thereafter, these captured snags are rectified, involving lots of rework cost, further project delays and shrinking profits and escalating budgets. Often final product finishes are not as desired, because those are not done right the first time. It has also been observed that same snags recur at several floors of tall buildings because of several different contractors engaged for different floors or poor quality of contractors involved. Poor skilled workers are another reason. Cost of such rework & extra delays can be 4-5% of the total cost of the project.

Most Construction companies practice or hire Quality Management Service. But they often fail to address the above-mentioned problems amicably. Although checklists, standard work methods for construction processes etc. are available as a part of ISO certification, but these documents are found not very user-friendly or useful. These are used more for 'marketing' or 'references. To resolve the aforementioned problems, a paradigm shift is required for construction projects involving tall Buildings (Hotels/ Apartments/ Commercial buildings) from the current practice of **"Finding & Fixing Snags"** to the next generation's practice of **"Predicting & Preventing Snags**".

This case refers to a premium Hotel-cum-apartments 30 stories Twin-Towers project, with 300 Guest Rooms & 100 Apartments (2 and 3 bed-rooms) in UAE. The total value of the project was around 500 million AED with a floor area of more than 5 million square feet. The project had already completed 30 months and was having 36% schedule overrun. Client had been allowed only further 6 months extension to complete huge pending work. A large amount of work was incomplete as only about 75% of Civil & MEP works and only 10% of Interior fit-out work was completed. Management wanted to leverage Lean philosophy for next 4 months' time, in order to cut down any further delay & cost. This was to be done by identifying, predicting and preventing SNAGS in the remaining phases of project, as a large number of snags are generated at Interior Fit-out stage. The expectation was to prevent more snags and rework, which could cause further delays in handing over, which also had a penalty clause for delays.

### Problems & Challenges Observed

A good deal of time was spent on the site for initial observations, discussions with key internal & external people including Contractors, involved at the project site, to understand the 'as is' condition of the working. Following were the key observations made:

- Repeated same Snags: in various guest rooms & apartment rooms at different floors as different contractors were working at different floors. There was no clear plan to avoid recurrence.
- No historical data available on the type of Snags: Past data on similar project not available with either Main Contractor or Other Sub-Contractors, for better learning.
- Lack of Qualified manpower for Snags inspection: No specially trained experts were available (except Architect) for Snags capturing was person dependent.
- No Culture of Predicting & Preventing Snags: Project Team considered proactive snag prediction & prevention as an overhead.
- Laborious efforts to find & close snags: prepared snagging list in Excel files, and tracking the progress of closure of those was very cumbersome, time-consuming and error-prone.

**Snags Capture:**

- Introduced pro-active Snag Observation Walks (SOWs) Schedules: Arranged key work-package focused proactive Snag Observation Walks thrice a week with relevant trades, to hunt the snags and capture. This was done for 3 weeks and later reduced to once-a-week for further 3 weeks. This resolved many simple problems on-the-spot and also gave opportunity to discuss plan of action for handling & preventing other snags.
- Agreed on the Criticality for prioritization: Used past projects knowledge, to identify work-packages having large potential of snags and rework (e.g. Civil-Finish & Washroom plumbing). Special focus was given on capturing snags in these work-packages to maximize preventing these during rest of the project.
- Observe the Site collectively: to capture the snags, and their criticality with the combined wisdom of SOW Team.

**Snags Prediction:**

- **Predict number of snags for the whole project:** This was based on Snags Density of the previous similar projects and "sampling" of five of the 30 floors of Hotel and Apartments separately. Some 100 possible snags were identified, including 43 over and above the previously snagging done by Sales Team from the mock-up Hotel Guest Rooms and Apartment. All these 100 would have led to more than 25000 snags (of different nature of Big, medium and small) for the total project, because of repetitions of snags at various floors of both towers.
- **Predict the Efforts, Cost, and Potential Delay due to snags:** Out of 100 snags identified, topmost 11 types of major snags contributing to 70% of the potential extra rework and delays estimation were chosen by the team. Efforts & Costs

were estimated based on discussions and some site observations for rework. (refer Table 7.6 below).

- **Conducted Root-cause-analysis:** for each of these 11 most critical types of snags.

**Table -7.6: Snags & Rework Efforts & Costs Predicted & Prevented**

| Sr. # | TOP SNAGS | # of Snags Predicted | # of Snags Prevented to occur | # of Snags prented from leakage to Client | Rework Efforts (Man-Days) Predicted | Rework Cost Efforts ('000 AED) Predicted | # of Days Delay predicted |
|---|---|---|---|---|---|---|---|
| 1 | Skirting in Rooms/ Lobby | 972 | 476 | 587 | 763 | 114 | 34 |
| 2 | Cistern-Tile misalignment | 220 | 118 | 102 | 1100 | 112 | 25 |
| 3 | Tub Misalignment | 235 | 102 | 65 | 470 | 94 | 30 |
| 4 | . | . | . | . | . | . | . |
| 5 | . | . | . | . | . | . | . |
| | . | . | . | . | . | . | . |
| 11 | . | . | . | . | . | . | . |
| | **TOTAL** | **5876** | **2523** | **3122** | **7806** | **1535** | **34** |

### Snags Prevention:

Out of 14 best practices of Defects prevention, we used the following ones:

### Some Popular ones like:

- (Dynamic) Inspection Checklists
- Root cause Analysis
- Self-certification of Quality

### Some new ones like:

- Create a Unique Snag bank or album for training. Checklists based on past project or Field visits
- Extensive use of simple Tools, like Jigs & Fixtures, for repetitive activities
- UABB: User Acceptance Based Build
- Mini mock-up for ensuring build-ability

We identified the most critical 11 type of snags, based on efforts and cost of rework, and final aesthetics, as the project was premium Hotel & Apartments. Some of these are shown in Figure 7.29.

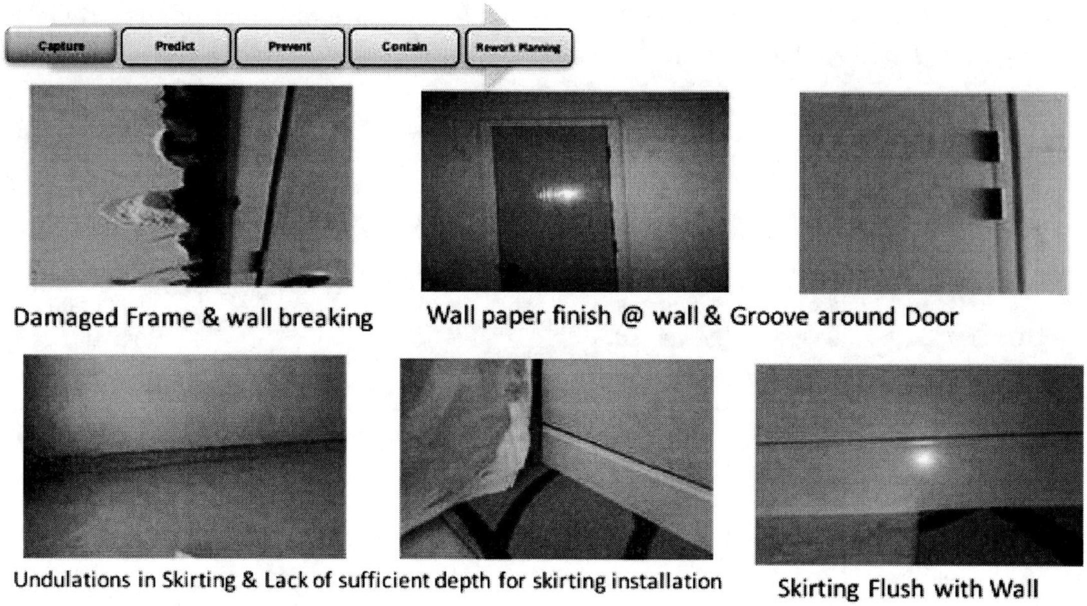


**Figure 7.29: Some of the Critical Defects**

These would have resulted in 5876 major and medium snags in the total building and accounting for about 70% of the rework efforts, suitable practices were agreed through collaborative workshops among the Main Contractors, PMC and Key Sub-Contractors. For instance, for Snag "Skirting in Room', mini-mock-up for build-ability was created to agree on the best method to construct and rectify. (Please. refer to Figure 7.29; bottom three pics)

The Rework Efforts and the Cost, including manpower efforts and material costs were estimated based on actual efforts and costs consumed by the agreed crew for one full floor, having 14 rooms. The estimated delay in the project due to these reworks were also estimated. The day-to-day work involving these 11 most critical snags above the 5th floor was stopped for 2 days, so that new snags are not created. First, the different contractors and site foremen were called at the mini-mock-up for the corrected snag. This was to educate on the use of right tools and right sequence of operations, along with working in team with other relevant trade. Then only they were allowed to start the work, in the 6th floor and above.

There was a 20 minutes Daily Stand-up meeting at site, among relevant trades for ensuring that new standard processes and the appropriate corrective actions are being followed to prevent further snags and rework. There was also random visit on different days and time at various floors, to ensure the practice of use of right Tools and Jigs & Fixtures and process is actually followed. We also supported the foremen and the workmen by providing on-the-spot training, wherever required, to ensure right execution.

A simple reward system was created, both financial (like free recharge on Mobile up to 100 AED) & non-financial (Certificate from the hand of Project Director with photograph), to motivate foremen/ workers to ensure implementation of agreed best practices. Table 7.7 below gives the summary of some of the benefits due to snags prevention and also prevention of potential leakage of snags to Client. Later the summary of all the benefits is also provided.

**Table 7.7: Summary of some of the benefits**

| SNAG TYPE | POTENTIAL (only South Tower) | | | |
|---|---|---|---|---|
| TOP SIX SNAGS | # of snags | REWORK EFFORTS | | POTENTIAL DELAY |
| | | MAN DAYS | COST ('AED) | |
| Skirting in rooms/ lobbies | 615 | 483 | 72,500 | 34 |
| Fitment & Groove in Interconnecting door | 205 | 307 | 53800 | 22 |
| Groove in Entry door | 190 | 250 | 37,500 | 18 |
| Cistern-Tiles poor alignment | 120 | 600 | 1,20,000 | 25 |
| Undulations in Ledge | 88 | 44 | 8,800 | 6 |
| Bath Tub | 150 | 300 | 60000 | 30 |
| TOTAL | 1368 | 1984 | 352600 | 34 |

**Table 7.7: Summary of some of the benefits**

| SNAG TYPE | POTENTIAL (only North Tower) | | | |
|---|---|---|---|---|
| (In Civil & MEP work only) | # of snags | REWORK EFFORTS | | Potential Delay (day)s |
| | | MAN DAYS | COST ('AED) | |
| Skirting in rooms/ lobbies | 357 | 280 | 42000 | 12 |
| Groove in Entry door | 140 | 180 | 27000 | 30 |
| Cistern-Tiles alignment | 100 | 500 | 100000 | 20 |
| Upstand – undulation | 119 | 535 | 107000 | 30 |
| Tub Misalignment | 85 | 170 | 34000 | 10 |
| TOTAL | 841 | 1725 | 320500 | 30 |

**Snags Containment:**

In this project, this step could not be completed for the lack of time and non-availability of appropriate data.

**Rework scheduling:**

Based on the estimated efforts and the manpower required for the 'look ahead' planning period of 8 weeks, all the rework was also included in the Primavera schedule, so that no further delay took place, due to shortage of resources. These measures resulted in several benefits. Some of the important ones are shown below.

**Benefits achieved**

In this project, although lifecycle management started at very last stage, that too for a very limited period of four months, the following benefits were derived from proactive capturing, prediction, prevention &closure.

- Improved Product Quality by 96%; preventing 43% of 5876 snags to occur & 53% leakage to client. Most of 53% would have also prevented if Lean initiative had started at earlier stage of project.
- Saved more than 800,000 AED Rework Cost by preventing 2523 of major or medium snags.
- Prevented further Project delay of 1.3 months by planning rework & preventing 2523 snags. This translated into saving of more than 10 million AED of cost because of penalty clause for project delay.
- Improved Tiling Productivity by 60%
- Provided learning & motivation for workmen
- Lesser conflicts among Contractors, although their savings were not quantified.

***Note:*** This project was carried out in 2015. Currently many Mobile Apps are available to automate these processes, avoiding manual work. Those will make the job even more productive and convenient.

*Make Ready meetings build foresight into safety. By aligning tasks, resources, and crews ahead of time, projects reduce risks, prevent delays, and create a SAFER ENVIRONMENT where planning protects people as much as progress.*

# CHAPTER 8

# MAKE-READY THE WORK: FOR SAFER AND SUSTAINABLE PROJECT

# 8 Make-Ready the Work for a Safer Project Environment

It is a proactive planning phase that ensures each task is both execution-ready and safe. This includes confirming the availability of materials, manpower, access, proper sequencing, and conducting safety checks. The goal is to minimize risks, prevent delays, and promote seamless coordination among trades, ultimately creating a hazard-free and efficient work environment.

## 8.1 Introduction:

Safety in construction is not an isolated requirement but integral part of the project delivery. It is essential now to stay focused on the environmental, health, safety, quality and continuous improvement aspects of project while delivering the scope of project within the time & cost define in the beginning. The figure 8.1 below depicts this aspect in brief.

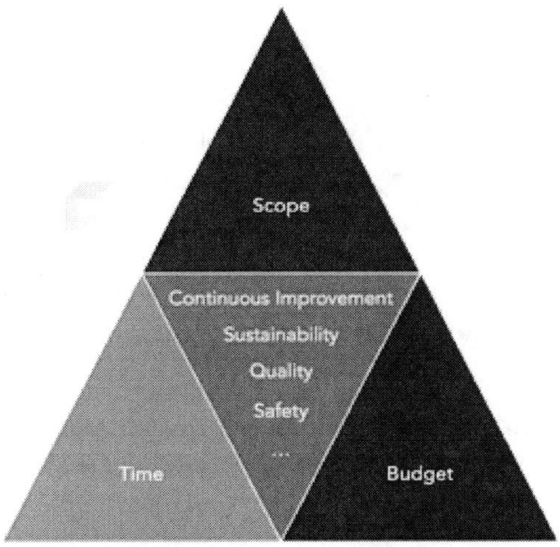


**Figure 8.1: Key requirements for a successful construction project**

The customers and clients are no longer willing to compromise the safety, quality and sustainability needs of the project for sake of reducing cost or delivering the project faster. In many countries, currently this is also a statutory and legal requirement. The competitiveness of organization and project managers are measured in the continuous improvement and enhancement they are bringing to the project while delivering the project. The lessons are expected to learn every day in today's project.

## 8.2 Need for Better Safety Management in Construction Projects

Occupational safety & health hazards are relatively high in construction industry, in comparison to other sectors. Workers suffer injuries, illness and even death in some cases, while engaged in construction activities with unsafe working conditions. Many die each year from construction accidents and also suffer from occupational diseases, due to dust, paints and other similar unhygienic work environment. "The real numbers are not known because often records are not kept and compensation is seldom paid", as per the 'Evaluation Report- Construction Action Programme (CAP)', by International Labour Organization (September 2006). Construction & extraction occupations with, transportation & material movement accounted for nearly half of all fatal occupational injuries (47.4 %) as per Occupational Safety & Health Administration (OSHA, 2020). The most shocking fact is that, 5 among top 10 most frequently occurring accidents violated standards reported by OSHA (2022) are from construction industry.

Safety is a critical parameter in civilized communities, prioritized by progressive clients and contractors over progress, cost, and quality. Unsafe environments and activities cause irreversible harm to individuals, organizations, and projects. Yet, safety is often treated as the responsibility of inspectors, with lapses corrected through reports or work stoppages. Instead of minimizing waste and controlling variation, many projects overburden people and machinery to meet milestones, pressuring CPS teams to emphasize productivity over safety.

Lean concepts help CPS create safe environments by removing constraints early, maintaining workflow, and eliminating recurring non-conformance through structured methods like first-run studies, observations, and PDCA. Safety management is no longer just compliance but a moral and economic necessity, requiring commitment from the entire project team. Five key reasons underscore its importance:

1. **Human suffering:** Injuries and accidents have severe impacts on the economic, social, physical, and psychological well-being of the affected person and their family. The long-term consequences and costs associated with safety mishaps are beyond monetary calculations. Better safety management can save lives, prevent injuries, and reduce human suffering to a great extent.
2. **Cost of accidents**: The costs associated with safety incidents and accidents can have a severe impact on the project. This impact is in terms of the Medical Expenses for treatment, Workers Compensation, Legal Cost, Property damage etc They can go beyond the project budget and have a cascading impact on project timelines. Improved safety management can save the project from cost escalations.
3. **Efficient work environment**: Unsafe working conditions are unproductive working environments. The downtime caused by the accident, including the time spent investigating the incident, cleaning up the site, and repairing damaged equipment or property can delay the project timelines and milestones. Besides this, the decreased morale among workers following accidents, can lead to decreased productivity, increased absenteeism, and higher turnover rates as workers may feel unsafe or dissatisfied with working conditions. A safe work environment enhances the confidence of workers and leads to improved productivity.

4. **Legal liability and obligations**: Both owner and contractor organizations are liable for providing a safer work environment and obliged to follow safety regulations. Liability and obligation may include the expenses related to legal proceedings, including lawyers' fees, court costs, and settlements or judgments resulting from lawsuits filed by injured workers or their families. Besides this, there may be fines imposed by regulatory agencies for violations of safety standards or failure to comply with regulations. These fines can be substantial and can vary depending on the severity of the violation and the jurisdiction. Better safety management provides an umbrella of protection to the project, people, and organizations from incidents and accidents.
5. **Reputation and branding**: Negative publicity and damage to the company's reputation can result in lost business opportunities, difficulty attracting new talent, and decreased investor confidence. Safety incidents, especially those resulting in loss of human life, damage the reputation of the organizations associated with the project and can impact future project opportunities. Better safety management can potentially safeguard the competitiveness of the organizations associated with the project.

In short, the better safety management in construction is the need of the hour. It's also Construction Team's commitment to society, obligation to fellow human, adherence to legal requirements, and part of project management efforts to reduce cost, increase productivity and deliver projects on time.

## 8.3 Integrating safety requirements in to Construction activities

Managing safety in silos complicates construction projects. The execution team often overlooks safety norms, leaving separate safety teams to enforce them, which disrupts productivity. This approach complicates the process, as safety is managed through additional enforcement steps, rather than integrated efforts. A separate department and inspection process are now required for quality, safety, and environmental standards, creating a sense that safety is someone else's responsibility.

Safety should be integrated into all project stages, from design stage to handover, in order to protect workers and ensure project success. Achieving this requires a well-planned strategy with active implementation and monitoring. A Lean mindset focused on continuous improvement is crucial. To embed safety into construction, we must focus on five key areas, managing both macro and micro aspects effectively.

- **Strategy and Planning**: Safety requires commitment from senior leadership. The plan should detail safety protocols, hazard assessments, emergency procedures, and training for workers. Regular risk assessments are essential, covering heavy machinery, working at heights, electrical hazards, and hazardous materials, with controls to mitigate risks. Embedding safety into strategy aids in seamless integration into risk management and planning, ensuring safety policies and procedures are actively and willingly practiced, not just enforced.

- **Integrate to Design & Procurement:** In the design phase, it's critical to assess safe execution of the project. Including safety in design and verifying it in procurement ensures materials, machinery, and workers are safeguarded. This approach supports the use of safe equipment and technology, prioritizing the well-being of all project participants.
- **Respect the People**: Lean Construction emphasizes "Respect for People." This requires comprehensive safety training, covering equipment use, hazard recognition, and emergency procedures. Regular safety meetings help address concerns, review incidents, and involve workers in safety discussions. Functional safety training, including participation of last planners like foremen, makes sites safer and more productive by valuing their input and experiences.
- **Collaboration and Gemba**: Safety implementation demands collaborative efforts from clients, consultants, and contractors, with regular on-site inspections (Gemba) to monitor and improve safety and quality. Routine inspections of scaffolding, ladders, machinery, and electrical systems are essential for hazard prevention. Session 4 further explores this process.
- **Continuous Improvement:** A Lean approach promotes continuous safety improvement without **waiting** for incidents to occur. Safety observations and near misses should be reviewed, using a PDCA approach to continuously evolve. Teams should learn from past incidents, and as some companies do, address safety issues in project reviews to prevent future risks.

This is also shown diagrammatically below in Figure 8.2.

**Figure 8.2: Safety Planning in Construction Projects**

## 8.4 'Make Ready' Meetings for Safer Project Environment

Safety is often seen as extra work due to limited understanding of safety readiness. While method statements and risk analysis are commonly used; dedicating 40-50 minutes monthly to review and mitigate upcoming risks can greatly enhance safety. Issues like client-driven last-minute changes, clearance delays, and progress pressures lead to a reactive "firefighting" approach. A collaborative Safety Planning System, similar to the Collaborative Planning System (CPS), is recommended. Developing a "make-ready" mentality across teams, including Design, Procurement, and Safety, can reduce hazards and boost productivity by making safety part of planning.

Like CPS, a Collaborative Safety Planning System should emphasize five habits: Trust, Collaboration, Variance Analysis, Lessons Learned, and Continuous Improvement. The focus here is on how to conduct "Make Ready" Meetings monthly and weekly after the initial planning for the next 4-6 weeks is complete.

Monthly "Make Ready" meetings help ensure a safe site environment. Safety Experts, Site Planning, and Engineers must attend to plan safe, efficient activities for the upcoming month. These meetings should set the agenda for safely preparing the site for scheduled work, as agreed upon by the team. This is shown as a diagram below in Figure 8.3.

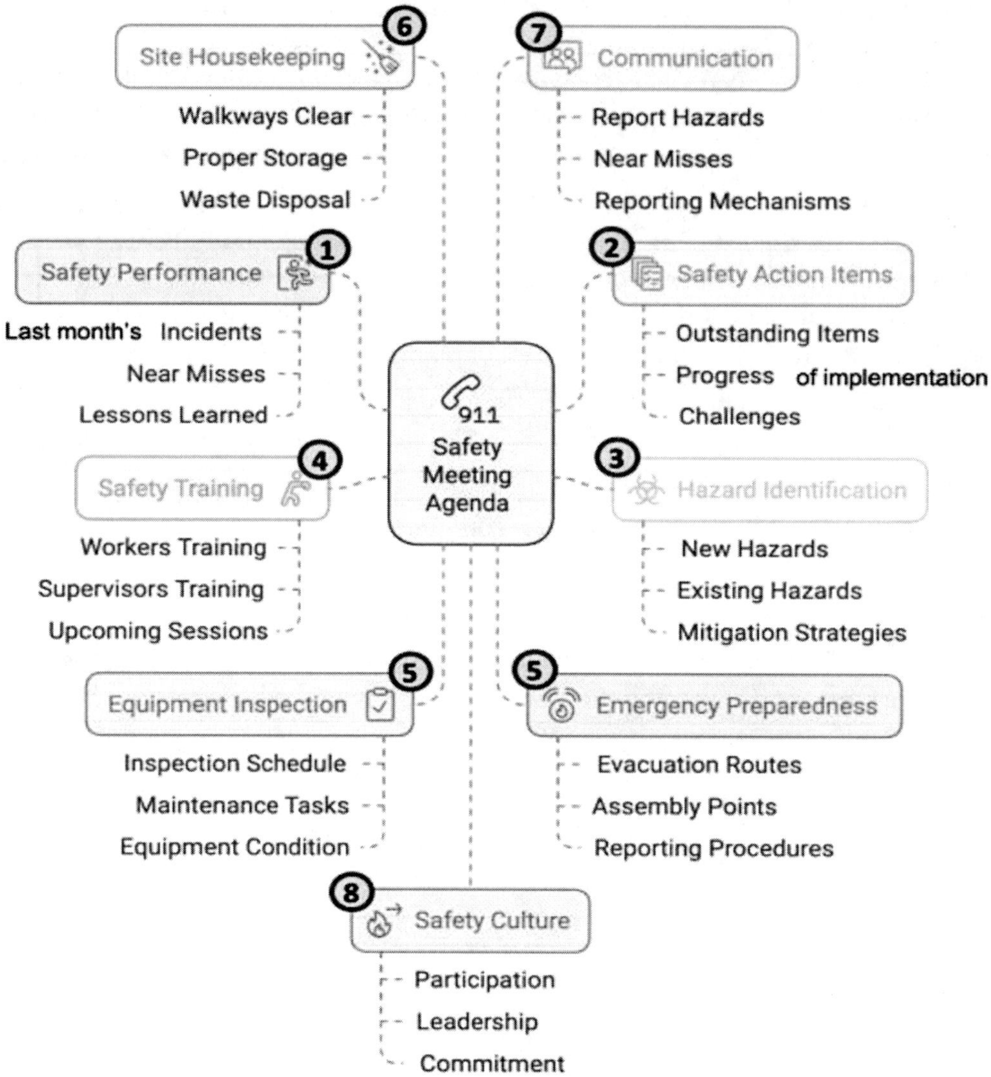


**Figure 8.3: Steps required to plan & conduct 'Make Ready' meeting on Safety**

Make-ready meetings are an invaluable tool for ensuring safety on construction projects. By promoting collaboration, identifying risks early on, and fostering a strong safety culture, these meetings can create a safer work environment for all involved, leading to improved project performance and overall success. The make-ready meetings can enhance teamwork, proactively identify and resolve potential risks in the project. It can also improve responsibility and accountability, providing an opportunity for everyone to have a common understanding of how to progress the work safely and efficiently. This ultimately leads to increased productivity in a safe work environment.

## 8.5 Proactively Capture Safety Risks with a Gemba Walk

Besides the monthly 'Make Ready' Meeting for the safety and the actions discussed in that, as explained in the section above, it is equally important to actually make a Gemba walk to the action points on the site. Please remember that the most practical and effective decisions are made at the work location, not in meeting rooms. Observing the work and work environment in person is five times more effective than listening to an analysis or observation report. The Lean concepts place greater importance on Gemba, the real workplace, for its effective results. Gemba walk is about going to the source of value creation to see things first-hand rather than relying on analysis, reports and assumptions. There are four major purposes of Gemba walk:

1. **Gain first-hand understanding of unsafe work conditions**, as identified in the Make Ready Meeting: Leaders can directly observe the process of performing activities and the challenges people are facing in real-world scenarios. This helps them to have a deeper understanding of proactive steps the organization and management can take for a safe work environment for all stakeholders.
2. **Identify improvement opportunities:** Understanding the ground reality in terms of inefficiencies, bottlenecks, and potential safety challenges helps leaders to identify and define improvement opportunities.
3. **Engage with the execution team:** Open communication with ground staff enables leaders to have insights into the problems faced by the team on-site and input for improvements from the working team there. Their suggestions must be welcome and evaluated with sincerity and preventive action should be taken.
4. **Continuous Improvement:** Gemba walks enable leaders to directly check countermeasures' effectiveness, experience their impact, first-hand, and practice the PDCA cycle. This allows either for standardizing improvements across projects or refining methods based on on-site observations and feedback from ground staff.

The Gemba Walk is a key technique in construction projects, where leaders visit the site to observe and understand work conditions. To maximize its impact, leaders should focus on observing, openly conversing with workers, and collaborating without imposing their views. Unsafe acts or situations observed can be turned into proactive safety measures. Gemba Walks connect leaders with site teams, fostering a safer, more efficient, and collaborative environment. This method supports not only safety but also boosts productivity, engagement, and Lean problem-solving. Through Gemba Walks, leaders can identify improvement areas and establish a culture of continuous improvement, benefiting the entire project team.

## 8.6 Predict the potential number, Types of Safety Risks, Their Impact and efforts to prevent for the rest of the project.

In the construction industry, success is achieved by the team's ability to identify, predict and prevent safety risks that could impact the project. To create a safe and efficient work environment, it's essential to take both proactive as well as corrective approaches to manage potential safety risks, and reduce their overall impact... We can take four primary proactive steps to identify potential safety risks in a project, which are listed below. All this can be done almost in the same way as we discussed about the defect's identification, prediction and prevention, in Chapter 7.

1. **Hazard Identification and Risk Assessment (HIRA):** This traditional method involves identifying potential hazards, evaluating their likelihood of occurrence, and the severity of their consequences. Then, we implement control measures to mitigate them. HIRA described here is like Proactively capturing (or identifying) the safety hazards and risks, similar to the defect's identification in 7.2.1 Two ways to do this effectively are through:

    - Based on Gemba Walk
    - Based on the Historical data of the similar projects of past.

    Like in the case of Defect density, we can define 'Hazard Density' per work-package or per work-Item. Then the prediction and prevention can be planned.

2. **Job Safety Analysis (JSA)**: This method breaks down each task into specific steps and identifies potential hazards and risks for each step. This approach helps to target the implementation of control measures. This step can also help to refine the measure of Hazard Density.

3. **Predictive Analysis and AI:** Employing technology such as Artificial Intelligence (AI) and machine learning helps to analyse historical data from past projects and industry trends. This approach helps to identify patterns and predict potential risk areas in the current project. By adopting these proactive measures, we can create a safer and more efficient work environment in construction projects. This can ensure the accuracy of Defect Density on a continuous and Dynamic way.

4. **Safety Planning, and Preventive Procedures:** Developing comprehensive safety plans and preventive procedures helps to address specific project risks and outline safe work practices. It's crucial to ensure everyone involved understands and adheres to these plans.

The following preventive methods provide insights into realistic work conditions on a construction site. Adopting the Lean Problem-Solving Model helps identify root causes, promoting a problem-solving culture among construction professionals to predict and prevent unsafe situations.

1. **Incident Reporting and Root-Cause Analysis**: Safety incidents or near-misses should be reported and investigated thoroughly. Establishing a trusted, collaborative environment ensures team confidence in management's support to identify and eliminate root causes. Concealing unsafe incidents often occurs but should be discouraged.

2. **Safety and 5S Audit**: Regular safety and 5S audits help identify potential lapses in safety. Corrective actions should enhance the overall project environment, not just specific areas, fostering a culture of problem-solving. Further examples are in Chapter 11 on 5S.

3. **Observations and Make-Ready Discussions**: Checking safety parameters during quality, progress, and technical inspections and in planning meetings is essential for site safety. Analysing consolidated observations from these discussions improves project safety conditions.

Identifying and predicting potential safety risks is vital, requiring ongoing review of risk assessments, practices, and safety plans as projects progress. This continuous process enables project managers to anticipate and mitigate risks, fostering a strong safety culture where all stakeholders are aware of safety measures in place.

## 8.7 Prioritize the Correcting of the Non-Conformance & Analyse the Causes to Prevent

Due to the dynamic nature of construction work environments, we always have to go for the reactive methods to ensure the compliances. It is important to stop the work and perform a corrective action to close the non-conformances identified during these interventions. Even in reactive approach, rather than repeatedly intervening to resolve issues, it is more effective to spend more time analysing the issues initially to predict and prevent their recurrence. The steps taken to anticipate and prevent unsafe conditions rely on a continuous improvement approach driven by the Lean Problem-Solving Methodology. This methodology is a process that enables a team to comprehend a problem, apply a solution, and embrace a culture of problem-solving within an organization or project (Refer to Figure 8.4).

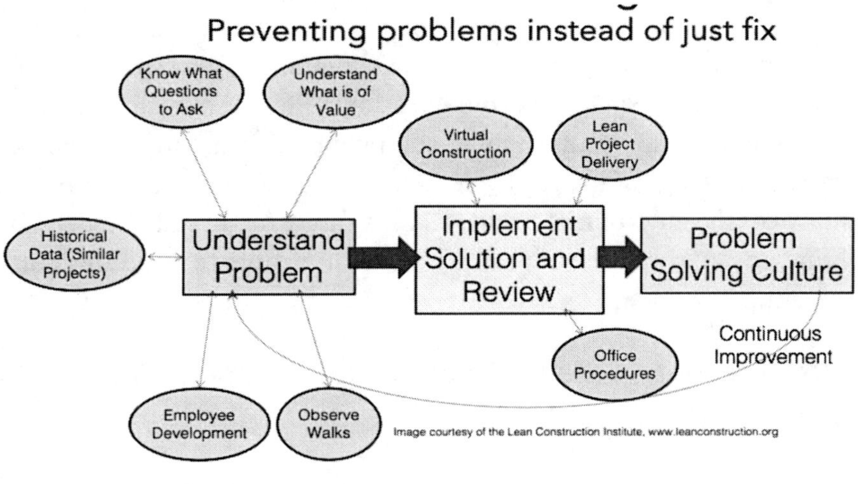


**Figure 8.4: Lean Problem-Solving Model**

Ensuring safety is a top priority for all stakeholders involved in construction projects. Whenever an unsafe activity is identified, it should be immediately halted, improved and verified before being redeployed. Although this approach can reduce the risk of unsafe work, it does not guarantee that the same problem won't occur again later in the day or in a different location. That's where the Lean Problem-Solving Model comes in handy. This model focuses on three key areas to effectively solve problems.

- **Understand Problem:** Construction professionals are experts in providing fixes or solutions for identified problems, whether they relate to quality, safety, or production. Any issues that are identified or highlighted will be resolved quickly with a "quick fix" or "jugad". Senior managers have likely resolved thousands of situations throughout their years of experience, enabling them to act and provide solutions immediately. While quick fixes may work instantaneously, they may backfire if the root causes are not resolved. A structured approach to analysing historical data using Pareto analysis to identify the vital few and finding the root causes can help teams understand the real problems. Engaging people during Gemba walks can also provide first-hand experience and make problems clearer for leaders. A deeper understanding of the causes that create unsafe situations or non-conformances can help teams deploy the right countermeasures, which can eliminate or minimize the recurrence of non-conformances. A detailed case study is elaborated in section 8.

- **Implement Solutions & Review:** There is no single solution that can completely eliminate or reduce non-conformance or unsafe acts in construction projects due to their constantly changing nature. To improve the situation, we must implement solutions using the Plan-Do-Check-Act (PDCA) cycle. Leaders and professionals must continuously strive for improvement and focus on achieving visible results in the project, rather than just documenting their improvement ideas.

A problem-solving mindset differs from fault-finding; in Safety and Quality, inspections and non-conformance notices often dominate, but true improvement requires trust, transparency, respect, and collaboration. KPIs and punishments may work briefly before people find shortcuts, while compliance-driven approaches produce documents without ensuring safety. A problem-solving culture eliminates root causes, continuously improves the environment, and sustains safety for all stakeholders. In short, beyond firefighting unsafe conditions, we must invest in people to build this culture for lasting improvement.

## 8.8 Ways to Prevent Safety Events to Happen and Avoiding Consequences

The following are eight most popular and effective methods to prevent safety events and avoid potential consequences. This area also shown in Fig 8.8 Below, giving little more information on the key activities in each of these.

- Safety Checklist Method
- 5S Method
- Job Hazard Analysis (JHA)
- Behaviour-Based Safety (BBS)
- Safety Incentive Programs

- Safety Culture Development
- Preventive Maintenance
- Incident Investigation and Root Cause Analysis

By incorporating these methods into the project's safety management system, stakeholders can enhance safety performance, prevent safety events, and minimize the consequences of workplace accidents and injuries. Indicatives safety hazards and their prevention has been provided in Figure 8.5.

## 8.9 Need for Sustainability in Construction

Sustainability in construction is an essential factor for the future of the built environment. As the global population continues to grow and urbanization accelerates, the demand for new buildings, infrastructure, and resources is at an all-time high. However, the environmental impact of construction activities has reached a point where traditional methods are no longer sustainable. Construction activities are responsible for a significant portion of global waste generation, energy consumption, and CO2 emissions.

**The need for sustainability arises from the following key factors:**

1. **Environmental Preservation**: Construction projects contribute heavily to resource depletion, waste generation, and pollution. Sustainability ensures that projects are designed and executed with minimal environmental damage, conserving resources for future generations.
2. **Economic Efficiency**: Sustainable construction practices often lead to cost savings in the long term through reduced energy consumption, efficient waste management, and improved resource utilization. While upfront costs may be higher, sustainable buildings and infrastructure have lower operating costs and longer lifespans.
3. **Social Responsibility**: Sustainable construction enhances the quality of life for communities. It provides healthier living environments, reduces environmental impact on local ecosystems, and ensures long-term access to essential resources such as water and clean air.
4. **Regulatory Compliance**: Governments worldwide are increasingly enforcing stricter environmental regulations, and green building certifications like LEED (Leadership in Energy and Environmental Design) are becoming more common. Adopting sustainable construction practices helps companies stay compliant with evolving laws and industry standards

**Figure 8.5: Different ways to prevent Safety Hazards**

Adopting sustainable construction practices helps companies stay compliant with evolving laws and industry standards.

Most of the Lean Practices discussed in this book, directly or indirectly support this need. This is explained below briefly.

## 8.10 Meaning of Sustainability in Construction

Sustainability in construction encompasses the entire lifecycle of a building or infrastructure project, from design and material selection to construction, operation, and demolition. The core principles of sustainability include:

1. **Resource Efficiency**: This focuses on minimizing the use of raw materials and energy during construction. Sustainable construction promotes the use of renewable resources, recyclable materials, and energy-efficient technologies.

   CPS, VSM, BIM and The Big Room support by ensuring that resources—both human and material—are utilized in the most efficient manner, reducing excess use and waste, which directly supports sustainability in terms of conserving resources.

2. **Energy Efficiency**: Buildings and infrastructure projects should be designed to minimize energy consumption. This includes utilizing energy-efficient systems, materials with high insulation values, renewable energy sources (e.g., solar, wind), and designing layouts that maximize natural light and ventilation.

   Safety management and predicting and preventing defects, drive energy efficiency by reducing energy waste through better planning, real-time defect prediction, and fewer disruptions during the project, which typically leads to optimized energy usage throughout the project lifecycle.

3. **Water Conservation**: Sustainable construction encourages the use of water-efficient systems, rainwater harvesting, and technologies that reduce water consumption. It is essential to minimize the environmental impact of water extraction and wastewater generation, particularly in water-scarce regions.

   CPS and The Big Room support this through Collaboration among the key stakeholders ensuring that water use is minimized and that construction processes are designed with water-saving mechanisms are in place, aligning with sustainability goals.

4. **Waste Reduction**: Construction waste is one of the largest contributors to landfill volumes globally. Sustainable construction emphasizes waste minimization during construction, as well as the recycling and reuse of materials. Efficient logistics and proper waste management strategies are key components of sustainability.

   Waste Reduction, Five S and predicting and preventing Defects, BIM etc. support this. These topics are foundational to reducing waste in the construction process by improving workflow efficiency, minimizing excess material use, and eliminating unnecessary motions or delays, which directly supports sustainable construction practices.

5. **Healthy Environments**: Sustainable buildings prioritize indoor environmental quality, ensuring healthy living and working spaces. This includes good air quality, use of non-toxic materials, and designs that promote occupant comfort and well-being.

   Safety management, Five S and CPS contribute to healthier environments by ensuring clean, safe, and well-organized spaces that minimize hazards and provide a comfortable and secure workspace for workers, directly enhancing their overall health and well-being.

## 8.11 'Making Ready' Meetings to Achieve Sustainability in Construction

Achieving sustainability in construction requires the adoption of specific strategies throughout the project's lifecycle. In figure below, we have mentioned some of the effective ways for maintaining sustainability. Figure 8.6 depicts this. This can be achieved the same way of 'Making Ready' meetings for Safety, we discussed in section 8.4.

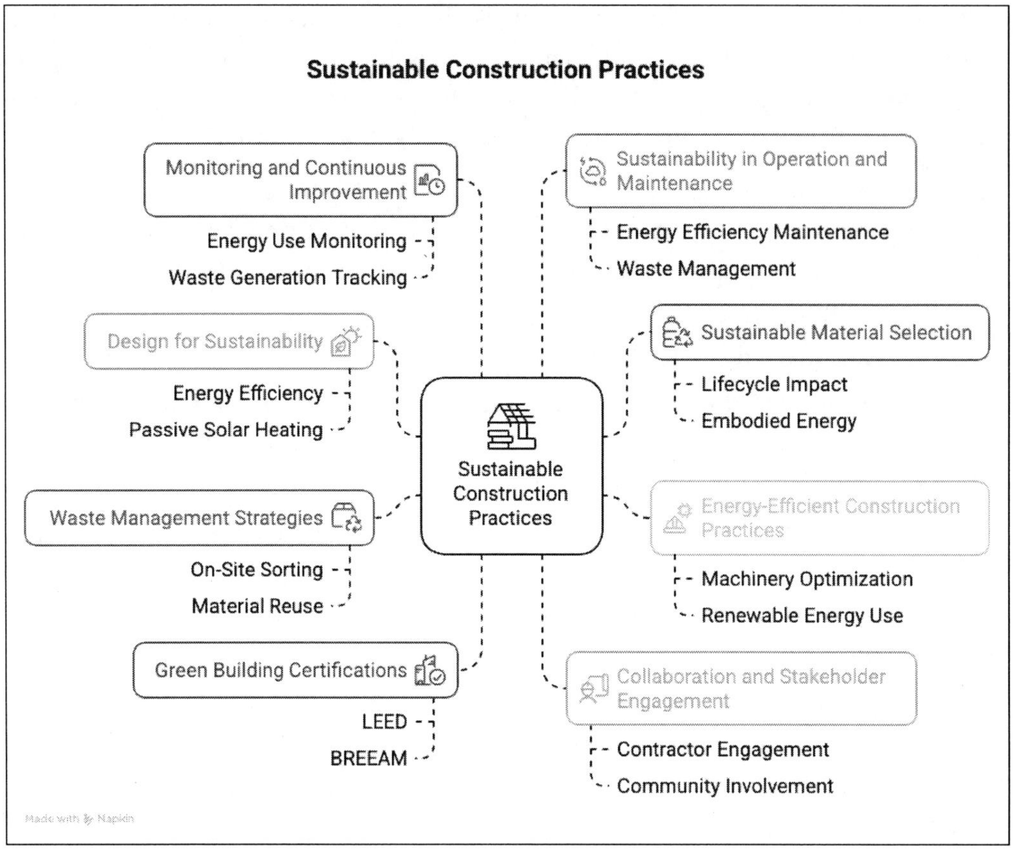


**Figures 8.6: Some effective steps and practices of achieving sustainability**

## Conclusion

Sustainability in construction is essential for the future. With global challenges like climate change and resource depletion, sustainable practices are more critical than ever. By merging Lean Construction with sustainability, we can achieve projects that are efficient, environmentally responsible, socially beneficial, and economically viable. The continuous improvement mindset of Lean Construction will ensure the construction industry is ready to meet 21st-century challenges and create lasting value for all stakeholders.

## 8.12 Annexure 8: A Case Study

We participated in an industrial project in a major Indian city, assisting a contractor in adopting Lean practices. During gap analysis and training, we found safety was a key concern, as the execution team felt restricted by the client engineers' strict zero-tolerance safety policy. The project faced around 250 NCRs weekly, requiring the execution team to halt work until each NCR was rectified and cleared, causing significant productivity delays. Using the Lean Problem-Solving Approach, we guided the construction team through improvements. Here is a brief summary of that journey, following the steps given in Figure 8.4 earlier.

**Understand Problem:** We collaborated with the contractor's safety team to analyse NCRs from the past three months, identifying frequent issues such as inadequate worksite barricading, poor housekeeping, unsupervised critical work, and unsafe access to work areas. These issues were observed during site visits and regularly discussed (Refer to the model in Figure 8.6).

**Classroom Training and Issue Identification**: Initial Lean training for site personnel introduced concepts and tools, revealing several factors behind high NCRs. Through observations and brainstorming, causes for missing barricading were identified: lack of barricade materials, safety team delays, and delayed information from safety engineers. A 12-month data review helped pinpoint recurring hazards.

**Root Cause Analysis Findings**: Using 5 Why analysis and Gemba walks, key issues were identified: barricade materials were hard to locate despite adequate stock, indirect communication delayed safety requests, and morning decisions on work location caused poor planning and coordination. The root causes traced back to inadequate planning and communication. Figure 8.7 shows different ways to prevent safety hazards.

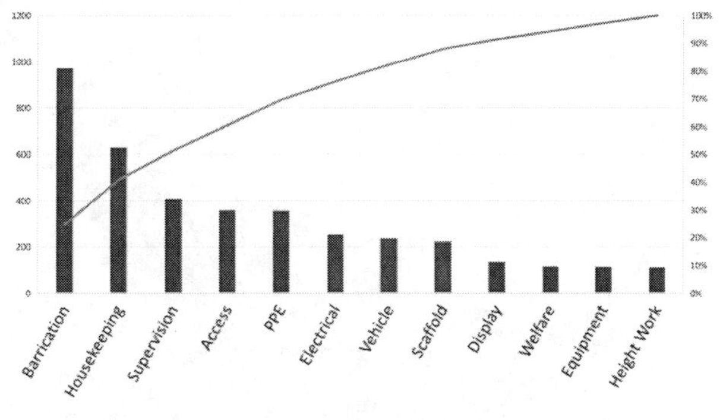


**Figure 8.7: Different ways to prevent Safety Hazards**

**Implement Solutions & Review**: Countermeasures included involving safety officers in evening meetings to pre-plan and confirm the next day's work locations. By analysing barricade completions against non-conformances, the team saw a 22% decrease in NCRs within two weeks, despite a 20% increase in number pf labour from 1700 to 1900 on site.

**Problem-Solving Culture**: The team embedded problem-solving in daily routines, regularly analysing NCRs, quality rejections, and approvals to find patterns. They used collaborative discussions and Gemba walks to identify root causes and apply targeted countermeasures.

# CHAPTER 9

# LEVERAGING DIGITAL TECHNOLOGIES FOR SUPERIOR PERFORMANCE

# 9 Leveraging Digital and Other Technologies

## 9.1 Digital & Other Technologies in Lean Construction 4.0 Era

In the fourth revolution of the global economy, new challenges in real estate and infrastructure construction are emerging. These Challenges were explained in Chapter 1. You should also read paper "Construction 4.0: New Generation Construction in a Hyper Connected world" [3, 8] for more details. You will recall some of these challenges influence the project performance measures, such as:

- Designing and executing new demand for more complex and iconic structural & buildings, leading to delayed projects, poorer Quality, and upwards revision of Budgets
- Providing better experience to new age customers, during the entire end-to-end process of apartment or villa booking to its owning or leasing.
- Focusing more on 'personalized' needs and mass-customization of designs of apartments & villas at affordable costs for the end-users.
- Providing world-class quality of infrastructure at a faster speed and at lower cost.

Therefore, in addition to LEAN philosophy discussed in earlier sections, the use of digital and other technologies is essential to attain superior performance of the Projects addressing above and others.

### 9.1.1 Three Types of Technologies:

Currently, there are three types of technologies that are transforming the construction industry. By effectively integrating these, companies can enhance the quality, productivity, and profitability of projects throughout their lifecycle. These technologies are categorized as follows:

- **Workflow Automation Technologies (WT):** This includes tools like ERP (Enterprise Resource Planning), CRM (Customer Relationship Management), e-Commerce, Portals, Project Management software, AutoCAD, and BIM (Building Information Modelling). These technologies handle structured data such as numbers and digits, and streamline business-related workflows. Many of you are likely already familiar with these systems.
- **Communication & Collaboration Technology (CT):** Examples include Virtual Reality, Augmented Reality, e-Commerce, social media like WhatsApp, mobile devices, Data Analytics, and Cloud platforms for collaboration among stakeholders. CT technologies support unstructured data, such as text messages, phone call audio, images, and video calls. With CT and WT technologies now on cloud servers, companies benefit from reduced capital investment. Most of these are explained in greater detail in this chapter.

- **Operational Technologies (OT):** This category includes IoT-based technologies, such as various sensors (like, motion, thermal, vibration, noise, RFID) and emerging construction methods like prefabrication, 3D printing, and robotic systems for tasks like painting and tiling. These technologies handle both structured and proprietary semi-structured data.

**Fig 9.1** provides the integration of these 3 technologies. Further, the integration of WT (ERP), CT (mobile apps, virtual reality) is also illustrated in a real-life case study in Annexure 9.1.

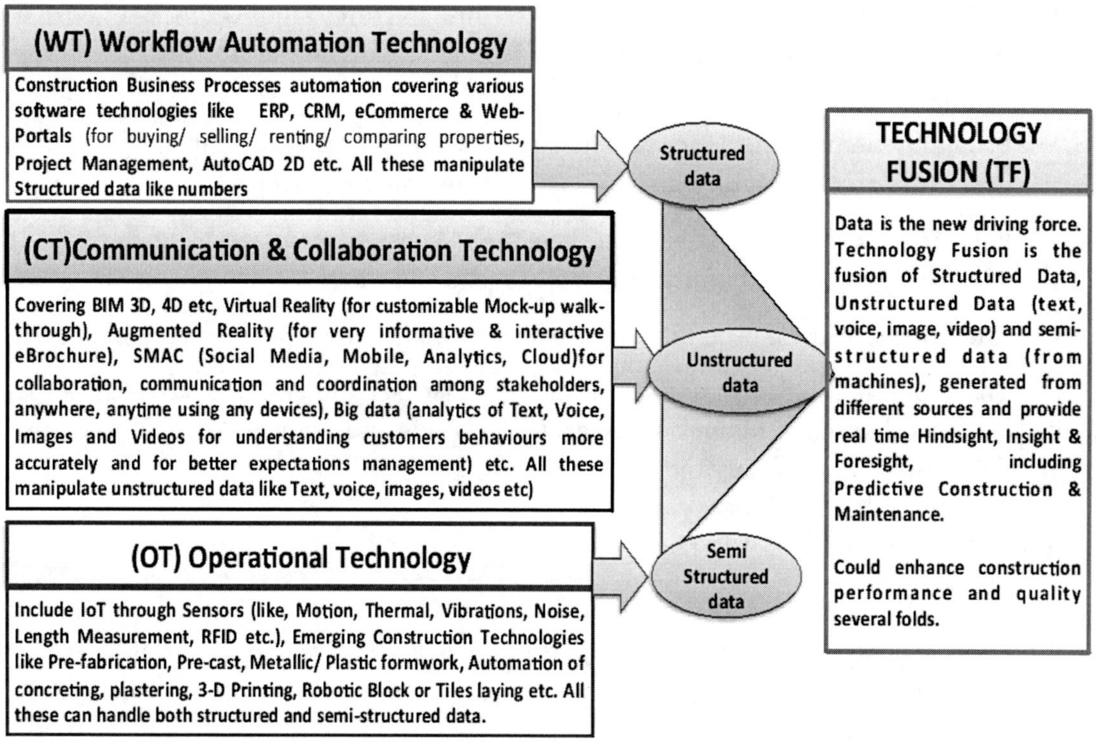


**Figure 9.1: Integration of three Technologies, giving multi-fold benefits**

These technologies significantly improve pre-construction, construction, and post-construction activities, such as:

- Architecture and design for buildings and infrastructure, including customization for interiors.
- Delivery of construction services, enhancing customer and stakeholder experience through real-time communication and coordination.
- Optimization of construction processes, project management, and safety, alongside sales and marketing.

- Accelerated development of just-in-time skills.
- Establishing a connected sales and supplier ecosystem for faster collaboration.
- Enabling innovation throughout the project lifecycle, from feasibility to handover and maintenance.

Tables 9.1 to 9.3 provide a summary of emerging OT and CT technologies relevant to construction, with further details in "Emerging Technologies in Housing Construction in India."

You will appreciate more the benefits of integration of WT (the ERP), CT (the mobile app, and Virtual reality & Gaming) after going through the real-life Case Study in Annexure 9.1.

**Table 9.1: Emerging Technologies related with Pre-Construction Processes**

| Emerging Technologies for Pre-Construction work | | | |
|---|---|---|---|
| **Phase** | **Sr #** | **Technology** | **Brief Description** |
| PRE-SALES | 1 | eCOMMERCE-Aggregation | Agrregating Sellers & Buyers at a common platform for business to get the best deal from various perspectives (e.g., MagicBrick.com; Housing.com etc) |
| | 2 | VIRTUAL REALITY | Preparing Finished building Mock-ups, customization and Walk-throughs to involve Customer in designing the Dream Projectand realizing it (e.g., SmartVizX.com) |
| | 3 | AUGMENTED REALITY | Preparing eBrochure, while simple is navigation and refering but with lots of information and pictures and visuals for easy understanding of the information users are looking for |
| | 4 | MOBILE-BASED VALUE ASSESSER | Comparison of various options of properties in terms of relevant parameters like Locations, Features, possible appreciation in next 5-10 years etc. |
| | 5 | CLOUD BASED CRM | Managing and engaging customers throughout lifecycle of sales to handover based on CRM usage, and not based on fixed cost licences (e.g. Salesforce.com) |
| | 6 | MASS-CUSTOMIZATION / CONFIGURATION OF FLAT & INTERIORS | Customize the floorplan of the Flat )or at least Interior design of the flats), for each user, and workout quotation and execution based on each customer's need (e.g., LivSpace.com) |
| ARCHITECURE & DESIGN | 7 | AUTOCAD | 2D Modeling for design and drawings |
| | 8 | BIM (Errorproofing) | 3D modeling of the various parts of the buildings, for finding alternative solutions to choose, finding design and possible executions defects (likepipe interferences) in advance and improving; Errorproofing & Cost reduction; 4D for cost management ; 5D for project scheduling |
| | 9 | SMART Buildings (IoT/ Sensor based) | Extensive use of various types of interconnected sensors for automatic switching on and off of light, opening and closing of doors , mood based room lighting pattern and temperatre etc |
| | 10 | IoT based Energy/ Water conservation | Sensor controlled AV, heating, Ventilation , water usage to minimize various wastes. |
| | 11 | LED | For decorative designers lighting |

### Table 9.2: Emerging Technologies related with Construction Work Packages

| Emerging Technologies for Construction work packages | | | | | |
|---|---|---|---|---|---|
| S.No | WORK-PACKAGES | Technololgy 1 | Technololgy 2 | Technololgy 3 | Technololgy 4 |
| 1 | Excavation | Mechanical Excavators | EPI Pulse Plasma Rock Fragmentation | Chemical Blasting | vacum excavation |
| 2 | Reinforcement | Smart Steel (pre- Cut & bent) in Plant | Pre fab Cage (Modularization) made in plant | COFFER/ METALLIC GRIDS, in place of Steel Bars | |
| 3 | Form Work | Plastic FORMS | Metal Alloys FORMS (like Myvan, Peri) | Climbing System of forming (like Peri) | Mettalic Grid instaed of Steel Rods (like Coffer) |
| 4 | Concrete Pour | Batch Mix | Ready Mixed Concrete (RMC) | Boom System of Concrete pouring | |
| 5 | Blockwork | Mortarless Blocks | Load Bearing RCC Walls | Pre-Fab Walls | Metallic Walls Like Steel |
| 6 | Electrical | Wireless/Sensors Technology | Cable-Conduit Repair Technology | Structured Cabling & Retrofit | |
| 7 | Fire Alarm (FA) | Linear Heat Detection | Flame Detection | Video Imaging detection | Control & Indicating Equipment |
| 8 | Security System | Digital Locks | Video Telephones | | |
| 9 | BMS | Programmable Logic Controllers(PLC's) | Internet of Things (IoT) | | |
| 10 | HVAC | Heat Recovery | Sensor based AC & ventilation control | | |
| 11 | CONSTRUCTION METHODS | Top-Down (Pre-Fab) Construction | 3-D Printing | | |

### Table 9.3: Emerging Technologies related with Construction Management

| Emerging Technologies for Construction | | | | |
|---|---|---|---|---|
| S.No | ACTIVITY being Managed | Technololgy 1 | Technololgy 2 | Technololgy 3 |
| 1 | MANPOWER Utization | Sensors Based Workers Tracking (including Pressure Sensors) | Mobile based Just in Time , Just Enough Daily Digital Training | Bionics based Time Logging |
| 2 | MACHINE Utization | Sensors Based Equipment Tracking | Analytics based Predictive Maintenance | Analytics based Spare Parts Ordering |
| 3 | MATERIAL Utization | Sensors Based Materials Tracking | Sensors Based Inventory Management | Drone based Survey Analysis |
| 4 | MONEY Utization | Mobile based Cash-Flow Management | Big Data Analytics | |
| 5 | PROJECT MANAGEMENT | Digital Camera based Remote Work Progress Monitoring | Drone based Project Progress Survey | Mobile based Social Collaboration Platform for Communication with Big Data Analytics |
| 6 | | The Last Planner Based Collaborative Scheduling and Execution Software (like vPlanner) | Cloud based Project management Softwares | Real Time and Predictive Scheduling |
| 7 | QUALITY MANAGEMENT | Paperless Quality Management of Processes & Products (like QEnhancer) | Prediction & Prevention of Snags Softwares | Machine Learning for Defects etc and alerting |
| 8 | PROCUREMENT MANAGEMENT | Cloud based Enterprise Resources Planning (like SAP, Oracle) | Sensor & GPS based Transport Vehicle Tracking Systems | |
| 9 | SAFETY HEALTH & ENVIRONMENT | Sensor based hazardous Location Safety Alerts | | |

### 9.1.2 Focus of Digital Technologies:

Many technologies are listed in the tables above. However, this book primarily focuses on digital technologies, their applications, and benefits, as other resources extensively cover operational and construction automation technologies. The remaining chapter focuses on digital technologies that are easily adoptable for construction sector for quick benefits.

### 9.1.3 Benefits of Digital technologies.

Based on extensive experiences of the authors, here are some guidelines to maximize the advantages of digital technologies, which can drive superior performance in construction projects and services.

**Guideline 1:** Digital tools eliminate distance barriers and enable real-time communication through collaborative platforms, fostering learning, innovation, and instant execution.

**Guideline 2:** Digital data, encompassing high-volume, varied types (like, text, images, voice, video), requires intelligent analytics to extract actionable insights. Using Descriptive, Predictive, and Prescriptive Analytics along with AI and Machine Learning helps automate recommendations and minimizes dependence on experts.

**Guideline 3:** Personalized customer experiences are vital. Focus on crafting unforgettable experiences, not only for customers but also to enhance interactions with business partners and employees.

**Guideline 4:** Digital businesses thrive on agility and a mindset geared toward exploration rather than stability. This environment promotes innovation, quick decisions, and timely results.

A case study in the chapter's annexure illustrates these principles further.

## 9.2 Emerging Digital Technologies for 'Pre-Construction Work

The following fast-developing technologies are widely adopted in the construction industry, helping solve new challenges, improve profitability, and foster long-term customer retention.

### 9.2.1 E-Commerce:

E-commerce leverages internet technology to buy or sell products online and integrates systems like mobile commerce, electronic funds transfer, supply chain management, online transactions, EDI, inventory management, and automated data collection. It spans online retail, electronic markets, and online auctions, and supports electronic business.

In real estate, E-commerce portals like www.MagicBricks.com, www.housing.com, www.99acres.com, www.Indiaproperty.com are commonly used for pre-sales of buildings, while in linear infrastructure projects, e-commerce is popular for online bidding and tendering, enhancing speed and transparency in government projects. For example, the Indian Government's e-procurement portal (https://eprocure.gov.in/eprocure/app) facilitates transparent public bidding. Real estate portals offer detailed pre-sales information, such as building location, campus layout, building appearance, floor plans, room interiors, and facilities like clubhouses and pools. For example, www.housing.com in the figure 9.2 shows building images, payment terms, project status, and area price trends, helping customers make informed decisions before visiting or engaging in sales discussions.

**Figure 9.2: Location, Campus, Building, Flat layout, inside rooms, facilities (Ref: www.Housing.com)**

### 9.2.2 Virtual Reality (VR)/ Augmented Reality (AR)/ Mixed Reality (MR)

We also recommend you read a small comprehensive article "Virtual, Augmented and Mixed Reality for the Construction Industry in a Nutshell" [1].

#### Virtual Reality (VR)

Through VR, the Customer can experience real-life like walk-through their complete dream house before actual construction starts. This will give them an idea about how it would actually look, after interiors are completely done. They can also have clear feel for spaces in real terms, and not a scaled down drawing version. They can feel, where is the congestion etc. VR requires following gadgets and technologies, shown in Fig 9.4 depicts, five of the pictures taken from a real project. This would minimize their surprises, after project completion

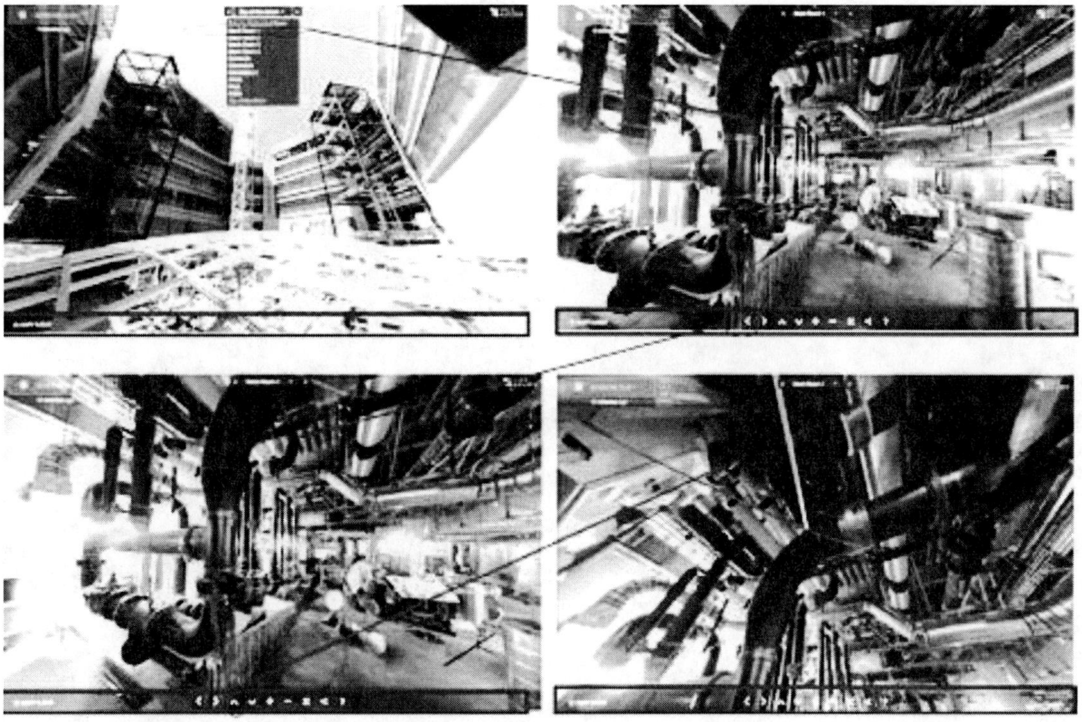

**Figure 9.3: Example of VR in Construction (Source: Reference [4])**

### Gadgets/ Hardware/Software Used for figure 9.4 :

1. Stereoscopic head-mounted display (providing separate images for each eye), like MS-HoloLens, Sony etc.
2. Stereo Sound
3. Head Motion Tracking Sensor
4. PC with NVIDIA RTX 1060 or AMD Redeon RX 480 Graphic Card
5. Processor: Intel i7 4770 or greater for Desktop or Intel Core i7-7700 or better for laptops
6. Software like Trezi (Gautam, Tithi, 2019, 21)

Figure 9.4: Example of VR Gadgets used

Figure 9.5: Example of Virtual Reality: Real-life walk-through

## Augmented reality (AR)

Augmented Reality (AR) is a virtual reality technology that merges real-world objects with computer-generated digital content. This overlay enhances how real surroundings might appear in different scenarios. An AR system layers virtual information over a live camera feed on smart glasses or mobile devices, allowing users to view three-dimensional images.

In real estate, a simple brochure may not suffice to sell an apartment. Sales teams can enhance the brochure experience by using AR to show far more "live" details about the property, which prospective buyers can view on their smartphones. An example is illustrated in Fig 9.6. Such dynamic, life-like experiences are beyond the capabilities of a traditional paper brochure.

**Figure 9.6: The AR application for a Sales Brochure for Apartment (source: https://www.unitear.com/)**

## Mixed reality (MR)

Mixed Reality (MR) combines aspects of both VR and AR, overlaying virtual objects onto the real world in a way that makes them appear naturally integrated. These virtual elements can be interacted with in real time within the actual environment.

For instance, as shown in Figure 9.7, a user wearing MR goggles can see real-time construction progress: when looking at a construction area, the user sees it is 91.5% complete directly through the lens. This virtual information aligns with the real object, eliminating the need for reports or feedback to assess project status.

**Figure 9.7: Two samples of M R applications for project progress and adding a sitting place in existing office. (source: Alice Bonasio Ref: [2])**

In another example of MR in Figure 9.8, additional construction is designed on top of an existing model to verify feasibility, aesthetics, and constructability by modifying virtual designs and equipment. Angus notes that MR uses headsets to project 3D objects into real spaces, blending VR and real objects. Unlike AR, where virtual objects are simply overlaid onto physical spaces, MR allows virtual objects to be anchored within them.

**A distinction between VR, AR, and MR is as follows:**

- **Virtual Reality (VR)**: A fully immersive experience that removes users from the real world into a digital environment.
- **Augmented Reality (AR):** Adds digital objects (e.g., text, images) to a live camera view on a smartphone, allowing additional information at specific physical locations.
- **Mixed Reality (MR):** Combines VR and AR, enabling not only the addition but interaction of virtual objects with real-world spaces.

**Figure 9.8: Sample MR Applications (Source: Angus, reference [1])**

These technologies help clarify project requirements, refine the client's vision, and enhance collaboration among stakeholders, reducing time and cost by minimizing rework and waiting. While applicable to infrastructure projects, linear infrastructure has lagged in adoption compared to real estate.

### 9.2.3 Cloud based Customer Relationship Management (CRM)

CRM (Customer Relationship Management) is essential in nearly all business sectors, including construction, addressing millennial stakeholders' communication and engagement needs. Key CRM features for the construction industry include:

a. **Core CRM:** Manages sales tasks, leads, contacts, pricing, document storage, and

b. marketing tools.

c. **Workflow automation**: Automates repetitive tasks, boosting productivity in sales and construction management.

d. **Collaboration tools:** Keeps teams synchronized with real-time data.

e. **Scheduling & calendar:** Manages schedules and activity logs.

f. **HR & staffing tools:** Provides access to HR data for project assignments.

g. **Proposal management:** Supports template-based proposal generation.

h. **Reporting & analytics:** Offers dashboards for regular reports and financial analysis.

i. **Customer support:** Includes customer surveys and support services.

Many companies use CRM beyond e-commerce expectations, now affordable through cloud-based SaaS packages like Salesforce. With cloud CRM, companies pay only for usage, eliminating initial CAPEX. Cloud-based CRM is ideal for construction businesses as it enables data access from any internet-connected device, anytime.

**Reasons for adopting cloud CRM like, Follow-up CRM, BuilderTrend, and CoreCon include:**

1. **Anytime, Anywhere Access:** Accessible on any device, aiding modern business flexibility.
2. **Ease of Use:** Easily learnable with maintenance handled by the provider.
3. **Affordability:** Cost-effective and scalable.
4. **High Security:** Enhanced web-based data security.
5. **Reliability:** High availability (up to 99.99%) and reliable operations.
6. **Scalability:** Easily adapts to more users and features.

### 9.2.4 Configurator Software for Designing Mass-Customized Homes and Interiors

Product Configurators helps to provide individual choices for personalization. This also bring acceleration in finalizing the Design. Simply put, a 'product-configurator' software can be defined as "a computer aided designing and manufacturing system to produce a customized product output." Two types of main configurator software are popular in in Construction Sector. One is for the building designs at the 'cold-shell' level and another for the 'interior design'. We are describing Product Configurator for Interior designs, after cold-shell is ready.

**Customized Interior Design Configurator:**

**A typical configurator for "interior designs" covers following choices for the users:**

a. **Civil Engineering Finish**: This can provide options like:
b. **Wall Finish:** Painting, Wallpaper, Fabric, Dado Tiling etc., and further choices of sub-categories in each of these.
c. **Room partitions**: Type: Sliding, Folding, Static, Removable etc. and also, it's material.
d. **Material:** Gypsum, Wood, Glass, Metal, Plastic etc. also their Sub-categories.
e. **Ceiling Finish**: Gypsum, Modular Grid, Wooden, Hardboard, Cardboard, POP, Tiles etc. and further sub-categories in each of these.
f. **Floor Finish**: Marble, Vitrified Tiles, Carpets, Wooden, Vinyl etc. and sub-categories

g. **Doors:** Wooden, metal, Plastic, Glass etc. and their sub-categories.

h. **Room Interior Furniture**: This can provide options like:

- **Apartment Portion**: Living Room, Bedroom, Dining, Guest Room. Figure 9.9 below shows a Kitchen & Figure 9.10 shows a living room, as examples.
- **For each room:**
    - Budget for the Interiors (Like Budget in INR Up to 50, 50 to 100, 100 to 150, 150 to 200, > 250 etc.)
    - Style: Looks & Feel of Contemporary, Classical, Futuristic, Heritage/Antic

**Figure 9.9: Kitchen Configuration: Styles to choose from, for same budgets. (courtesy: LivSpace Pvt Ltd, Bangalore, India)**

**Figure 9.10: Living Rooms: Different style to choose from, for different budgets. (courtesy: LivSpace Pvt Ltd, Bangalore, India)**

## 9.3 Digital Technologies for Architecture & Design

The digital technologies have influenced the way Architecture & Designers function today. This is in terms of making the Architects more creative as well as more productive. This is achieved by automating several processes involved in the development of final Architecture drawing and also detailed shop-floor or construction-ready drawings. The most important technologies available are described below with some illustrations.

### 9.3.1 Computer-Aided Design (CAD)

CAD, familiar to many, was introduced in the 1960s and is now essential for most designers and architects. It not only aids in drafting but also allows engineering calculations, blending roles of draftsmen, designers, and engineers. CAD software ranges from 2D vector-based systems to 3D solid modelers, with capabilities like rotating designs in three dimensions for comprehensive views, including from inside an object. Some CAD tools support dynamic mathematical modelling.

Initially used in manufacturing, CAD is now integral to designing construction projects, including small residences to large industrial structures and Linear Infra projects like bridges and highways. CAD assists throughout the engineering process, from conceptual design to strength analysis, and is crucial in Real Estate Construction for MEP systems (AC ducting, electrical wiring, etc.) and large machinery projects like oil, gas, and power plants.

CAD software helps architects and designers to manage construction projects with an integrated interface on personal computers. 3D modelling and rotations reveal design defects or interferences, reducing rework and project delays. Leading 2D and 3D CAD software providers include Autodesk (AutoCAD), Dassault Systems, and Siemens PLM. Figure 9.11 illustrates 2D and 3D samples from these packages.

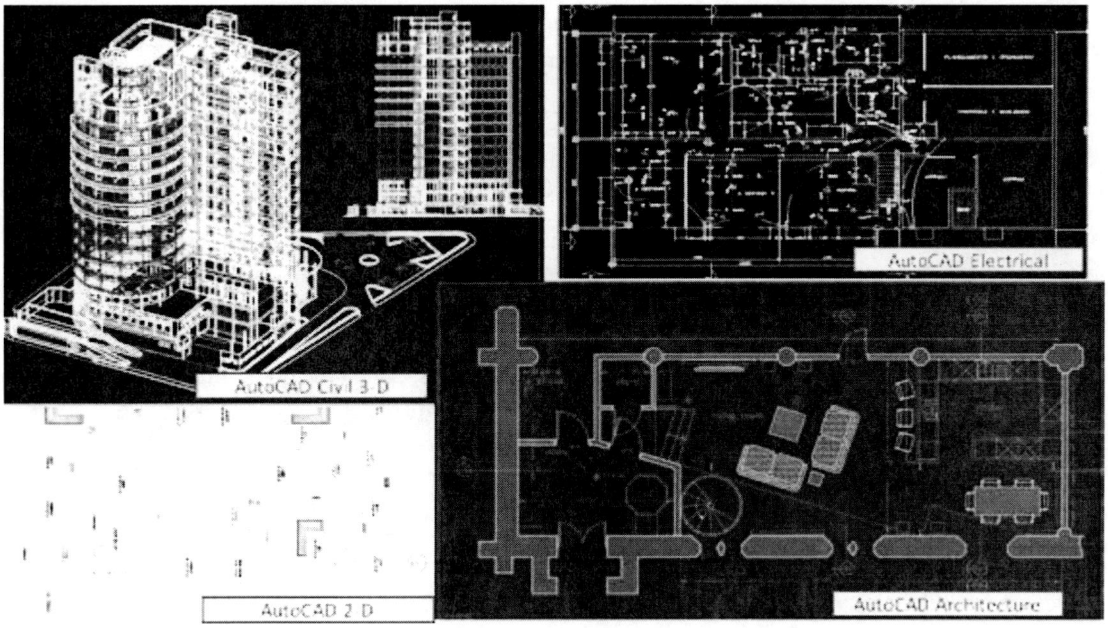


**Figure 9.11: Sample outputs from various packages of AutoCAD (source: Wikipedia)**

### 9.3.2 BIM (Building Information Modelling)

We would like to acknowledge here that most of the material used in this section for explanations has been taken from one of global leaders in BIM domain and product, Autodesk. Their descriptions are very precise, yet very comprehensive. This is available on the link https://www.autodesk.com/solutions/bim/benefits-of-bim [4].

Building Information Modelling (BIM) is a process encompassing several key functions like creation of an intelligent 3D model, document sharing and management, coordination and simulation during the entire lifecycle of a project, covering planning, designing, building, operating and maintaining. BIM is specially used to design and document real estate and infrastructure projects. Every detail of a building is modelled in BIM. The model is also used for analysis to explore various design options in order to choose the best.

**a.** For instance, this can be used for trying different designs of Water drain system and other Pipelines layouts in the same building, to see the implication of the costs and the aesthetics, as a part of Value Engineering exercise.

**b.** It can create 3-D visualizations that help different stakeholders understand the look and feel of the building, to choose the best, before these are actually built. These may not be as good experience as Virtual Reality, but much better than typical 2-D drawings.

c. It can also check various possible defects in advance specially, interference of various MEP piping, cabling and AC Conduit etc. These possible defects can be resolved by changing pipes layouts before the construction starts.

When everything is finalized, the model is also used to generate the design documentation for construction. BIM is a smarter, more efficient ways to resolve their problems of optimum design and build. BIM is also very useful for Infrastructure design projects of roads, highways, bridges etc. You can learn more on this from reference [8]. In construction industries, the use of BIM is quite common now in following six areas.

- **Architecture:** Making better design decisions, improving building and infra projects performance, collaborating more effectively during project lifecycle.
- **Civil:** Using intelligent, connected workflows to help improve predictability, productivity, and profitability.
- **Construction:** Digitizing the construction site and connecting project information from design through construction and handover.
- **Plant:** Managing the design and construction of intelligent piping, structures, and processes more efficiently and collaboratively throughout project lifecycle and at hand-off.
- **MEP:** Improving MEP (mechanical, electrical, and plumbing) design quality, cost and convenience; collaborating in real time to support the project delivery process.
- **Structure:** Exploring structural designs and detailing software for winning new business and supporting project delivery.

Figure 9.12 depicts some example of BIM in different areas of Architecture.

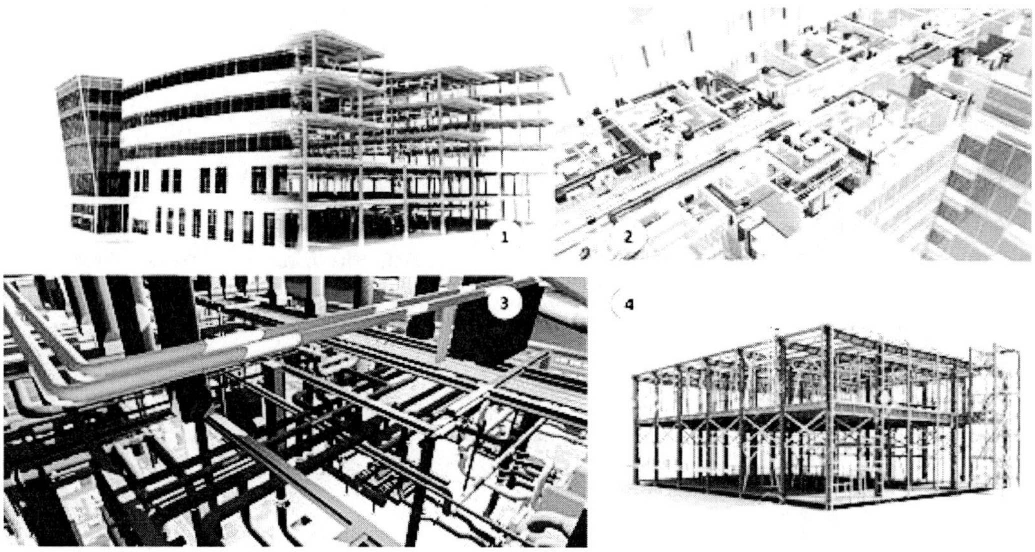

**Figure 9.12: Some Examples of BIM for Building Architecture & MEP Design (source: www.autodesk.com, www.bimcrunch.com)**

Unfortunately, the capabilities of BIM have got to be leveraged far more than what is being done currently. Moreover, now you have BIM 4-D and BIM 5-D versions as well, and people are still not yet ready with the capability of BIM 3-D. However, the meaning of 4-D and 5-D versions are explained very briefly below:

**BIM 4-D: (incorporating Construction sequencing)**

4D BIM adds an extra dimension of information to a project information model in the form of scheduling data. This data is added to components which will build in details as the project progresses. This information can be used to obtain accurate program information and visualizations showing how your project will develop sequentially.

Time-related information for a particular element might include information on lead time, how long it takes to install/construct, the time needed to become operational/ harden/ cure, the sequence in which components should be installed, and dependencies on other areas of the project.

**BIM 5-D (incorporating Costs)**

Incorporating and extracting accurate cost information is the 5th dimension of BIM. Once the BoQ (Bill of Quantity) is prepared with 3D model, in 5D the considerations might include capital costs (the costs of purchasing and installing various equipment and items like Chillers for AC), its associated running costs and the cost of renewal/replacement down the line, for calculating Life Cycle Costs. These calculations can be made on the basis of the data and associated information linked to particular work-Items within the graphical model. This information allows cost managers to easily extrapolate the quantities of a given component on a project, applying rates to those quantities, thereby reaching an overall cost for the development. This would also help computing the cash-flow requirements at different point of times. Although, BIM 4D and 5D looks quite promising, its use is still at very nascent stage.

### 9.3.3 IoT (Internet of Things/ Sensors) for Smart Buildings

In the era of 'smart' Buildings and 'smart' Cities, the role of Internet of Things (IoT) is extremely important in designing and buildings. This is also getting popular in linear infra projects. For instance, when you design and install Lighting across a long road, the basic components like metallic pole can also be designed to have digital Camera for managing traffic and crime around it, Wi-Fi booster for networks, to give constant network to the vehicles and people passing through it etc.

Unfortunately, we have observed that the knowledge of the Architects and Designers is very limited in this aspect. There is a big need for them to understand the fundamental concept of IoT and collaborate with digital technology experts who can help architects and designers design an excellent customer experience from the construction of the 'smart' product. The paragraphs below provide some brief introduction to this technology.

IoT is a system of networking or interconnection of various 'non-living things', like computing devices (like Laptops, Mobile phones, machines) and gadgets (like cars, refrigerators, microwave Owen), any other physical objects (like trees, furniture etc.) and 'living things' like, animals or people. These things can be identified uniquely (UIDs) and have the ability to communicate (transfer data over a network) among themselves, without requiring 'human-to-human' or 'human-to-computer' interaction.

IoT is driven by combination of three forces, namely, Sensors and Actuators, Network Connectivity, and People and Processes.

### a. Sensors & Actuators

Sensors and Actuators: Besides our physical world, which we perceive through our five senses, we are also surrounded today by the cyber world, with GPS sensors. These sensors provide location data, and sense like eyes and ears using cameras, mobile phones and microphones. Moreover, these are several other types of sensors that can measure practically everything, such as temperature, weight, and pressure. Typically, different types of digital sensors have been designed to measure things such as:

- Location, Proximity, and physical presence of anything
- Motion, Velocity and displacement
- Temperature
- Humidity, Moisture
- Acoustic, Sound, Vibrations
- Chemicals & gases of different types
- Flow of different Fluids like water, oil
- Force, Load, Torque, Pressure
- Leakages and levels of Liquids
- Electric, magnetic forces
- Acceleration, tilt
- Machine vision, ambient light etc.

Since all this data can be measured and stored, it can also be shared with other sensors, if proper network connectivity and communication protocols are available.

### b. Network Connectivity:

These can take the digitalized data as inputs from one sensor or device and can transfer to other sensor or device through any of following networks or technologies:

- PAN (Personal Area Network)

- LAN (Local Area Network)
- MAN (Metropolitan Area Network)
- WAN (Wide Area Network)
- Bluetooth
- Wi-Fi
- 2G/ GSM/ CDMA/
- 3G/ GPS/ GPRS
- Cellular 4G/ LTE/ 5G etc.

For details on these networks, you should conduct the Google search and get more information about these, about the basic technology and the utility and functions/ features.

### c. People & Processes

These networked inputs can be combined into bi-directional systems that integrate data, people, processes and system for better decision making. Some common processes used by the people are listed below.

- Location and tracking Objects
- Customer Relationship Management
- Business Analytics (data, text, voice, image, video etc.) & Cloud
- Upgrades and Configurations
- Remote monitoring & Maintenance
- Automation & Control
- Supply Chain management
- Mobile Devices & Apps
- Security & Apps
- Financial management

Interactions among these three entities are creating and can create several new types of innovative 'smart' applications and services. In the context of the Real Estate and Infra Projects design and construction, you may consider the following applications, as illustration. Many of these can be activated or controlled through smartphone or tablet, or touch-buttons.

1. Smart dimming & multicolour LED Bulb for mood management

2. Smart Gadgets (Voice command operated Air Conditioners, curtains, TV & appliances like refrigerators, microwave ovens, dishwashers, washing machines, stoves etc.)
3. Smart SECURITY (Biometric authorization at main door, Video door phone, Smart Lock with digital camera linked to smartphone, Panic button in Bedroom)
4. Smart Smoke & Gas leakage ALARM in kitchen
5. Smart Energy monitoring (HVAC, Room Light and AC & Fans etc.)
6. Smart Irrigation Control (for in-house plants and landscaped trees)
7. Smart Water Grid/ Outlets for distribution of water in a city or gated community.
8. Smart traffic control on roads, on long congested bridges
9. Smart Infotainment
10. Smart Predictive Maintenance of various Buildings, Roads and Equipment
11. Smart Emergency Alerts (Theft Alarm etc.)

As a Smart home example, as one shown in figure 9.13 you may like to read and see about a major Building project, the EMAAR's Digi Homes coming in north India. (www.emaardigihome.co.in).

One interesting example in Infra is about a 'smart' water distribution system in Vijayawada district of Andhra Pradesh. This is shown below in Figure 9.16, and was developed by Efftronics Systems Pvt. Ltd, Vijayawada. Their IoT covered and monitored a total of 59 Water reservoirs, 5 Water Treatment Plants, spread over about 58 Square Kilometres of area in a city with population of about 10 Lakhs people. This almost removed the manual intervention, and reduced the water consumption and its wastage. Such is the power of IoT.

A total of 59 Water Reservoirs, 5 Water Treatment Plants, spread over about 58 Square Kilometers

**Figure 9.13: Smart Homes with a sample of smart door lock with various opening options (source: www.emaardigihome.co.in)**

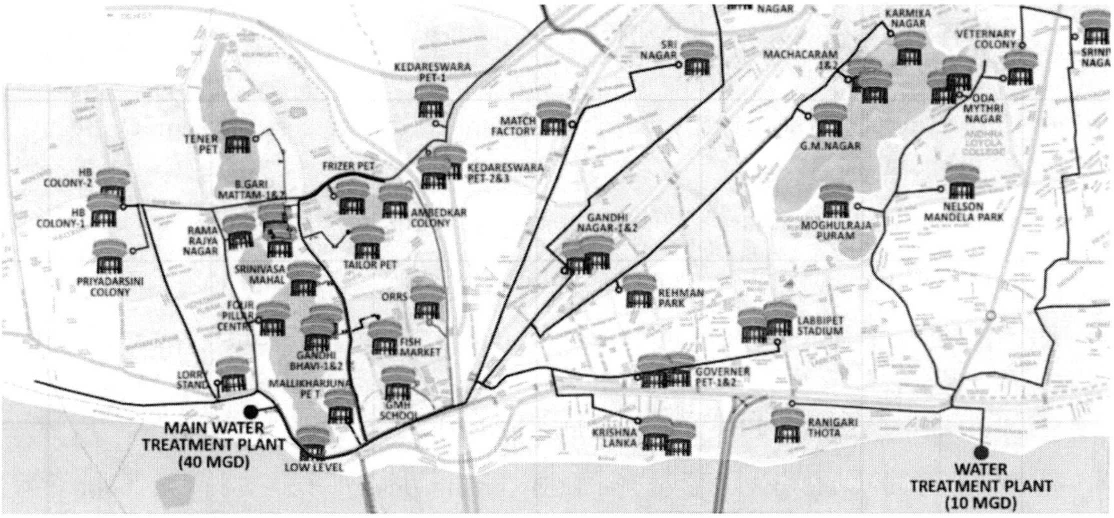


**Figure 9.14: Smart Water Distribution System in Vijayawada (courtesy Efftronics Systems Pvt. Ltd)**

## 9.3.4 LED (Light Emitting Diode) Technology

LED, a semiconductor light source that emits light when current flows through it, is familiar technology to most. LEDs can be programmed to emit various colours based on energy requirements. In 2014, the Nobel Prize in Physics was awarded to its key inventors from the U.S. and Japan. LEDs offer substantial benefits, especially in building and infrastructure design, saving 80-90% power compared to conventional sources and boasting a long lifespan.

A major LED initiative in India was the distribution of 30 crore LED bulbs under the Ujala scheme, saving over 15,000 crore rupees annually. Many streetlights are now LED, delivering more light at lower cost. Key LED advantages include high efficiency, customizable colour without filters, compact size, instant illumination, durability under frequent cycling, dimming capability, minimal heat output, long lifespan, and resistance to shock.

**Advantages of LED:**

- **Efficiency:** LEDs emit more lumens per watt than incandescent light bulbs. LED lighting fixtures is not affected by its shape and size, unlike fluorescent light bulbs or tube-lights. Therefore, any shape can be created out of it very easily and at low cost.
- **Colour:** LEDs can emit light of a desired colour without using any colour filters as traditional lighting methods need.
- **Compact Size:** LEDs can be smaller than 2 mm.

- **Warmup time:** LEDs light up almost instantaneously
- **Cycling:** LEDs can be subject to frequent on-off cycling without failure, unlike tube lights or fluorescence lights
- **Dimming:** LEDs can easily be dimmed hence can give stroboscopic effect, if required.
- **Cool light:** LEDs radiate very little heat and are more eco-friendly
- **Slow failure**: LEDs mainly fail by dimming over time & not abruptly
- **Longer Life:** LEDs have 5 times longer life than fluorescent tubes & 40 times the incandescent light bulbs.
- **Shock resistance:** LEDs, are not damaged with external shock.
- **Focus:** The solid package of the LED can be designed to focus its light in a particular way.
- **Ambience temperature Independent:** Now LED lights have been designed and tested to work between -40 to +100 degrees Celsius.

**Disadvantages (but most of these have been resolved now)**

- **Voltage sensitivity:** LEDs works with a voltage above their threshold voltage and a current below their rating.
- **Area light source:** LEDs are difficult to apply to uses which need a spherical light field.
- **Light pollution:** white LEDs used for outdoor lighting cause more sky glow.
- **Impact on insects:** LEDs are attractive to insects more than sodium-vapor lights
- **Thermal runaway**: Running two or more LED strings from a single current source may result in LED failure.

Despite some limitations, like voltage sensitivity and potential impact on insects, LEDs are valuable for building lighting, offering:

1. Creative flexibility for low-cost, unique fixtures
2. Long-term power savings despite higher initial costs
3. Enhanced building aesthetics at night
4. Reduced need for backup power sources

Figures 9.15 and 9.16 show examples of LED fixture designs and building aesthetics at night. LEDs are now widely used in process industries and solar-powered street and highway lighting.

**Figure 9.15: Illustration of easy unique designs possible using LED (Courtesy: LivSpace Pvt Ltd, Bangalore, India)**

**Figure 9.16: Illustration of external LED lighting designs giving different looks & feel to the same building (source: www.Oneindia.com)**

## 9.4 Emerging Digital Technologies for Construction Work

Following are some of the main digital technologies, which are influencing the Construction

work in Construction 4.0 environment, in a big way. Industry has to adopt and adapt these fast, in order to sustain and achieve superior performance.

### 9.4.1 3-D Printing

3D printing, or Additive Manufacturing, is a computer-controlled layering process creating three-dimensional shapes, especially useful for unique prototyping and complex components. Its applications span product design, complex parts, unavailable equipment spares, tool-making, consumer electronics, aerospace, medical fields, customized baking, and now construction.

Construction 3D printers enable the creation of building components or even entire structures. Combined with BIM technology, 3D printing can speed up construction, reduce labour costs, minimize waste, and allow work in hazardous areas. For example, Shanghai's WinSun Decoration Design Engineering used large 3D printers to build 10 small demonstration houses in under 24 hours for less than $5,000 each. In India, IIT Madras and TVASTA Manufacturing Solutions recently built a 600 sq. ft. 3D-printed house in Chennai in just five days. This technology has potential to meet India's housing needs faster and more affordably.

3D printing offers benefits like speed, cost-effectiveness, complex design capabilities, and zero waste, despite high material and equipment costs. However, limitations persist, including the need for fast-setting, strong, low-porosity binders, durable and flexible construction equipment, risks of cement setting within equipment, and issues with layer binding and porosity. Figures 9.17 and 9.18 depicts some real examples of 3D printing.

Ref: https://www.3dnatives.com/en/3d-printing-construction-310120184/ (for 1, 2, 3)
https://www.designingbuildings.co.uk/wiki/3D_Printed_Office_Dubai (for 4)

**Figure 9.17: Some sample creation of Construction 3-D printing**

**Figure 9.18: First 3D-Printed house, developed in India by IIT-Madras & TVASTA**

In conclusion, it's essential to understand and adopt 3D printing technology in construction to harness its advantages as research continues to overcome current limitations.

### 9.4.2 Robotics

Currently, the construction industry is among the least automated, despite challenges like labour shortages and low productivity. India faces even greater hurdles. However, with the successful integration of robots in manufacturing, they are now slowly entering construction as well. For instance, Tesla's E-Vehicle plant in the U.S. uses 5,000 human workers and 1,000 robots. Similarly, robots are beginning to assist in new construction, renovation, and demolition, indicating an upcoming robotic transformation in this traditionally manual industry. Robots typically excel in repetitive, controlled tasks, unlike most construction sites. Yet, advancements have led to robots for specific construction tasks:

1. Brick laying and masonry for buildings (Ref fig 9.19 below)
2. 3-D printing robots for complex shapes, buildings, even small concrete bridges
3. Earthwork of excavation and levelling Robots for roads
4. Finishing work robots, like tiling, plastering, putty application and painting (getting made in India too. Please refer fig 9.20)
5. Concreting components and assemblies

For example, a brick-laying robot can lay 5,000 bricks per day compared to a human team's 350, and a robot-assisted plastering team can cover 3,000 square feet per day versus a human team's 200 square feet. Robots don't face issues like fatigue, distractions, or labour disputes, providing consistent productivity and 24/7 service.

As construction companies look to automate more and more tasks for the sake of efficiency and productivity, demand for construction robots will grow steadily.

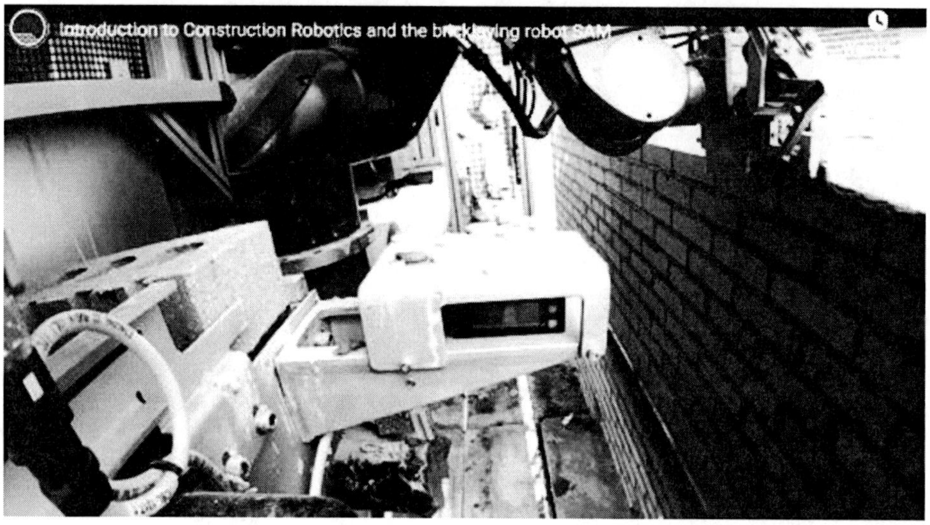


**Figure 9.19: An example of Brick-laying Robot (source: reference [22])**

**Figure 9.20: An example of Plastering & Painting Robots (Applying Putty) (Courtesy: Pace Robotics, India)**

Robots also enable rapid creation of complex designs with high precision, reducing time, costs, and waste, particularly with 3D printing. This allows quicker planning and development and easy adaptation to custom designs. Robots can also enhance human abilities, for instance, by allowing workers to lift heavier loads safely and focus on creative tasks while robots handle repetitive or hazardous jobs.

During health crises like the COVID-19 pandemic, robot usage increased due to their immunity to social distancing needs. As the construction industry evolves, adopting robotics early will likely yield significant advantages in efficiency and safety.

In essence, following are the key benefits of robotics for Construction Industry, hence it is highly recommended:

1. Same set of human labour can produce several folds more output
2. Customized design and complex shapes
3. Working in the hazardous environment
4. Robotics automation can be augmented with other devices

### 9.4.3 GPS and Drones Technologies

Surveying land for construction of linear infrastructure projects, like bridges and roads, has long been a major challenge. The "Total Station" has improved accuracy over traditional theodolites, but road surveys still take significant time. For instance, surveying a 100-kilometer road with measurements every 20 meters requires mapping 5,000 positions and takes 3-4 months. For bridges over water, such as in the sea, finding precise pile and pier locations is even far more harder.

GPS and drone technologies have revolutionized this process. GPS, managed by the US Air Force, provides time and location data globally. For better understanding how it works, you may like to know that a GPS requires at least 4 satellites to find receivers 3-D position (latitude, longitude, and altitude, namely the X, Y and Z dimension). Fig 9.21 below depicts briefly how it works.

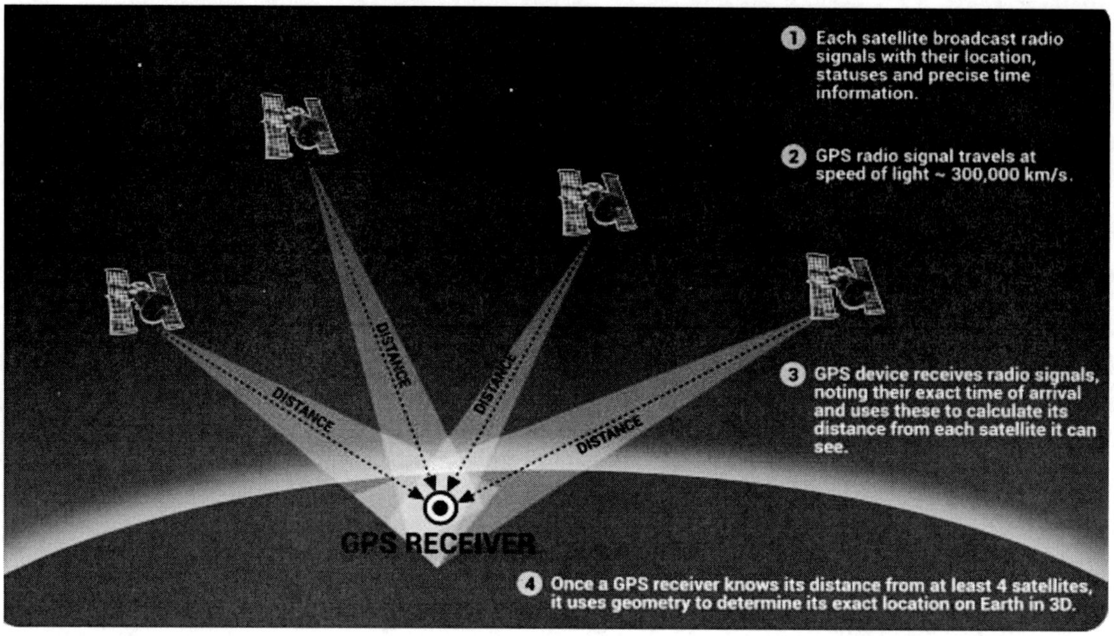


**Figure 9.21: Illustrative visual of GPS system. (source: reference [23])**

Once the position of a receiver is calculated with respect to four satellites, the GPS device can easily mathematically calculate, the speed and location of the object with GPS receiver and also the distance to destination. Since these satellites have atomic clocks for timing, it is extremely accurate in computing distance and speed etc., despite being so far away from the moving objects.

Drones, originally UAVs, are remote-controlled aircraft used in various industries, now including construction for surveying, capturing images and videos from a height. In construction, GPS and drones are valuable for:

1. High-level geography surveys for designing long roads and highways.
2. Building surveys for identifying defects in hard-to-reach places.
3. Mapping and inspecting highways for road defects, divider alignment, and expansions.
4. Precision surveys to map earth contours, aiding in road construction planning.
5. Pinpointing pile and pier locations for projects like L&T's Mumbai Trans Harbour Sea Link.
6. Remote site inspections, covering risky areas efficiently.
7. Enhancing health and safety through regular site monitoring.
8. Virtual site walkarounds

A typical Drone is shown in the Figure 9.22 below. Currently this technology has matured quite a bit and is quite accurate.

**Figure 9.22: An illustration of Drone flying with a remote control**

Although these technologies are extremely useful and very cost-effective, there are a few technical limitations currently, but research and development are going on these at many academic and research institutions in the globe to address these. Some of these issues are:

- Time synchronization between individual satellites and the GPS receiver
- Real time update of the exact location of the GPS satellite
- Interference with other signals

Recently he Indian GPS system was also developed the Indian Space Research Organization (ISRO). This is known as NavIC (Navigation with Indian Constellation). NavIC consists of a constellation of seven satellites in geostationary and inclined orbits. Soon it will be available to people for common commercial use.

We strongly recommend exploring the usage of these technologies in your construction projects. Use lots of innovation to get excellent benefits.

## 9.5 Emerging Digital Technologies for Project & Quality Management

The digital technologies are also influencing in a big way the quality management, project management and construction management at the work-site. By adopting these, the industry can achieve far superior performances compared to currently.

### 9.5.1 New Technologies for better Resources Utilization

One of the main problems faced in Project management of Construction Projects is low utilization of the resources, especially the manpower and the equipment. In

Construction projects, the work locations at the project-site are spread over a long horizontal or vertical area. In these widely spread areas, the work is performed by various teams of workmen or expensive Equipment are deployed. It is difficult to monitor those dispersed working areas at project-site for a better supervision and performance. This impacts the productivity and the output of the resources adversely.

The new digital technologies, like biometrics, RFID, various type of sensors, and IOT are now available to address these issues amicably. These are explained very briefly below, with some examples describing how could these be useful in construction project management.

a. **Biometric Technology:** this generally refer to the use of **technology** to 'identify' a person based on some aspect of their biology, like face recognition, fingerprint, hand geometry, Iris recognition and DNA. Fingerprint recognition is one of the first and original **biometric technologies** that have been grouped loosely under digital forensics.

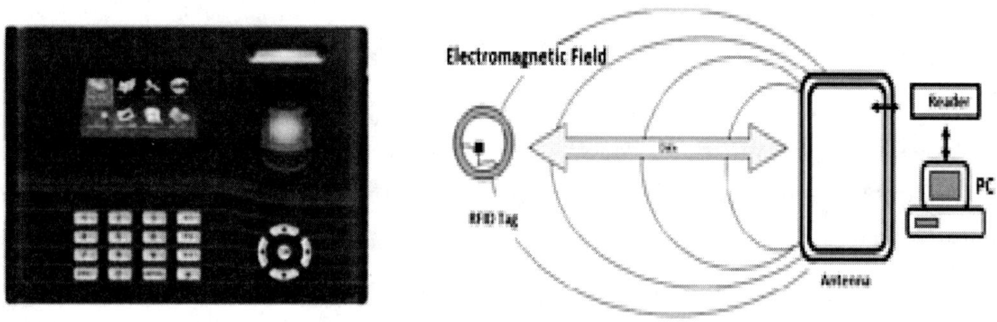


**Figure 9.23: Some components of Biometric technologies**

b. **RFID Technology:** Radio-frequency identification **(RFID)** uses electromagnetic fields to automatically identify and track **RFID** tags attached to objects. The tags contain electronically stored information about the product etc. Passive type of tags collect energy from a nearby **RFID** reader's interrogating radio waves. These are very popular in Retail Malls and Hypermarket products. Unlike a barcode, the tag need not be within the line of sight of the reader, so it may be embedded in the tracked object. Therefore, there are active type of tags also which have a local power source (such as a battery) and may operate hundreds of meters from the RFID reader, used in transportation and logistics business for vehicle location tracker etc.

c. Digital Sensors: Sensors are sophisticated devices that are frequently used to detect and respond to electrical or optical signals. We have discussed about these in the IoT technology section also. A Sensor converts the physical parameter (for example: temperature, blood pressure, humidity, speed, etc.) into a signal which can be measured electrically. As an illustration, let us talk about the labour utilization on the construction site.

Most organizations are very happy to use Bionic system of attendance, in the name of newer technologies. This is good for sure, but not good-enough. Since this is a centralized system installed at the entry gate, there is usually a big queue of workmen, waiting for their turn to check-in and register their presence. A picture below shows this. Our observation is that as much as 35-50 minutes, every day, might be wasted in this, out of 450 minutes of their work, for each worker, as shown in Fig 9.24. But after entering the site, there are no ways to track the workmen about where they actually are, and whether they are working at a given moment.

**Figure 9.24: Queue of Workmen, waiting for their turn for punching their attendance after lunch hours, using Bionics.**

If the planned number of workmen in a given working location is 50, and workmen working there for the most of the time are only 43, there will be a backlog of work for the next day, and it will keep on accumulating. This is because you would come to know about this, only "after" the harm is done. There is better technology now which is mobile and RFID based for this, which could produce better results. For example, a solution based on Mobile phone would:

- Allow attendance from Workmen's mobile
- The workmen will be tracked, whether and where exactly they are working. This will use the motion sensors and Geo locational sensor in the mobile
- Attendance can be based on geo-fencing. That is at a closed given location, number of workmen working can be tracked remotely, and if they are less than a certain pre-defined number, an alert can go to project manager automatically for taking corrective action.

Such system will resolve one of the major problems of construction project management, both in multi-story tall buildings, horizontally spread gated communities, or a linear Infra Project. You could find some solutions for reference from companies called "factoHR" and "Efftronics". You can explore in Google for this.

### 9.5.2 Mobile enabled Quality Management

You will recall that in Chapter 3, we talked about achieving excellent Quality, by focusing on four dimensions, namely Products, Services, Processes and People.

For the easy implementation of the concepts of achieving excellent Quality of Products, services, Processes and people, several mobile soft-wares are available now. Some of the common features of these mobile-based packages are given below. Fig 9.25 also shows a sample of one of such App.

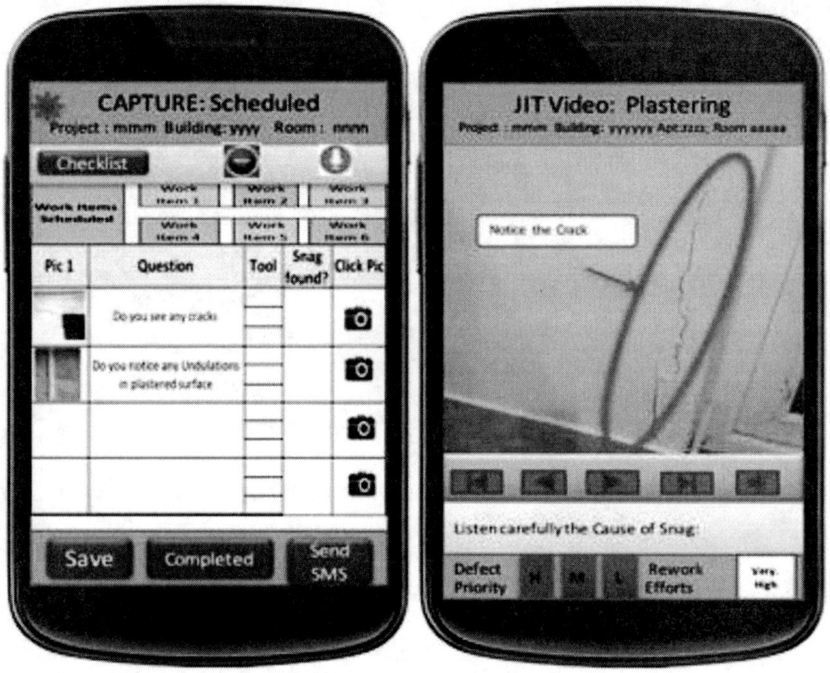


**Figure 9.25: Sample Quality Management Mobile App, showing Checklist & Training Video**

#### 1. Mobile-based Product Quality Audit/ Defects Checklists Management

- Pro-Active Defect finding during various stages of completion of the project on the site, based on Dynamic Checklists (regularly updated, based on regular checks) based on master data created in past
- Issuing the Defects Reports with Snag / Defect pictures
- Managing the whole life of Defects report, starting from capturing, allocating, fixing, approving the defect corrections, closing defect and also alerting for long delays. Paperless operations

- Preparing various Inspection Reports

**2. Mobile-based Checklists Management for construction processes and supplier's material**

- Inspection reports of the suppliers' material specially tested those tested internally by them.
- Inspecting the workmanship on the site, based on Pre-Created Checklists, which are updated regularly
- Issuing the Non-Complaints Reports (NCRs)
- Tracking the NCR till closed, and alerting for long-pending ones
- Preparing various Inspection Reports

**3. Mobile-based People "just-in-time, just enough" training for present day's work**

- Playing 8-10 minutes videos for the workers at the Toolbox meeting, for the work they are supposed to do on that day.
- Videos cover the best practices, common mistakes, tools & instruments to be used for the construction process, impact of poor workmanship on project

**4. RFI and Complaint management**

One such App available is 'CQaaS -Construction Quality as a Service' [9]. There are other emerging software systems which have the following features as well:

- Predict and forecast the number and type of defects & snags in a given project or at during particular duration of the project.
- Predict and forecast the impact of defects and snags in terms of the extra manpower, extra material and extra cost required for rework. These can also estimate the extra delays of completion of the project, for defect corrections.
- Recommend preventive measures to avoid those defects/ snags, in order to save money, efforts and extra delay due to rework involved.

Usage of technologies like Artificial Intelligence, Machine Learning and Analytics in Construction can make GPS, Drone and IoT technologies even more powerful. We can very briefly define these terms and possible usage.

- **Artificial Intelligence:** This is the science and engineering of developing Computer Systems to act like intelligent machines, capable of performing tasks normally requiring human intelligence like visual perception, speech recognition, decision making, and translation between languages.

- **Machine Learning:** This is the application of Artificial Intelligence hat provides systems the ability to automatically learn and improve from experience without being explicitly programmed. This focuses on the development of computer programs that can access data, recognize the repetitive patterns of behaviour and use it learn for themselves.

- **Analytics:** This is systematic computational analysis of data (number, text, voice, images etc.) or statistics, using relevant mathematical algorithms.

In the software like 'CQaaS' we discussed above, innovative use of AI, ML and Analytics technologies can be thought of. For example, you would know that almost same type of defects/snags take place in similar projects. Like, a crack in the tile, tile layout alignment line not matching with the next tile, poor tiles corner joint etc. With the collection of such data in pictures, using image mapping technology from the various projects on a company, the software can 'learn' the type of defect/ snag over a period of time. Then by using machine learning, the defect / snag can also be automatically captured (or approved) by the software, while capturing at the new site. This will eliminate the need of skills of expert knowledge of Snagging, at the same time ensure that no snag is missed by the person.

Just imagine that before handing over the hotel to its owner, a drone is sent to perform snagging in the entire building in each of the guest rooms. Drone has the camera and the AI, ML software already loaded. It would do the total snagging in may be 2-3 hours and identify all the snags without missing any. It will also generate the Snags reports. All this would be done without any human intervention. This is the kind of scenario; you can think of in Construction 4.0 era. Of course, we are still far from this scenario.

While capturing a defect or a snag (Figure 9.26), artificial intelligence can automatically convey through a mobile alert message, about the impact of that defect / snag on the cost and time of repairs, and also the impact on the delay of the project, without any expert being available there. Almost same system can be used for finding the surface defects of long roads and railway tracks.

**Figure 9.26: Snagging using a Quality Management Software**

**We can summarize the benefits of these software Apps:**

- Total Paperless operations,
- Making documentation error-free and accessible from anywhere, any time
- Enabling Predictive Maintenance of the key Equipment
- Reporting Progress of the Project
- Ensuring no defects/ snags are missed, even though the people are not trained.
- Taking care of non-availability of skilled manpower in the world, at least partly.

If you refer our earlier discussions on the, 'Role of PMCs in 21st Century' to take care of Quality Issues, you will realize that all these new technologies are a big support to the Quality Manager of coming era to meet the new Age customers' expectations. Table 9. 4 below show as an illustration, how each of the new role of Project Managers can be supported by new technologies. In fact, in our opinion, sky is the limit for innovation of using AI, machine learning, IoT and other technologies in Construction. The good news is that these are available more easily now and are affordable. Millennials are particularly very excited about this.

### 9.5.3 Collaborative Distributed meetings and training webinars Platforms

One of the most popular collaborative communication platforms, which you must already be using are software like Skype, Cisco-WebEx, Zoom and Microsoft Team Meeting.

Table 9. 4 Role of Project Managers for Ensuring Quality in 21st Century

| SR. NO. | FOUR FOCUS AREAS | ROLE of PROJECT MANAGER for ENSURING QUALITY | Whether Lean Digital Thinking Helps ? |
|---|---|---|---|
| 1 | DOMAIN KNOWLEDGE & PROJECT LIFE CYCLE PHASES | Must have or acquire the knowledge of the finer requirements of the type of project being taken up (For examples Hospital Building, Resort or Hotel Building, Malls, Commercial Office Workplace, Residential for different economic status of people etc) | Yes (Historical Data, AI and Data Analytics) |
| 2 | | Basic Methodologies and Processes best suited for the Project (like which ones from 49 processes of PMBOK, or whether lean Construction Practice like The Last Planner. VSM etc, or PRINCE 2 framework etc) | Yes (Mobile enabled Portals/ Apps) |
| 3 | | Knowledge about the Geography and the local culture , Food practices, Work-Culture etc of the Work-Site (India itself is having quite different work culture in North, South, East , West and North-East; Project in Singapore, UK, UAE and Srilanka will be quite different etc.) ) | Yes (Mobile enabled Information on different ciuntries and culture) |
| 4 | | Construction Technologies chosen or to be practiced ; Also use of Value Engineering | Yes (Mobiled enabled eLearning) |
| 5 | | Curiosity for knowing and learning new technologies like Virtual/ Augmented Reality, BIM etc | Yes (Mobile Enabled Jst in Time , Just enough training): Search Engines) |
| 6 | PROJECT MANAGEMENT LIFE CYCLE | Knowledge and awareness of the detailed activities to be done, specially the most crucial ones from Quality and Safety point of view. | Yes (Quality management Apps like QEnhanser) |
| 7 | | Focus on the P-D-C-A cycle continuously (Plan-Do-Check-Act Cycle) | Yes (Prescriptive Big data) |
| 8 | | Focus on Risk management ; Identification of critical Risks in advance and plan for their management | Yes (AI and Data Analytics) |
| 9 | | Optimal Utilization of the Resources like Workforce, Equipments (specially Critical and Expensive), | Yes (RFID, AI and Data Analytics) |
| 10 | | Correct calibration of all those machineries and Instruments | Yes (Enterprise Video / Text Chats, QM Software)) |
| 11 | | Constant focus on elimination of wastes and continuous improvements | Yes (Just-in-Time, Just Enough Training, eLearning)) |
| 12 | | Curiosity for knowing, learning and using new digital technologies and softwares (like Qenhancers; Cloud based ERP & CRM , PM etc.) for improved outcome of PMCs | Yes (Just-in-Time, Just Enough Training, eLearning)) |
| 13 | | Regular Communication with the team , and play the role of Enabler/ Problem Solver in the Governance / Project Reviews | Yes (Collaboration on Video Chats,, AI and Data Analytics) |
| 14 | | Ensure Just-in-Time, just enough short trainings for the relevant people , to be used immediately | Yes (Just-in-Time, Just Enough Training, eLearning)) |
| 15 | | More focus on Defects/ Snags prediction and prevention, rather than capturing and correcting | Yes (QM Software with Predictive & prescriptive Data Analytics) |
| 16 | | Leaning the Concepts of Lean Construction like The Last Planner, Value Stream Mapping, Gemba Walk for non-value Adding waste elimination | Yes (Just-in-Time, Just Enough Training, eLearning)) |
| 17 | PROJECT LEADERSHIP | Close involvement of Clients or End-Users at the early stages | Yes (Collaboration on Video Chats) |
| 18 | | Good Rapport and relationship with the Client and Users | Yes (Cloud based CRM) |
| 19 | | Good Rapport and relation with Key Sub-Contractors and Suppliers | Yes (Cloud based CRM) |
| 20 | | Quick Decision making capability | Yes (Historical Data, AI and Data Analytics) |
| 21 | | Ability to motivate the Team and partners | Yes (Using communications digitally and personally) |
| 22 | | Good listening capability to understand the problems from the team and finding their Innovative solution | Yes (Using communications digitally and personally) |
| 23 | | Knowledge of Cultural, Social, Political environment at the Project Site. Ability to manage those | Yes (Mobile enabled Information on different ciuntries and culture) |
| 24 | | Emotional Maturity | Yes (Mobile enabled Self-Development on Leadership) |
| 25 | | Reserving some time per week to think for a repeated or bottleneck problem and try to find better solution with or without the help of the team | Digital Calendar and Reminders |
| 26 | | Be Pro-Active rather than Reactive, managing the Crisis | Digital Calendar and Reminders |
| 27 | | Asking the right questions , so that people may themselves find / suggest solution | Yes (Mobile enabled Self-Development on Leadership) |
| 28 | PROJECT ENTERPRENEURSHIP INNOVATION & RISK TAKING | Focus on the BIG PICTURE: the real VALUE to the Client and the End-Users | |
| 29 | | Basic Knowledge of general management areas like Human Resources Management, Finance Management, Marketing , Procurement management and Legal (specially statutory and environment | Yes (Mobile enabled Self-Development on eLearning) |
| 30 | | Use of Innovation and Capability of taking Risk in trying out new solutions | Yes (Mobile enabled Self-Development on eLearning) |
| 31 | | Try to be self-motivated and passionate to make it happen | Yes (Mobile enabled Self-Development on eLearning) |

Using these platforms will generally provide you with free access to the following functions:

- Discuss with the people, physically located at different places, using audio, videos, camera, sharing screens etc.
- Conduct Webinars and training to the people located at different places, with provision of asking questions, taking survey polls etc.

- Videos record the whole session to upload in the portal, to see by the authorized people, who did or did-not attend the webinars or training sessions.

We are not providing much information on this, as this is already well-known to most people, like the use of WhatsApp. However, we are briefly describing a new technology of Virtual Collaboration, called Teleporting.

### 9.5.4 Teleporting (distributed VR based Collaboration)

After Covid 19 global crisis, most people are familiar with many popular collaborative communication platforms, like Cisco-Webex, Zoom, Microsoft Team Meeting, WhatsApp Video calls etc. Through these, people located at different places can also do video-conferencing for any desired communications and collaboration. Generally, people can do the following using these collaboration platforms:

- Discuss with the other people physically located at different places, using audio, videos, camera, sharing screens etc.
- Conduct Webinars and training to the people located at different places, with facilities for asking questions, taking survey polls etc.
- Videos record the whole session to upload in the portal, to see by the authorized people, who did or did-not attend the webinars or training sessions.

However, now there is much better platform emerging for more realistic communication and collaboration. In construction, the stakeholders like Designers & Architects, Clients, Contractors, Materials Suppliers and Consultants can use this new platform in real time, almost as if they are located physically at the Construction site. This is especially good during Design & Build phases. This platform is called "teleportation", a kind of VR Collaboration. In this, using a VR software like Trezi [10, 11], different stakeholders can log-in from different locations. Some of them must be having the VR gadgets like shown in Figure 9.6. Others can have only desktops with the required specifications for RAM and Processor to manipulate high-resolution images and videos. Below are some more capabilities of this new real-time close-to-real VR collaboration.

- VR based 'Teleport' Software lets hosts or guests sign-in via MS- Outlook option. However, for non-VR mode (i.e., in Desktop mode) the SW will require an i3-sixth Generation processor, 8GB ram and Intel integrated graphics card to run.
- This will let people invite even a not-having this Software into the collaboration with an installation link. Teleport software reflects its back in the teleport session with different coloured icons. Each collaborating person has a different colour code and his/ her name attached to that, for easy identification. Users without this software will also be able to edit access will have the ability to insert objects in the scene. These objects are created with Revit, or FBX
- Some Software (like 'Trezi') Teleport will have an option for X-ray vision for looking through the design, specially, useful for electrical systems, ducts, and

plumbing etc. Users will also be able to interact with design models like 'Lego' blocks.

The case-Study in the annexure actually shows the use and benefits of such a VR based collaboration, for any discussions, changes required to be made, RFI clarifications etc.

### 9.5.5 Digital Technologies for Project, Procurement & Cost Management

One of the key points in chapter 1 is that construction is a major sector of the global economy, employing about 7% of the labour force and involving $10 trillion spent annually on construction-related activities. While many sectors have embraced AI and digital technologies to boost productivity, the construction sector has progressed slowly.

The construction sector can greatly benefit from AI, ML, and similar technologies for quality and cost management. Sumana Rao [9] has discussed this in detail. Here are some key insights:

- **Prevent Cost Overruns:** Analytics of historical data, considering factors like project size and contractor history, enable AI (like, Artificial Neural Networks) to predict cost overruns. Predictive models help establish realistic timelines for future projects. Implementing Just-in-Time training for workers and highlighting safety measures can reduce defects and enhance productivity. Additionally, crowd-sourcing ideas for site issues can also help minimize cost overruns by leveraging best practices.
- **Mitigate or Minimize Risk**: Construction projects face risks related to quality, safety, time, and cost. Larger, complex projects carry greater risks, often due to poor communication among subcontractors. Adopting the Lean Planning System, such as "The Last Planner or Collaborative Planning System" for scheduling, along with daily huddles, can help mitigate risks. General contractors can utilize AI to monitor and prioritize risks, allowing teams to focus on significant factors. AI can also rate subcontractors based on risk scores from past performances.
- **Improve Project Planning:** "The Last Planner" revolutionized collaborative scheduling, making it more realistic. AI and robots can address issues of delays and budget overruns. For instance, some companies use robots for 3D scanning of sites, feeding data into deep neural networks to assess project progress. If deviations are detected, management can address problems before they escalate. Future algorithms will leverage "reinforcement learning" to optimize project planning. Technologies like Augmented and Mixed Reality can also enhance information visualization.
- **Enhance Productivity at Job-sites**: Robots excel at repetitive tasks, such as pouring concrete and bricklaying, which saves time and allows human workers to focus on improving construction processes. Project managers can remotely track job site work in real time.

- **Compensate Labour Shortages with AI**: The global shortage of skilled labour is exacerbated by insufficient investment in skill development programs. Solutions discussed in section 9.4 suggest that construction firms should invest in AI and data science to enhance productivity and lessen reliance on skilled labour. AI can optimize labour and equipment allocation, allowing project managers to balance workloads effectively. Experts anticipate that construction robots will become increasingly intelligent and autonomous.
- **Construct Components Off-Site and Assemble On-Site**: Many construction firms are now using off-site factories with autonomous robots to produce pre-cast or prefabricated building components. These components are assembled by human workers on-site. Autonomous machinery can efficiently complete structures like walls, allowing workers to focus on plumbing, HVAC, and electrical systems. This can involve simpler pre-fabrication or complex 3D printing.
- **AI and Big Data in Construction**: AI systems can leverage vast amounts of data generated daily, turning every job site into a data source. Information from mobile devices, drone videos, and sensors provides opportunities for professionals to analyse insights. It is estimated that robotics, AI, and IoT can reduce construction costs by up to 20%. Engineers might use VR goggles to direct mini robots in monitoring construction progress, while AI plans routing for electrical and plumbing systems and enhances safety by tracking real-time interactions.

Despite fears of job losses due to AI, it is unlikely to replace the human workforce entirely. Instead, AI will reshape business models, reduce errors, improve safety, and increase efficiency. Construction leaders should prioritize investments where AI can have the most significant impact. Early adopters will shape the industry's future and gain short- and long-term advantages.

### 9.5.6 New Technologies for Safety, Health and Environment

Occupational Health and Safety is one of most important aspects of construction, as 85% of the accidents take place due to behavioural aspect of the workmen. In most construction sites, regardless of location and type, some risk factors do exist. These include risks of falling down, crashing with Equipment, electrocution, fire outage, short circuit, dealing with heavy-weight machinery, and even casualty in the worst possible scenario.

While deploying surveillance cameras and fire alarms are not new, using AI and Machine Learning technologies-based solutions takes the level of safe-working conditions to a new height. The solution can be used to look for risk factors and any escalation can be instantly identified and reported. These can study the Safety climate, Safety culture and behaviour. This helps minimize the scope of injury and accident at such project sites. These can

- Recognize the Objects like People, Tools and Equipment
- Track the objects like People, Tools and Equipment
- **Recognize their actions like:**
  - People- Where located, movement, walking and at what speed, postures like Standing, sitting, bending crawling, picking, ladder climbing etc.
  - Equipment (Dumper Truck, Hydraulic Excavator, Tower Crane Concrete Mixer Truck etc.); where located, moving and at what speed, Idle etc., also, the level of Vibrations, Temperature, Noise.
- Assess if the object is a potential hazard
- Assess if the behaviour is a possible hazard
- Predict how the object will behave
- Predict if and when the next incident will occur
- Alert the through possible warning

Real-time recognition of safety hazards and corrective actions enhances protection of people and equipment without human intervention. NCRs raised by the security team can be tracked digitally through their full lifecycle, eliminating paperwork and mirroring Quality Management's NCR tracking. See Chapter 8 on Safety, Health, and Environment for details.

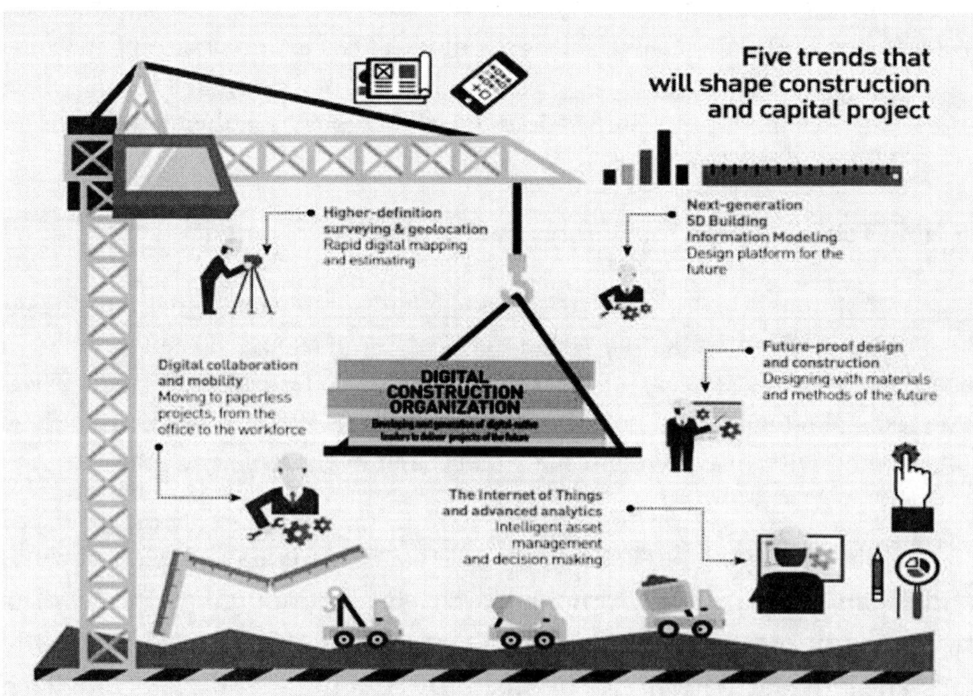


**Figure 9.27: Some key trends in Construction (source: reference [25])**

In summary, considering the immense challenges of rising demand in the Construction Sector, limited capital, and high expectations from modern customers and stakeholders, a 4th revolution in the industry is essential. Other factors, such as new-age entrepreneurs, low productivity and profitability, increased MNC presence, FDI, and low worker skill levels, further drive this need. This revolution should adopt the Construction 4.0 framework, integrating Workflow, Communication & Collaboration, and Operational technologies, along with the LEAN Philosophy, to meet the demands of a developed India and enhance project performance. Figure 9.27 depicts such a scenario.

## 9.6 References


1. HoloBuilder. (2016). *Virtual, augmented and mixed reality for the construction industry in* a nutshell. Medium. https://medium.com/holobuilder/virtual-augmented-and-mixed-reality-for-the-construction-industry-in-a-nutshell-2bcd190548bc
2. *Forbes, L. H., & Ahmed, S. M. (2019). Modern construction: Lean project delivery and integrated practices. CRC Press, Taylor & Francis Group.*
3. *Rastogi, S. (2019, April). Construction 4.0: New generation construction in a hyper-connected world. The Masterbuilder.*
4. *Prasad, A. M. (2018). Emerging technologies in housing construction in India. Ministry of Housing and Urban Affairs. https://mohua.gov.in/upload/uploadfiles/files/1-%20IIT%20Chennai.pdf*
5. *Robotics Business Review. (2019). 5 ways robotics will disrupt construction industry in 2019. https://www.roboticsbusinessreview.com/news/5-ways-robotics-will-disrupt-construction-industry-in-2019/*
6. *Kumar, M. (2019). How GPS works. GeeksforGeeks. https://www.geeksforgeeks.org*
7. *Rastogi, S., & Tewari, G. (2018, September). Lean Digital Thinking™ framework for addressing problems of Construction 4.0. In Proceedings of the International Conference ICIEIND 2018 on Industry 4.0 (SOA University, Bhubaneswar, India).*
8. *Rastogi, S. (2017, July 28–29). Construction 4.0: The 4th generation revolution. In Proceedings of Lean Construction Conference (ILCC 2017). Chennai, India.*
9. *Rao, S. (n.d.). The benefits of AI in construction. Trimble Constructible. https://constructible.trimble.com/construction-industry/the-benefits-of-ai-in-construction*
10. *Tewari, G., & Tewari, T. (2019). Trezi. [Video]. YouTube. https://www.youtube.com/watch?v=GNqzduDMtjA*
11. *Tewari, G., & Tewari, T. (2021). Trezi collaboration meeting. [Video]. YouTube. https://www.youtube.com/watch?v=K5fvmpyvfTc*

## 9.7 ANNEXURE I: VIRTUAL REALITY-AIDED-DESIGN, BUILD and HAND-OVER: A CASE STUDY

Here is a case study, which incorporated three emerging Digital technologies, which resolved Customers/ End-users issues all the three stages of the project, Design Stage, at Build Stage and also at Handover Stage. This resolved many of the problems at different stages of the project, as mentioned earlier in the Chapter. Following three technologies were engaged:

- Virtual Reality
- Product Configurator
- Teleporting (distributed VR based Collaboration)

**The Case-Study: Experiences and benefits:**

Based out of Siliguri and Kolkata, 'Raamesh Singhal Design' is a leading Architects & Designers company who has been focusing on interior designing and architectural projects across North-East India and Bhutan for the last 20 years. The firm specializes in design & build of residential bungalows, residential apartments and star hotels. Being in secluded and remote place with a difficult terrain for travel and communications, they always faced problems for bringing satisfaction to their clients and users, in that region. They also had to do several iterations of designs and taking the colored printed copies to the clients for discussions. This consumed big amount of time for creating renders and communicate their designs effectively to their customers. Therefore, for the design & build of a 100 guest rooms, star-hotel in Gangtok, they tried to find, digital technologies to feel if those could bring wow to the customer for this mega project. Some details the project are listed below:

- **Project Name:** Raamesh Singhal Design's Interior Design of a Gangtok Hotel
- Resort surrounded by mountains
- **Project Size:** 100 Rooms Executive Hotel Design & Build, 70,000 SFT carpet area in Gangtok,
- **Project Cost:** In the range of about INR '100' crores (USD 12.5 Millions)

**Design & Architecture stage:**

In such projects, the rendering was earlier done using various printouts of 3-D designs through AutoCAD's Revit, by physically visiting the client several times. This used to take a few hours and consume unnecessary time at each stage of the design. The changes suggested by clients also were to incorporated in Architect's office, and again visit the clients for their approval. Therefore, they engaged the VR technology, with Product configurator, using Trezi. There were two experts from Trezi, who helped them in this. Raamesh Singhal Designs were looking for a way to improve customer satisfaction.

Fig 9.28 below are actual pictures from the 360 Walk-through designed & shared using Trezi platform, without physically visiting the Client.

**Figure 9.28: Project Pictures taken from 360-degree VR Walk-Through**

Then, for any changes suggested the discussions were done using 'Teleport Trezi' where people from Design & Architecture team, the Client, the Main Contractor and the appropriate Supplier came together for understanding the changes required, and incorporating immediately, in real time. Figure 8 below shows one such change of changing the Switch material and colour through Product Configurator, with the various options available for the floor. Similarly, floor-material was also changed. Trezi team let the clients make the changes on-the-spot based on client feedback, at design-stage itself, also telling possible cost differences.

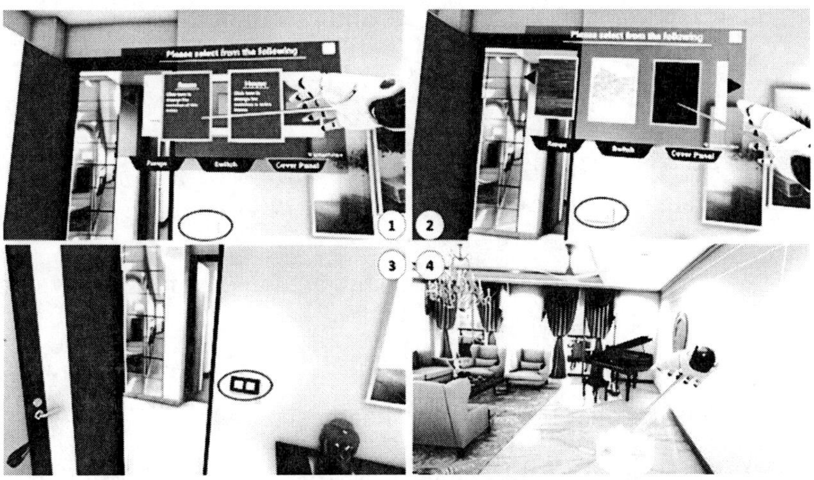


**Figure 9.29: Project Picture for Workplace**

**Build Stage:** During build stage also, many changes were suggested and incorporated with the help of Teleport collaboration of relevant stakeholders, giving real 'wow' to the client. Figure 9.29 depicts the VR-based teleport collaboration for incorporating the changes in the chart displays and the chair-type and location in their office room.

**Handover Phase:** In this phase, although the service of VR-Based 360 walk-throughs of the actual hotel constructed was not prepared, as the clients were already very happy with the results. However, it was possible to do that very easily. If done, that would have been very useful for making hand-over very smooth. Following were the key benefits actually achieved in this project:

**For Architects and Designers:**

- Better design visualization close to real-life experience, bringing speed and WOW
- Real-time Multiple design options & iterations, reducing the delays in design finalization
- Multiple design tools, without much expenses.
- Better communication and coordination with other project stakeholders, leading to having the identical mental and visual impressions in their minds.

**For Builders and Suppliers of finish work-items:**

- Saved cost & wastage associated with product display & physical mock-ups. With
- VR, no physical mock-ups were required.
- Faster decision-making by architect and customer, for having the visualization close to reality. Cognitive assumptions and variation in visualization will no more be there.
- Better understanding of finish products and its quality.

**For Customers & End users:**

- Better ability to visualize the dreams and understand the design intent
- Effective participation in the conceptual and detailed design, build & handover process
- Almost no gap between expectation and actual experience, even in build & handover

The client was really excited about the outcome, and the ease of reaching there. They actually wrote a letter of appreciation to Raamesh.

In **Construction 4.0**, digital tools such as Design Configurators, Virtual Reality (VR), and VR-based "Teleport" platforms (e.g., Trezi, Figure 9.30) enable architects to involve clients directly in interior design with real-time visualization and collaboration. These technologies provide immersive experiences for clients and stakeholders — designers, contractors, consultants, and suppliers — at every stage, helping clients make informed decisions and real-time changes. At handover, a VR 360° walk-through of finished interiors allows final adjustments before site visits.

**Figure 9.30: Making real-time changes using Teleporting Collaboration**

Such tools enhance communication among stakeholders, align perceptions, and deliver a "wow" experience to clients, while supporting faster completion and reduced project costs.

# Chapter 10

# Lean Construction in Government Sector

# 10 Lean Construction in Government Sector

## 10.1 Need for Embracing Lean Construction in the Government Sector

Adopting Lean Construction is crucial for the government and public sectors as they primarily utilize taxpayers' money. Poor project execution often results in sub-optimal use of investments, effectively wasting taxpayers' money. The government, being the largest investor in this sector, underperforms compared to private-sector projects. Around 70-80% of projects face delays, which lead to revised budgets, poor-quality facilities, and public dissatisfaction. The following details underscore the importance and investment levels of the Indian Government in this sector.

In Chapter 1, Figure 1.1, two pillars of the construction sector are discussed: Real Estate and Infrastructure. Real Estate includes Commercial (hotels, malls, office buildings, hospitals, educational facilities, etc.) and Residential (apartments, villas, affordable housing). Infrastructure includes Transport (roads, railways, bridges, etc.), Energy & Utility (oil, gas, power plants), and Industrial (manufacturing, defence). The government promotes Foreign Direct Investment (FDI) differently, grouping real estate Infrastructure with urban development (water supply, sanitation, schools, healthcare) and treating roads & highways, renewable Energy, defence manufacturing, retail, and tourism as separate sectors.

**Key industry data (source: https://www.investindia.gov.in):**

- India's construction investment is expected to reach $1.4 trillion by 2025, becoming the third-largest global market.
- The sector employs 51 million people, second only to Railways.
- In 2022, it became the sixth largest FDI recipient, with $32 billion in inflows during the COVID period (2000–2023).
- By 2030, urban centres will require 25 million additional housing units, with 600 million urban residents expected.
- The outlay for PM Awas Yojana is being enhanced by 66 % to over 79,000 Crore
- Urban Infrastructure Development Fund: INR 10,000 Cr Outlay per year to create urban infrastructure in tier 2 & 3 cities.
- India has the second-largest road network globally, spanning 63.73 lakh km.
- In 2023, capital investment outlay increased by 33% to ₹10 lakh crore.

Despite these investments, 67% of government projects face delays, often resulting in cost overruns and quality issues. For example, delayed bridges increase commuting times, fuel costs, and inconvenience. Additionally, delays can alter project specifications due to various changes, inflating costs further.

**A 2023 KPMG global research report highlighted:**

1. 41% of projects exceeded time and cost by over 20%.
2. 72% identified rigorous planning as the key to overcoming challenges.
3. 52% emphasized having standardized risk management processes.
4. 90% saw a need for innovative project execution approaches.
5. 97% highlighted the need for accurate material and equipment estimation.
6. 86% stressed innovative workforce training programs.
7. 62% pointed to leadership and cultural shifts for safety improvements.

These inefficiencies are estimated to cost $150 billion annually, causing public frustration and delaying critical facilities. Addressing these issues through Lean Project Management 4.0 can significantly reduce risks and improve performance. Embracing this philosophy is an urgent necessity for the Government sector, presently and at least in next 2 decades.

## 10.2 Key Issues & Risks for Delays, Over-Budgets, and Poor Quality in Government Sector Projects

To understand these challenges, we interviewed around 45 senior officials from State Public Works Departments and private corporations and engaged with 10 contracting companies. Based on these discussions and past experiences, we identified key issues which lead to delays, overspending, and poor quality. Overspending often results from project delays, design changes, and poor initial estimates. Using a 100 Km State Road Project as an example, we categorized these risks into six types for better clarity, applicable to projects like real estate, bridges, and highways.

Delays occur in pre-construction, construction, and handover phases, with most delays in linear infrastructure arising from pre-construction. Roads often traverse multiple districts, municipalities, and villages, involving diverse "External Stakeholders" such as political, social, and government groups. Managing these stakeholders is challenging, as State Works Departments often lack appropriate authority and skills for this. Poor collaboration, political agendas, and personal gains frequently are the key reasons for delays, disputes, and cost escalations. Some of these are mentioned below in brief.

### 10.2.1 Social Risks

- Encroachers, supported by political or local influencers like religious leaders and media, demand unrelated CSR tasks (e.g., repairing schools or temples). Some of these influencers include Religious Leaders like Temple Pandits, Masjid Maulanas, and others like School teachers, local Social Reformers etc., general Public and even, Media/ Press. Resistance by them often leads to work stoppages.

- Contractors face pressure to buy overpriced, low-quality materials and labour skills from local powerful and influential people and, resulting in delays.
- Land encroachment disputes and public demonstrations, often influenced by local leaders and influencers, disrupt progress.

### 10.2.2 Political Risks

- Local leaders delay permissions for quarries or sometimes demand corrupt practices.
- Contractors are pressured to employ poor-quality equipment, materials, or labour from politicians' businesses.
- Politicians demand CSR-like tasks (e.g., building schools) for image-building, adding costs outside the contract.

### 10.2.3 Government-Departments Risks

- Routine delays from approvals by overburdened offices like District Collectorates and Municipalities.
- Electricity Boards delay shifting poles due to outdated designs or unclear budgets.
- Water and Sewerage Departments face design, capacity, and budget issues, causing delays.
- Forestry and Environment Departments often exhibit power struggles and delays.

### 10.2.4 Contractors and Suppliers Risks

- Collusion among contractors during bidding leads to higher costs and poor quality.
- Changes during projects result in cost disputes; some contractors inflate quotes or submit fake invoices.
- Contractor over-commit resources across projects, causing delays and quality issues.

### 10.2.5 Consultants and Domain Experts

- Poor consultant selection and lobbying lead to sub-optimal results.
- Insufficient time or fees for preparing Detailed Project Reports (DPRs) lead to inaccurate estimates and risks.
- Under-quoting for contracts causes resource and quality issues during execution.

- Poor Bill of Quantities (BoQ) and estimation leading to inflated pricing and delays.
- Inadequacy of capacity and capability of the resources (manpower, Equipment, Finances etc. They take up additional projects beyond their capacity and keep rotating their available resources in different projects.

### 10.2.6 Client Risks (Bureaucratic Hierarchy)

- Internal silos among bureaucrats and external funding agencies delay approvals.
- Issues like leaked bidding information, fake invoices, and delayed payments affect project cash flow.
- Poor contract management, lack of risk management plans, and inadequate IT adoption lead to inefficiencies.
- Selection of favourite EPC Contractor, who might have under-quoted, and later trying for revised budget approvals
- Lack of proper due diligence on consultants and contractors' selection
- Non-compliance with safety regulations and poor community engagement disrupt timelines and public relations.

Despite these persistent issues, innovative solutions and guidelines are urgently needed to improve outcomes. The next section explores these possibilities in detail.

### 10.2.7 Vigilance in Governance: Balancing Oversight and Initiative

This is an important area in the Government Sector projects, which needs to be relooked in the current scenario of Lean and Agile Projects works in the Government. In order to ensure Effective governance, Government relies on **vigilance systems** to prevent corruption and misconduct. This is specially so because they are using public money. However, in practice, these systems often struggle to balance ethical oversight with encouraging initiative due to constant innovation required in the Construction Designing, planning to execution. This is especially true in the Indian administrative context.

Vigilance is necessary because some officials are corrupt, and inaction would erode trust. The key is to find the right balance between diligent oversight and overly restrictive scrutiny, so that Government Officers can also take some innovative steps for better projects and facilities for the public, at the same time, minimize unnecessary procedures and processes of Vigilance, mostly caused by vested interest of the public representatives, like unprofessional local journalists, some corrupt politicians and social activists.

#### 10.2.7.1 The Paradox of Administrative Action

**Officials often face a dilemma:** "One who 'works' makes a thousand mistakes; one who 'doesn't work' makes only one." Those who take initiative for bringing changes in design, materials specifications, construction processes etc, risk scrutiny, while those who avoid action face little criticism, fostering a culture of passivity.

This paradox makes governance risk-averse, hindering progress and innovation. Proactive officers risk punishment, especially from unverified or anonymous complaints. This leads to decision paralysis, low morale, and a general reluctance to take charge.

#### 10.2.7.2 Issues Undermining Vigilance Integrity

**Several structural problems weaken vigilance systems:**

- **Unfiltered Anonymous Complaints:** Acting on anonymous complaints without proper filtering creates a defensive culture. This stifles innovation and reduces leadership to risk aversion. A good system must differentiate honest mistakes from malicious intent. Anonymous complaints need a structured vetting process, like an independent panel, to prevent misuse. The culture should also celebrate courage and constructive governance, not just the absence of corruption. Plus, 360-degree feedback (peer, subordinate, and citizen insights) should be used to provide a more holistic view of an officer's conduct.
- **Overreliance on Quantitative Metrics:** Success is often measured by volume—like the number of traps laid or suspensions. This makes vigilance a spectacle rather than a tool for systemic improvement. This metric-driven approach rewards statistical output over genuine cleansing, promoting **performative vigilance**.
- **Erosion of Natural Justice:** Often, suspicion is treated as guilty until proven innocent, especially when records are reinterpreted with hindsight. This shifts the **burden of proof** onto the accused, forcing honest officials to jump through hoops to prove their innocence, undermining the principle of "innocent until proven guilty."

These issues collectively create a vigilance system that punishes intent, hinders initiative, and fails to distinguish between good-faith errors and deliberate misconduct.

#### 10.2.7.3 The Legitimate Role of Vigilance

Despite these challenges, vigilance is crucial. Without it, unchecked power, especially in areas like procurement and approvals, can lead to widespread decay. Effective and impartial vigilance acts as the state's **immune system**, identifying and neutralizing threats to integrity.

The goal should be not to weaken vigilance, but to **refine its structure** to make it fair, sharp, and constructive.

#### 10.2.7.4 Towards a Balanced Framework

The future of vigilance lies in transforming it from a punitive tool into a **trust-building institution**. This requires blending ethical supervision with motivational leadership, fostering a culture where integrity is rewarded, initiative is protected, and misconduct is addressed justly. Government has to relook on the vigilance system with a fresh new approach and change the regulations and clauses in the light of 21$^{st}$ Century requirements of agility, and use of technology, as against only manual systems. This also required some judiciary reforms as well.

**Key elements of this reformed approach include:**

- **Credibility Screening of Complaints:** All anonymous or unsubstantiated complaints should go through a triage process with independent panels. This saves resources for genuine cases and protects officials from malicious targeting.
- **Redefining Success Metrics:** Vigilance performance should be measured by multi-dimensional indicators like structural improvements, whistle-blower protection, and deterrence. We should incentivize problem-solvers, not just fault-finders, and train vigilance officers to mentor departments rather than just investigate punitively.
- **Incentives for the Proactive and Honest:** Officers showing integrity and results should receive safeguards and recognition. Peer reviews, citizen feedback, and performance reviews can balance isolated complaints.
- **Mentorship within Vigilance Cadres:** Vigilance bodies should become enablers, advising and training departments on risk mitigation, compliance, and ethical dilemmas.

#### 10.2.7.5 Conclusion: From Watchdog to Guardian

A truly vigilant state upholds justice while respecting its honest officers. When vigilance instils fear instead of fairness, it damages governance. The solution is to transform vigilance into a **constructive institution** that mentors, motivates, and monitors equally.

By combining analytical intelligence with human judgment and shifting from fault-finding to value creation, vigilance can usher in an era of purposeful governance. It's true success will be measured not by how many it punishes, but by how many it **protects, inspires, and enables.**

### 10.2.7.6 A Case Example:

Ramesh, a brilliant Executive Engineer, once brimmed with an infectious zeal for innovation. His eyes would light up discussing new materials for roads or smarter drainage designs. Early in his career, he spearheaded a project using a novel, cost-effective interlocking paver for a rural road. It was quicker to install, required less maintenance, and saved the government significant money. The villagers lauded him; the project was a resounding success.

**The Unforeseen Storm**

Then came the storm. An anonymous complaint, fuelled by a disgruntled contractor Ramesh had previously blacklisted and fanned by a local journalist seeking headlines, triggered a vigilance inquiry. Despite the project's undeniable success and cost savings, every decision Ramesh made was dissected with microscopic precision. His innovative choice of material was questioned, procedures were reinterpreted, and even his motives were cast under a shadow of suspicion. He spent months battling accusations, providing endless documentation, and enduring humiliating interrogations. The mental and emotional toll was immense. He was eventually cleared, but the experience left an indelible scar.

**The Burden of Caution**

Years later, Ramesh is a respected, sensible, and undeniably honest officer. Yet, the spark of innovation has dimmed. A new project comes up, offering an opportunity for a ground breaking, environmentally friendly waste management solution. The old Ramesh would have jumped at it. The current Ramesh, however, hesitates. He remembers the sleepless nights, the public humiliation, the fear. He opts for the tried-and-true, albeit less efficient and more expensive, conventional method. "It's safer," he sighs, "No one can question 'standard practice."

**The Cost of Fear**

The project finishes on time and within budget, but it's just another project, not the transformative one it could have been. The community misses out on a better solution, and the exchequer bears a higher cost. Ramesh's experience is a stark reminder: when vigilance, however well-intentioned, fails to differentiate between genuine malfeasance and bold initiative, it fosters an environment of **risk aversion** and **decision paralysis**. The true cost isn't just to individual careers, but to public progress itself. Government vigilance must strike a delicate balance, acting as a guardian against corruption while simultaneously nurturing the courage to innovate.

## 10.3 Recommendations to Resolve above Issues

Based on over 100 years of combined experience and discussions with government officials, contractors, and consultants, the following recommendations aim to resolve the because of above six categories of risks discussed above. Some of these are already proven solutions, while others are innovative ideas requiring more piloting for wider acceptance. You will appreciate that the Chronic problems may demand non-traditional solutions.

### 10.3.1 Managing Social Risks

To mitigate risks of delays, profit loss, and inefficient resource use, following steps are recommended. Lean Construction 4.0 can actually reduce project costs by 5-10%, enabling reinvestment to increase profits and goodwill.

Allocate up to 0.5% of the project cost for CSR activities like repairing temples, building school washrooms, or low-cost sports facilities. These actions can reduce delays and build far better local support.

To manage local trouble-makers and naysayers, allocate another 0.5% of budget to engage capable and positive-thinking local groups. which has been effective in a few states in India, actually expedites project progress by addressing these troublemakers.

For encroachments, establish ties with senior government officers to assist in land clearance and coordinate with municipal bodies for rehabilitation. Tailored project-specific solutions are crucial for managing these types of risks.

### 10 3.2 Managing Political Risks

Minimizing political risks also benefits from the assumption of a 5-10% cost-saving through faster project completion. This is because of saving the Site Cost, and Delay penalty, and interest on the working Capital for the duration by which the project is completed early. In some cases, there are early completion bonus as well. Reinvesting these savings enhances profit margins and stakeholder relationships.

- Regularly engage with local influencers to secure quarry access and involve them in events to foster goodwill, like project inauguration.
- Consider acquiring their equipment and labour, accounting for only 10% of the promised productivity to ensure steady income and prevent disruptions.
- Address political CSR requests, invite stakeholders to site visits, and manage branding carefully without aligning exclusively with one party. Follow legal guidelines for party fund donations.
- Use intelligence agencies to profile key stakeholders and assess CSR needs, compensating for state governments' limitations in handling these aspects effectively.

In summary, the main EPC contractors should allocate 1.5-2% of project costs for managing external factors based on intelligence reports. This can save 6-10% of the project cost through faster project completion by controlling key risks. The benefits include:

- Site Establishment Cost savings for the early completed period
- Inflation cost savings,
- Early completion bonuses,
- Avoiding delay penalties,
- Reduced interest on working capital
- Resource availability for new projects,
- Improved stakeholder satisfaction, and
- Enhanced reputation for timely delivery
- Sometimes, also the status of Preferred Vendor

### 10.3.3 Recommendations for Managing Government Department Risks

Section 10.2.3 outlines key government stakeholders in most projects, which include District Collectorates, Municipal Corporations, Forestry, Environment, Revenue, PWD, Electricity, Water, Sewerage, and Police Departments. Delays due to inter-departmental silo working can be mitigated with the following measures:

- **Quarterly Review Meetings:** Chaired by senior officials such as the Prime Minister or his/her representative, Chief Minister, or relevant minister, these two-hour meetings should include all stakeholders, including EPC contractors, to address major delays and budget or prioritization issues promptly. The actions for individuals must be agreed and listed with responsibility allocated and target date. If some government policy is necessary, that can also be looked into, after the meeting.
- **Monthly Review Meetings:** Led by the Chief Secretary or Minister, these ensure follow-ups on quarterly commitments, resolve inter-departmental issues, and involve site visits for critical projects.
- **Weekly/Fortnightly Meetings:** Conducted at the Chief Engineer level, these meetings address immediate issues or escalate unresolved matters to higher-level reviews.

These practices, pioneered by Nitin Gadkari, the central Minister for Road Transport & Highways, have nearly doubled infrastructure project progress in India, significantly enhancing government efficiency and public services.

### 10.3.4 Recommendations for Managing EPC Contractors and Suppliers Risks

Main EPC contractors handle engineering, procurement, construction, and project delivery, involving numerous suppliers and contractors. Risks during pre-construction, construction, and maintenance phases can be addressed as follows:

- **Contractor Selection:** Choose contractors based on proven experience, Current and future construction capacity and capability of the similar sized projects, technical expertise, financial stability, and adherence to legal and ethical practices. Avoid L1 contractors who may under-quote; instead, consider L2 contractors with proper justification.
- **Streamline Suppliers:** Limit manpower suppliers to the top-performing ones. For example, in one project, reducing 123 suppliers to the top 20 enabled better quality and reliability. A short-term loan kitty helped suppliers expand capacity, resolving cash-flow issues and boosting revenue.
- **Resource Reallocation:** When a contractor manages multiple projects, prioritize high-priority projects by reallocating resources temporarily. This ensures faster completion of critical projects while delaying lower-priority ones without impacting overall timelines.

For example, if contractors are managing six PWD projects, they can discuss with the main contractors the importance of each project. They can suggest reducing work on two low-priority projects for a few months and reallocating resources to four high-priority projects. After quickly completing these four projects with extra support, resources can be redeployed to the two delayed projects for faster completion. This approach enables all six projects to be completed on time, rather than leaving each in varying stages of completion.

- **Fraud Prevention:** Impose strict penalties for fraud and fake invoicing. Simplify CAG laws with in a period of 1-2 years to deter malpractice.
- **Improved Estimation Models:** Develop a parametric estimation model using historical data and a digital app, supported by central and state governments, to improve accuracy in time and quantity estimations. If such Digital App is not available, invest into developing one.

By adopting these measures, project execution can be optimized, benefiting the country, states, public, and all stakeholders.

### 10.3.5 Recommendations for Managing Various Consultants and Domain Experts

Key risks and issues for clients reflect poor project performance, therefore our recommendations include:

- Address frauds like, information leaks to bidders, fake invoicing, and exaggerated claims with better due diligence and strict penalties. Simplify laws for the Comptroller and Auditor General (CAG) of India, over a period of 1-2 years.
- Monthly reviews by the Chief Secretary or Minister of Construction should address delays in payments and approvals, focusing on resolving major roadblocks.
- Prevent under-quoting by EPC contractors by improving BoQ quantities and project time estimation using parametric models based on past projects, developed with support from central and state governments.
- Conduct due diligence when selecting consultants and contractors to assess resources and capabilities. Address capacity shortages by reallocating resources from lower-priority to high-priority projects.
- Enhance contract management by dividing contracts into two parts: critical, actionable specifications (the 'real' Contract Data) and routine legal clauses (General Conditions of Contract).
- Quarterly reviews by the Chief Secretary should handle high-level risks involving departments outside the PWD, such as Environment, Electricity, and Forestry.
- Educate stakeholders about construction methods, emerging technologies, and quality management through short videos and events.
- Provide training on quality standards and defect prediction to prevent issues across the project lifecycle.
- Enforce strict compliance with safety and health regulations.
- Encourage EPC contractors to allocate 0.5% of project costs to CSR activities, generating goodwill and societal benefits.

### 10.3.6 Recommendations for Managing Client Risks

Our recommendations for client's mirror those above and include:

- Tackle fraud with severe penalties and due diligence. Simplify laws for better compliance and oversight.
- Address payment and approval delays with monthly reviews led by senior officials to ensure timely action.
- Develop a parametric estimation app for accurate project budgeting and timelines, leveraging historical data.
- Improve contract clarity by separating critical project clauses from routine legal terms.

- Manage interdepartmental risks through quarterly high-level reviews for quick decisions and resource allocation.
- Facilitate workshops for contractors, suppliers, and engineers to standardize materials and specifications, ensuring resource levelling and project alignment.
- Train staff on quality standards, safety regulations, and emerging construction practices to prevent defects and ensure compliance.
- Foster innovation and stakeholder engagement to address social concerns and ensure timely project completion.

The following innovative best practice has been developed by the authors and piloted for a big project and is recommended for others to adopt. This is equally true for the private sector as well. In this case, which involved the construction of 89 sports infrastructure facilities across the State with higher specifications and tight timelines, a Facilitation Workshop was organized by the Client. The key stakeholders for Construction, namely, the Contractors, the Suppliers, and Engineers participated for direct communications and understanding each other's requirements and the problems to arrive at agreed solution. This led to several advantages:

- **Pre-approved Brands:** Alternate brands were pre-decided for each product, allowing Contractors to choose among approved options. Suppliers showcased their products, enabling Contractors to assess alternatives based on cost and local availability.
- **Supplier Capacity and capability Assessment:** Supplier capacities and capabilities were evaluated beforehand, facilitating equitable distribution of orders based on production capacity, proximity, and existing commitments, ensuring resource levelling.
- **Visual Aids:** New products and facility models were displayed, helping stakeholders visualize the completed sports facilities.

The workshop streamlined material selection, approval, and purchase for all structures, enabling uniform specifications and adherence to the project timeline. Figure 10.1 provide some of the glimpses of such a Workshop in Odisha.

**Figure 10.1: Some pictures from the Suppliers' Meet for accelerating Procurement Process**

**Based on the success of this, our key recommendations include:**

1. **Streamline Regulatory Processes:** Simplify and make regulations transparent to avoid delays.
2. **Robust Budget Planning:** Conduct feasibility studies, hold DPR Consultants accountable for risks (especially if they have missed to anticipate), and develop contingency plans for budget constraints. We also suggest to pay DPR Consultants more and give them more time to gather information. This will reduce future surprises and the delays. Also develop contingency plans to manage budget constraints and uncertainties effectively.
3. **Depoliticize Projects:** Engage independent experts to manage political interference and ensure continuity despite government changes.
4. **Clear Scope Definition:** Define detailed project scopes and manage changes, in terms of specifications and materials changes innovatively to stay within budget.
5. **Efficient Procurement:** Use pre-qualified vendor lists and standardized bidding to streamline contractor selection. Develop some key vendors over time, who can be associated and maintain a long- term business relations productively
6. **Stakeholder Engagement:** Maintain transparent communication with political, social, governmental, contractors and suppliers as stakeholders.
7. **Workforce Development:** Invest in training to bridge skill gaps and ensure an available workforce.
8. **Contract Management:** Simplify contracts with clear terms and dispute resolution mechanisms.
9. **Technology Adoption:** Phase in technology with training and infrastructure upgrades for better project efficiency.

10. **Collaborative Planning:** Incorporate Lean planning and innovative methods to address delays in schedule, quality, or budget. Many of these are explained in this book
11. **Quality Management:** Predict and prevent defects with strict quality controls and regular inspections.
12. **Safety Compliance:** Enforce robust safety protocols and adhere to health standards.
13. **Continuous Improvement:** Learn from past projects, document lessons, and implement best practices.

Adopting these strategies and Project Management 4.0 methodologies, as outlined in this book, will lead to successful project outcomes.

## 10.4 Selection of 'Right' Project in Government Sector for adopting Lean

Adopting Lean Construction starts with selecting a pilot project that guarantees success, demonstrating its benefits and encouraging wider adoption. Guidelines include:

- **Public Interest:** Choose project, in which public is keen, and any success there will be known to all like a wild-fire through Social-Media. For example, the construction of the 22-kilometer Atal Bridge, the longest Sea Bridge in Mumbai was a of great public interest t impact and visibility. (Please. see Appendix 10, for details)
- **Measurable Results:** Choose projects with visible public impact, such as completing facilities ahead of schedule or under budget, to showcase Lean principles. For example, the 2.8-kilometer River Bridge mentioned in Annexure 6 was inaugurated by the Chief Minister of Odisha three months earlier than publicly announced. This became a big news on TV. (Please see Appendix 6, for details)
- **Strategic Importance:** Select projects with substantial public impact to highlight government commitment. Moderate complexity ensures manageable implementation. For example, the case study about the 650-bed Medical College Hospital, demonstrating how Lean principles can enhance outcomes and public perception. (Please. see Appendix 3, for details)
- **Management Commitment:** Strong leadership buy-in is essential for resource allocation and visible support. Past performance metrics should guide project selection.
- **Risk Profile:** Pick projects with manageable risks to showcase Lean's effectiveness in mitigating challenges.
- **Positive Leadership:** Ensure the project leader and team have an enthusiastic, problem-solving mindset and a collaborative attitude.

In summary, a well-planned pilot project demonstrating Lean principles can establish a strong foundation for their adoption in future government projects.

## 10.5 Right Method of Project Progress Review in Government Sector (Focus on making up the 'Lost Time' due to Avoidable & Unavoidable Reasons)

In Government infrastructure as well as other projects, delays often stem from present ineffective Project Progress Reviews. Current practices primarily focus on comparing planned versus completed quantities, which are useful for making invoice payments and 'S' Curve preparation. This is because the Invoices and payments are based on the Quantity completed during a period. But this fails to reflect actual project progress. This approach is more beneficial for financial officers than for business heads or project managers seeking to understand real-time delays. Key limitations include:

- **Lack of Tangible Progress Assessment:** Quantity-based reviews don't adequately reflect physical progress. While tasks may be completed, they may not significantly advance overall building completion. A tangible progress report could provide better mental satisfaction and a clearer picture of reality. This is the reason, why most Project managers want to visit the work-site to getting the 'right' feel of progress.
- **Disconnection from Real-world Progress:** Relying solely on quantities often overlooks bottlenecks, delays, and corrective actions needed to stay on schedule. This makes it difficult to get clear 'focus' for the 'corrective actions' to be taken in immediate next period to catch-up with the delays.
- **Inadequate Quality Assessment:** Quantity-based metrics also overlook the quality of work completed. Simply measuring the quantity of tasks accomplished may lead to neglecting quality control and assurance processes, potentially resulting in rework, defects, or safety issues.
- **Limited Integration Insights:** Physical reviews offer a clearer view of task coordination, such as plastering, putty application, and painting, highlighting potential clashes among subcontractors. All these three activities, normally done by different sub-contractors are interdependent, which is not reflected correctly in the planning CPM/PERT Chart.
- **Neglecting Site-specific Factors:** Quantity reviews fail to consider on-ground realities like weather, site access, or unforeseen challenges, which physical reviews can address.

Combining quantity-based and physical building completion reviews provides a more comprehensive understanding of project progress, aiding better decision-making.

Annexures 10 below, provides two case studies from real estate and infrastructure projects that illustrate these concepts in detail.

## 10.6 How Government Officials Can Support Main EPC Contractors

To enhance efficiency and foster goodwill, Government officials can implement the following strategies:

- **Allocate CSR Budget:** Advise contractors to reserve up to 0.5% of project costs for CSR activities, such as repairing local infrastructure like mini-sports, religious or educational public facilities. This builds local support and reduces delays.
- **Manage Local Challenges:** Dedicate another 0.5% of the budget to engage local groups for handling issues like land encroachment or local leaders and influencers, as proven effective in many states in India.
- **Facilitate Approvals:** Build relationships with professionally dedicated government officers to expedite land vacation and permissions. Use municipal and police support to rehabilitate displaced individuals.
- **Engage Stakeholders:** Regularly involve local stakeholders for quarry access or CSR activities, leveraging events and contributions for political goodwill.
- **Address Resource Productivity:** Consider the poor-quality equipment and labour resources at reduced productivity rates (say only 10%, instead of 70-80%) so that there are no interferences from influencers in order to maintain the good progress by the other resources. This will minimize disruptions, by the social and political influencers as their interests has been taken care of.
- **Streamline Reviews:** Conduct quarterly meetings chaired by Prime Ministers (for mega National level projects or Chief Ministers (for Mega and critical projects at State level). These meetings should address delays in payments, approvals, and deviations. Monthly meetings by Chief Secretaries or Ministers should ensure commitments are met. Fortnightly or weekly reviews at the Chief Engineer level can resolve site-specific issues.
- **Select Experienced Contractors:** Avoid reliance solely on L1 contractors. Prioritize track records, technical expertise, and past project success, and if required choose L2 contractors with proven records, with justification.
- **Improve Subcontractor Engagement:** Ensure contractors use fewer, reliable subcontractors and top suppliers for better quality and cost management.
- **Enhance Resource Balancing:** Use portfolio management techniques to redistribute resources across projects, mitigating capacity and capability risks.

- **Address Fraud:** Impose strict penalties for fake invoicing and simplify laws for the Comptroller and Auditor General (CAG).
- **Improve Estimation Models:** Develop parametric estimation models and free digital tools for better time and quantity predictions.
- **Strengthen Contract Management:** Simplify contracts into two separate parts, namely, short critical specifications and bulk routine documents for better clarity.
- **Focus on Training:** Provide frequent education on best practices, emerging technologies, and quality management through videos and events.
- **Ensure Safety Compliance:** Adhere to stringent safety and health regulations to prevent disruptions, penalties, and reputational risks.

These measures, when combined with the methodologies in this book, can ensure faster project completion, enhanced quality, and significant cost savings.

## 10.7 ANNEXURE I: Two Case Studies/ Examples clarifying the Concepts

### Case Study 1- Infrastructure Project: India's longest Sea Bridge

This case is about constructing a 22-kilometer-long bridge over the sea in Maharashtra state. The bridge will connect three critical locations in Mumbai and Navi Mumbai, including an international airport and a seaport. This infrastructure will reduce the current travel time from approximately 150 minutes to just 20 minutes for a substantial number of vehicles commuting daily. The bridge's cost is estimated at sixteen thousand crores of Indian Rupees (about 2 billion USD). To expedite construction, the entire work was divided between two main EPC contractors, with one responsible for constructing about two-thirds of the bridge and the other for the remaining one-third. Relevant details are illustrated in Figure 10.2.

By mid-2019, a total of 20 months had elapsed since the project's commencement, but it was already delayed by about 10 months. The senior management team's primary focus was to minimize further delays. They sought to foster a different mindset among stakeholders to recover a portion of the lost time and complete this project milestone as close to the originally committed target date as possible.

**Figure 10.2: The Complex Construction Project: Sea Bridge in Maharashtra of India**

**Some key information about this bridge and the Pile Foundation are given below:**

1. Total length of this project = 22 km; Out of this about 17 km inside the sea, and rest outside connecting roads/ land
2. About 85% of the Spans involved were of 60 meters, another 10% were of 90 meters.
3. Pile Diameter about 2.2 meters; 30-40 meters in length and about 15-20 meters below the Sea-level
4. Key Equipment involved in construction process: Finger Platform, Guide frames, Reverse circulating Drilling machine, Pile Foundation Liners, Pre-Fabricated steel Cage, Concreting the Pile cage
5. One Pile cap on Four piles for 60 meters span
6. Two Piers on the Foundation Cap (1 on the set of 2 Piles)
7. Pier Cap in each of the Two Tiersspan
8. Twenty Pre-casrt Girder Segments per span of 60 meters
9. Super structure consisting of Crash-Barrier, View Barrier, Sound Barrier, Hand Rails etc

Our initial step was to understand how the project's progress was being measured and reviewed and how action plans were decided and executed. This understanding was crucial for assessing the project's actual health and determining the areas where efforts were needed to catch up. Table 10.1 provides an example of how progress was reported and monitored at the time.

**Table 10.1 Project Construction Monthly Plan and Review (Current Sample for one major part of the Project)**

| STRUCTURE | AREA | SCOPE | PLAN FTM | | FTD | FTM | CUM | UPTO MAY 2019 |
|---|---|---|---|---|---|---|---|---|
| PILE | Side 2 | 485 | 4 | | x | 5 | 54 | 49 |
| | Side1 | 224 | 12 | | 1 | 9 | 53 | 44 |
| | Side 3 | 509 | 19 | | 2 | 15 | 69 | 54 |
| | TOTAL | 1218 | 35 | | 3 | 29 | 176 | 147 |
| PILE CAP | Side 2 | 101 | 1 | | x | 1 | 3 | 2 |
| | Side1 | 56 | 3 | | x | x | 6 | 3 |
| | Side 3 | 157 | 2 | | x | x | 2 | x |
| | TOTAL | 314 | 6 | | x | x | 11 | 2 |
| PIER | Side 2 | 198 | x | | x | x | 5 | x |
| | Side1 | 113 | 2 | | 1 | 3 | 5 | 2 |
| | Side 3 | 229 | 6 | | x | x | x | x |
| | TOTAL | 540 | 6 | | 1 | 3 | 5 | 2 |
| PIER CAP | Side 2 | 198 | x | | x | x | x | x |
| | Side1 | 113 | x | | x | x | x | x |
| | Side 3 | 229 | x | | x | x | x | x |
| | TOTAL | 540 | x | | x | x | x | x |

- **Side 1**: Piling starting from mid-sea towards land (Side 1 and 2 are being constructed in parallel)
- **Side 2:** Piling starting from land towards mid-sea, finally joining with Side 1
- **Side 3**: Piling on the Land for the Flyover joining the Sea Bridge
- **FTD**: Finished till Date (in current month)
- **FTM:** Finished till Month
- **CUM:** Finished till Now, from the very beginning

The committed date for completing all the piles, as shown in the review table, was 18 months away. However, achieving this within the committed timeframe seemed impossible, a fact that was neither obvious nor visible in the current project progress review to either the client (the government) or the EPC main contractor. Furthermore, the corrective actions required to meet the committed date were unclear, apart from expediting all activities.

To address these issues, the review method was first modified to a span-wise (a physical and visible segment of the project) review, complemented by a visual display of the project plan and progress. The revised tabular format for project progress review is shown in Table 10.2, and the visual display part is depicted in Table 10.3 As suggested earlier in this chapter, the key point was to focus on the most repetitive and critical part of the project, which in this case was the 60-meter bridge span. This segment was to be repeated 200 times to complete the bridge. Consequently, the progress of pile foundations, pile caps, and piers was measured not in absolute numbers but in terms of the number of spans planned and completed. Partial completion of a span was not considered; a span was either fully completed or not. This method was easier for all stakeholders to comprehend. Additionally, the actual cycle time achieved for each completed activity for one span was recorded against the required cycle time, providing insights into the progress relative to the planned target date.

**A quick analysis of the data in the revised table revealed the following approximately (not accurate analysis, only for illustration):**

1. Remaining number of piling foundations on Side 2 = 100
2. Remaining number of spans for piling on Side 2 = 100
3. Current average cycle time for piling per span on Side 2 = 23 days
4. Number of months required to complete at the current speed = 38 months (at 23 days per span)
5. Number of months available to complete this piling work = 21 months (as per the committed date)
6. Expected project delay = 17 months (about 80%)
7. Average cycle time required to meet the commitment = 13 days

Achieving this reduction from 23 days to 13 days per span was extremely challenging due to the complexities of sea-based operations. However, it was critical to meet commitments and maintain credibility with both the main contractor and the client.

The revised month-wise progress was also visually represented in Figure 10.4. The centreline depicted the bridge, with span numbers allocated to the completed portions. Above this line, the planned completion of various key construction activities (such as piling) was shown in different colours. Below the line, the actual completion of these activities was displayed in lighter shades of the same colours for easy correlation. This method provided a clearer understanding of progress and directed efforts toward reducing the cycle time for repetitive tasks, thereby helping to recover lost schedule time in upcoming planning cycles.

Despite the challenges of reducing the piling foundation cycle time from 23 days to 13 days per span, the team approached this with determination and guidance from a Lean Construction expert. Due to the complexity of drilling machines and other equipment, engineering modifications were not feasible. Instead, management solutions were implemented to enhance productivity and reduce cycle times using Lean Construction concepts like Value Stream Mapping, Collaborative Planning Systems, and the elimination of the Eight Wastes

**Table 10.2 Project Construction Monthly Plan and Review (Recommended Sample for an Infra. Project)**

| Structure | Area | Scope(# of Spans) | Plan FTM | Planned Cycle Time | Actuals | | | Target Date | | Remarks (KEY ACTION being Taken to Recover Lost Time) |
|---|---|---|---|---|---|---|---|---|---|---|
| | | | | | Cycle Time | FTM | CUM | PROMISED | FORECAST | |
| Piling | Side2 | 80 | 0 | 14 | 23 | 0 | 13 | 44228 | | |
| Pile Caps Fitting | Side2 | 80 | 0 | 18 | 26 | 0 | 3 | 44287 | | |
| Piering | Side2 | 162 | 0 | 16 | 0 | 0 | 0 | 44317 | | |
| Pier Cap Fitting | Side2 | 162 | 0 | 7 | 0 | 0 | 0 | 44348 | | |

| Structure | Area | Scope(# of Spans) | Plan FTM | Planned Cycle Time | Actuals | | | Target Date | | Remarks (KEY ACTION being Taken to Recover Lost Time) |
|---|---|---|---|---|---|---|---|---|---|---|
| | | | | | Cycle Time | FTM | CUM | PROMISED | FORECAST | |
| Piling | Side3 | | | | | | | | | |
| Pile Caps Fitting | Side3 | | | | | | | | | |
| Piering | Side3 | | | | | | | | | |
| Pier Cap Fitting | Side3 | | | | | | | | | |

| SUMMARY | TOTAL PILES | TOTAL PILE CAPPED SPANS | TOTAL PIERED | TOTAL PIER CAPPED SPANS | REMARKS(KEY ACTION Being Taken to Recover Lost Time) |
|---|---|---|---|---|---|
| Side1 | | | | | |
| Side2 | | | | | |
| Side3 | | | | | |
| TOTAL | | | | | |

**Table 10.3 Project Construction Monthly Plan and Review (Recommended Sample for an Infra. Project)**

**(SAMPLE) STATUS of SCHEDULED & COMPLETED WORK as on (say) July 11, 2019**

| | | TOTAL COMPLETION DATES | 1 | 2 | 3 | 4 | 5 | 6 | 7 | 8 | 9 | 10 | 11 | 12 | 13 (Review Week) | 14 | 15 | 16 | 17 | 18 | 19 | 20 |
|---|---|---|---|---|---|---|---|---|---|---|---|---|---|---|---|---|---|---|---|---|---|---|
| PLANNED SIDE 1 | Pier Caps | pqr | | | | | | | | | | | | | | | | | | | | |
| | Piers | abc | | | | | | | | | | | | | | | | | | | | |
| | Pile Caps | xyz | | | | | | | | | | | | | | | | | | | | |
| | Piles | 15-03-2019 | | | | | | | | | | | | | | | | | | | | |
| | Fingers | cde | | | | | | | | | | | | | | | | | | | | |
| | Access Bridge | pac | | | | | | | | | | | | | | | | | | | | |
| SPAN # | | | 1 | 2 | 3 | 4 | 5 | 6 | 7 | 8 | 9 | 10 | 11 | 12 | 13 | 14 | 15 | 16 | 17 | 18 | 19 | 20 |
| ACTUAL SIDE 1 | Access Bridge | | | | | | | | | | | | | | | | | | | | | |
| | Fingers | | | | | | | | | | | | | | | | | | | | | |
| | Pile | ??????? | | | | | | | | | | | | | | | | | | | | |
| | Pile Caps | | | | | | | | | | | | | | | | | | | | | |
| | Piers | | | | | | | | | | | | | | | | | | | | | |
| | Piers Caps | | | | | | | | | | | | | | | | | | | | | |

The arrival of the monsoon season, which halts construction due to turbulent weather, presented a unique opportunity to focus on preventive and corrective measures. During these three months, the following actions were undertaken:

1. The remaining project was conceptualized as 200 identical mini-projects of one span each, to be completed in 13 days without significant engineering changes.

2. A detailed critical review of the current 23-day cycle for one span was conducted, including:
    - Root cause analysis to identify delays and key issues.
    - Development of a new detailed PERT chart for one span, incorporating overlapping activities and proactive problem-solving.
    - Detailed Value Stream Mapping to identify and mitigate delays and inefficiencies.
3. Special attention was given to process improvements for heavy equipment, such as drilling machines, to minimize waste and increase output.
4. A simple day-wise Gantt chart was developed for daily planning, execution, and monitoring.
5. Collaborative planning sessions were implemented, including a 15-minute daily stand-up meeting for key personnel to discuss the next day's schedule and resource readiness. Weekly 60-minute "Make Ready" meetings ensured proactive resolution of constraints.

After the monsoon, the piling foundation cycle time gradually decreased from 23 days to 18 days, then 15 days, and finally stabilized at 13–14 days for the remainder of the project. This remarkable improvement effectively doubled the project's progress rate for such complex operations, representing a significant achievement for the team.

This case highlights the critical importance of defining an accurate method for project progress review. A realistic review system not only provides a clear picture of progress but also guides stakeholders toward actionable solutions for achieving target completion dates. We must also highlight that is extremely important to define the 'correct' method of project progress review, so that the course of action is also known for achieving the target completion time realistically.

**Case Study 2- Real Estate Project:**

This is another significant project in the state of Odisha[2], focusing on the construction of 89 multi-utility indoor sports complexes across the state to promote sports at the grassroots level. The total project cost was approximately 680 crore INR. Each of the 89 facilities was scheduled for completion within 12 months. Figure 10.3 below provides high-level details about the building design and layout.

The ambitious project of 89 sports complexes faced a tight timeline, so over 70 contractors were engaged to work simultaneously across sites, aiming for completion within 12 months. Larger agencies handled 2–3 contracts, while most managed one. To

---

[2] *The authors acknowledge the contribution of Er Jeevanananda Nayak, Chief Engineer, and Er. Trinath Behera, Superintending Engineer for providing the necessary data and adapting the Lean Construction 4.0 practices.*

save time, only two standardized designs were used: a Ground+2 structure for big cities and a Ground+1 for smaller towns. While plinths and superstructures were identical, foundations varied due to soil, land use, and geological conditions, creating challenges..

**Below are some illustrative cases:**

- In some instances, one or two very old trees existed on the land, requiring the Forest Department's permission to remove them and plant a large number of trees as per the norms to maintain environmental balance.
- Certain sites had water bodies that needed to be filled before achieving the construction level.
- Some locations were on highlands with rock outcrops, necessitating significant time and effort to level the ground.
- Existing structures at two sites had to be demolished after relocating the facilities.
- Overhead transmission lines were another issue that limited site activities.

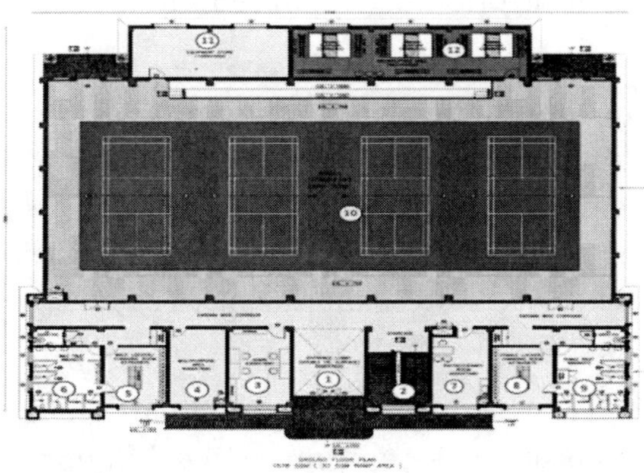

**Figure 10.3: The building external design and internal layout for a sample Indoor stadium**

**Some key information about this project are given below:**

1. Entrance Lobby size = 5000 x 7020 mm
2. Administration office size = 4000 x 7690 mm
3. Multi-purpose space size = 4000x 7690 mm
4. Male Locker Room = 4378x 5875 mm
5. Female Locker Room = 4378x 5875 mm
6. Male Toilet = 4860x 6375 mm. DA 2200x1700 mm
7. Female Toilet = 4860x 6375 mm. DA 2200x1700 mm
8. Arena Space = 40000x 22470 mm
9. Physiotherapy Room = 4000x 7690 mm
10. Equipment Sores = 11099 x 5000 mm
11. Weight lifting Hall = 15514 x 5000 mm

**12.** Staircase

Due to varying local conditions and different dates for awarding contracts, the project progress for different buildings was at varying stages, ranging discreetly from ground level to roof level. Contractors were informed of the overall target dates for completion and handover during the tendering process, with a strong emphasis on adherence. Consequently, projects awarded later required swift and efficient action by the agencies due to the shorter available timeline.

Based on this information, the selected main contractors submitted month-wise, work-package-wise quantity completion plans. Observations on project plan statuses and suggested improvements were made to enhance performance and achieve the tight deadlines. Observations were restricted to the top 4-5 issues to ensure manageability. Key observations and recommendations are detailed below:

**a.** The basic activities for each building were not consistent across contractors' project plans, despite the buildings being identical. To address this, a half-day workshop involving about 15 main contractors was organized to identify identical activities for all buildings.

**b.** Contractors had initially prepared only quantity-wise plans to track progress via an "S" Curve. They were instructed to create Gantt Chart Plans for execution and review by physical location, using the common activities identified earlier. This approach aimed to provide clarity for the entire project, highlight dependencies between subcontractors, and facilitate more accurate commitments. They were given two days to complete this task.

**c.** Another half-day workshop was held to compare the contractors' Gantt Chart Plans. Since the superstructures of the 15 projects were identical, the durations of similar construction activities were expected to be uniform. Given the similar types of resources for all the agencies, like manpower skills, machinery, and equipment sourced from the same state, **"critical path activities"** durations were compared among their plans to assess estimation accuracy. However, significant anomalies were observed, likely due to variations among estimators and schedulers. Examples of these variations are shown below (also illustrated in Table 10.4):

- Ground Floor Slab Casting: 7–10 days to as much as 25 days.
- Truss Erection for the Roof: 10–20 days to as much as 32 days.
- Wooden Flooring: 15–20 days to as much as 50 days.
- Finishing Civil Work: 30–40 days to as much as 60 days.
- Water Supply & Sanitation, Electrical Work, and Handover: 40–50 days to as much as 95 days.

**Table 10.4: Sample Variation for the duration of the identical activities by different Contractors**

COMPARISON of DIFFERENT (NAC Smller one TYPE) PROJECTS using GANTT CHART

Work Name: Construction of Biju Pattanaik Indoor Stadium(MPH) at Bellaguntha in the District of Ganjam for the year 2021-22.

| | | Proj SITE: GANJAM ---Agency: Anup Kumar Dash | | | Proj SITE: KODALA ----Agency : Sarika Civcon | | | Proj SITE: CHIKITI ---Agency: Rabindra Kmr | | | Proj SITE: BELGUNTA----Agency: Prashanta Patnaik | | | Proj SITE: GOPALPUR --Agency: Chitta Ranjan Swain | | | Proj SITE: POLSARA. Agency : Prasanta Patnaik | | | Proj SITE: KS NAGAR. Agency: Prashanta Patnaik | | |
|---|---|---|---|---|---|---|---|---|---|---|---|---|---|---|---|---|---|---|---|---|---|---|
| | ACTIVITY/Task Name | START DATE | FINISH DATE | ACTIVITY DURATION | START DATE | FINISH DATE | ACTIVITY DURATION | START DATE | FINISH DATE | ACTIVITY DURATION | START DATE | FINISH DATE | ACTIVITY DURATION | START DATE | FINISH DATE | ACTIVITY DURATION | START DATE | FINISH DATE | ACTIVITY DUARATION | START DATE | FINISH DATE | ACTIVITY DURATION |
| | OVERALL PROJECT DURATION PLANNED | 28-03-2022 | 27-12-2022 | 270 Days | 06-Jan | 05-Oct | 270 Days | | | 270 Days | 27-Dec | 26-Sep | 270 Days | 11-Jan | 10-Oct | 270 Days | | | 270 Days | 29-Dec | 26-Sep | 270 Days |
| 1 | Upto Lintel Level | | | | | | | | | | 01-06-2022 | 20-06-2022 | 20 D | | | | | | | 09-06-2022 | 03-07-2022 | |
| 1a | Upto Roof Level | 05-06-2022 | 30-06-2022 | 25 D | 05-06-2022 | 30-06-2022 | 25 D | 05-06-2022 | 30-06-2022 | 25 D | | | | 05-06-2022 | 30-06-2022 | 25 D | | | 7 D | | | |
| 2 | G.F Slab Casting of Gym, Yoga Room, Equipment Room | 21-06-2022 | 15-07-2022 | 25 D | 21-06-2022 | 15-07-2022 | 25 D | 21-06-2022 | 10-07-2022 | 20 D | 21-06-2022 | 30-06-2022 | 10 D | 21-06-2022 | 10-07-2022 | 20 D | | | 7 D | | | 20 D |
| 3 | G.F Slab Casting of Male & Female Toilet, Locker Room, Stair Case | 05-07-2022 | 20-07-2022 | 15 D | 05-07-2022 | 20-07-2022 | 15 D | 05-07-2022 | 20-07-2022 | 15 D | 01-07-2022 | 10-07-2022 | 10 D | 05-07-2022 | 20-08-2022 | 15 D | | | 7 D | | | 20 D |
| 4 | First Floor Slab Casting and Arena Area Column upto elevation 11.05 Mtr. | 23-07-2022 | 03-08-2022 | 10 D | 23-07-2022 | 03-08-2022 | 10 D | 23-07-2022 | 03-08-2022 | 10 D | 11-07-2022 | 31-07-2022 | 20 D | 15-07-2022 | 10-08-2022 | 25 D | | | 25 D | | | 15 D |
| 5 | Mumty Slab Casting at Elevation 11.05 Mtr. | 07-08-2022 | 20-08-2022 | 13 D | 07-08-2022 | 20-08-2022 | 13 D | 01-08-2022 | 20-08-2022 | 13 D | 01-08-2022 | 10-08-2022 | 10 D | 01-08-2022 | 25-08-2022 | 25 D | | | 15 D | 01-08-2022 | 26-08-2022 | 25 D |
| 6 | Truss Erection and Roof Treatment | 23-07-2022 | 25-08-2022 | 32 D | 23-07-2022 | 25-08-2022 | 32 D | 23-07-2022 | 25-08-2022 | 32 D | 11-08-2022 | 20-08-2022 | 10 D | 06-08-2022 | 31-08-2022 | 25 D | | | | | | 20 D |
| 7 | Wooden Flooring | 26-08-2022 | 10-09-2022 | 15 D | 26-08-2022 | 10-09-2022 | 15 D | 26-08-2022 | 10-09-2022 | 15 D | 21-08-2022 | 20-10-2022 | 60 D | 26-08-2022 | 10-09-2022 | 15 D | 22-08-2022 | 13-10-2022 | 20 D | 26-08-2022 | 15-10-2022 | 50 D |
| 8 | Finishing Work of Civil Work, Driven Way | 16-08-2022 | 15-09-2022 | 30 D | 16-08-2022 | 15-09-2022 | 30 D | 16-08-2022 | 15-09-2022 | 30 D | 21-09-2022 | 30-11-2022 | 60 D | 21-09-2022 | 30-10-2022 | 40 D | 15-08-2022 | 20-11-2022 | 95 D | 26-08-2022 | 22-11-2022 | 85 D |
| 9 | PH Work, El Work & Handing over the Building | 26-08-2022 | 05-10-2022 | 40 D | 26-08-2022 | 05-10-2022 | 40 D | 26-08-2022 | 05-10-2022 | 40 D | 11-11-2022 | 31-12-2022 | 50 D | 01-09-2022 | 10-10-2022 | 40 D | 23-08-2022 | 27-11-2022 | 95 D | 31-08-2022 | 29-11-2022 | 95 D |

**d.** Progress on long-duration activities such as wooden flooring, finishing civil work, plumbing, and electrical work could not be tracked weekly. Delays in these activities could not be flagged in advance by either the Works Department or contractors, necessitating their breakdown into smaller tasks.

**e.** Risks were not identified in advance, resulting in unnecessary delays, stress, and repeated excuses throughout the project.

**Actions Suggested for Improvement:**

1. **Standardize Work Methods:** Conduct detailed discussions among contractors about the Work Method Standards (WMS) to be followed for various processes. Share knowledge to identify best practices and agree on optimal durations for activities. Update the Gantt Charts accordingly to reflect these standards. All the activities on critical path were taken up and standardized.
2. **Break Long-Duration Activities into Sub-Milestones:** This allows weekly progress reviews and timely corrective actions. Make project progress 'visual' through Gantt Charts and other pictorial reports.
3. **Risk Identification and Management:** Anticipate risks related to drawings, approvals, equipment, manpower, materials, weather, absenteeism during major religious functions (e.g., Dussehra, Durga Puja), funding, and local issues. Prepare a dynamic constraint list to track key issues over the next six weeks, outlining resolution plans and responsibilities.
4. **Focused Project Reviews:** Limit reviews to 60 minutes with all key stakeholders present. Address the following questions:
    - What Activities and Tasks were planned for the last month and week?
    - What was achieved?
    - What was not achieved, and why?
    - How will these issues be resolved, and what is the plan to make up for 'lost time'?
    - What support is needed from the Works Department?
5. **Refine Processes:** Collaborate to improve any bottlenecks, even in the best practices identified by contractors.
6. **Suppliers' Summit (a new Best Practice):** Organize workshops to address bulk supply risks for materials like wooden flooring, roofing materials, sports equipment, and furniture etc. Key outcomes included pre-approved alternate brands, capacity assessments of suppliers, and visualization of completed facilities.

The Facilitation Workshop was organised by the Client, wherein all the Contractors, Suppliers, Engineers were present. Advantages of the workshop were manifold, like:

- For any single product, alternate brands were pre-decided so that the Contractors have the option of selecting one of the approved brands. Different Suppliers exhibited their products with specifications and the Contractors got an opportunity to choose between the alternatives considering the commercial aspect vis-a vis its availability in their region.
- Capacities of the Supplying Agencies was assessed beforehand so that the entire quantity could be distributed according to their production capacity, orders in hand, proximity to the work sites; thus, ensuring resource levelling.
- All the new products to be used in the facility were displayed along with the models of the facilities for convenience of all the three entities. This helped the Contractors, Engineers and representatives of the Client to visualise the completed view of the sports facility post construction.

The Facilitation Workshop helped in selection, approval and purchase decisions in respect of these materials in one stroke for all structures. By the time the workshop was over, the project had the advantage of adoption of uniform specifications with approved materials while everyone knowing to stick to the target date.

**Benefits Derived:**

1. Of the 89 buildings, 40 were completed within the tight deadline. This was sort of record for success in PWD.
2. Except for four buildings stuck in judicial disputes, all others were completed with only a two-month delay.
3. This achievement marked a significant milestone for the state government and its people.
4. Valuable lessons were learned for future repetitive projects, with potential for broader adoption of these practices and greater efficiency.

## 10.8 References

1. *KPMG (2023), "Familiar Challenges, New Approaches: Global Construction Survey 2023-India Edition", https://kpmg.com/in/en/insights/2023/07/global-construction-survey-2023-india-edition.html*

# CHAPTER 11

# FIVE S: ELIMINATING STRESS & TIME WASTE

# 11. Five S

## 11.1 What is Five S

Five S is an important part of Lean Construction 4.0. However, in some cases, it is treated as the starting point for Lean Construction, while in other cases, it is just ignored. Five S is a systematic approach to a better and cleaner workplace. It is also an attitude of pride and care about the place where one works, and is a practical way to initiate employees' participation in the elimination of some of the eight construction wastes.

The basic principle governing 'Five S' is good housekeeping. If integrated with the work culture, it significantly improves productivity, safety, quality and morale by eliminating delays, various hassles, unnecessary Inventory and also unnecessary motion and transportation. Basically, it organises the workplace to keep it neat and clean and to maintain standardised conditions so to attain the discipline that is needed to do a good job. We have seen some people say that if you can find man, material, machine, tool, drawing or other information etc in less than a minute, you have a good Five S system.,

In fact, you can call Five S a philosophy, a religion or simply a way of life in our daily work. Developed by Japanese using five Japanese words, all starting with S and written in the brackets, it can be simply described as in the next long sentence. Five S begins with discarding the things not required in the workplace (Seiri); arranging all the necessary things in an orderly manner (Seiton); maintaining a clean environment removing distractions (Seiso); standardizing the locations and processes (Seiketsu); and finally maintaining this system on a continuous basis (Shitsuke) as a part of the habit. All employees must follow this discipline of such standards and rules in their daily work. Each of these Five steps associated with S, represents a specific aspect of the methodology, which collectively contribute to enhancing productivity and reducing waste, besides reducing the mental stress of the people associated with keep on searching things of their regular use.

This concept can be applied at Site, in the office and also in the homes as well. Some of the areas of applications are listed below, where have large number of things accommodated causing problems of wasting a lot of time searching for the things mentioned there on a routine basis. Also having excess of the things listed below making the place cluttered, which can be de-cluttered using Five S.

1. **At the FACTORY/ CONSTRUCTION Site:** Floors, Staircases. Stores, Lofts, Filing Room, Shelfs, Top or Behind Shelves, Aisles, Work tables, Corners, Walls etc.
2. **In the OFFICE:** Printers area, Desktops, CPUs/ Computer Racks, Photocopier, Phones, Cable/ Wires, Cupboards & Mouse, Crockery, other Tools & Equipment etc.

3. **In the FURNITURE**, both at office and Construction Site: (Not-in-use or damaged) Chairs, Tables, Desks, Cabinet (for Filing & Books), Display Board, White Boards, Carpets, Trash cans, Wall hangings, decorative Items etc.
4. **In the STORES:** Materials: Stationery; Manuals, Collaterals, Books, Charts/ Graphs, Drawings, Bulk Materials in Open Yard etc.

Sometimes 'safety' is added as the sixth 'S'. However, in this book, we have covered important area like safety as a separate Chapter. Sometimes, people also use names like, Five S campaign or movement, Can Do, or just simply housekeeping for this Five S initiative.

## 11.2 Why do we need Five S:

**In short, the following benefits can be derived from Five S:**

1. Increased efficiency and productivity: Five S reduces time wasted in searching for
2. tools, materials, or information, resulting in improved workflow and productivity.
3. Improved quality: Five S helps in identifying and addressing defects and issues
4. promptly, leading to improved Asset or service quality.
5. Employee morale and engagement: A well-organized and visually appealing
6. workspace fosters employee satisfaction, morale, and engagement.
7. Cost savings: Five S reduces waste, eliminates unnecessary inventory, and optimizes
8. space utilization, resulting in cost savings for the organization
9. Enhanced workplace safety: A clean and organized workplace minimizes safety
10. hazards, reduces accidents, and promotes a safe working environment.
11. Reduces machinery down time and improves operation time.
12. Opens up more work and storage space, ensuring better utilization
13. Helps to bring to attention abnormalities, cracks, mistakes and missing parts at a
14. glance, and thus eliminates accidents and defects.
15. Encourages better discipline at the work place.

**This is summarised in the diagram given below:**

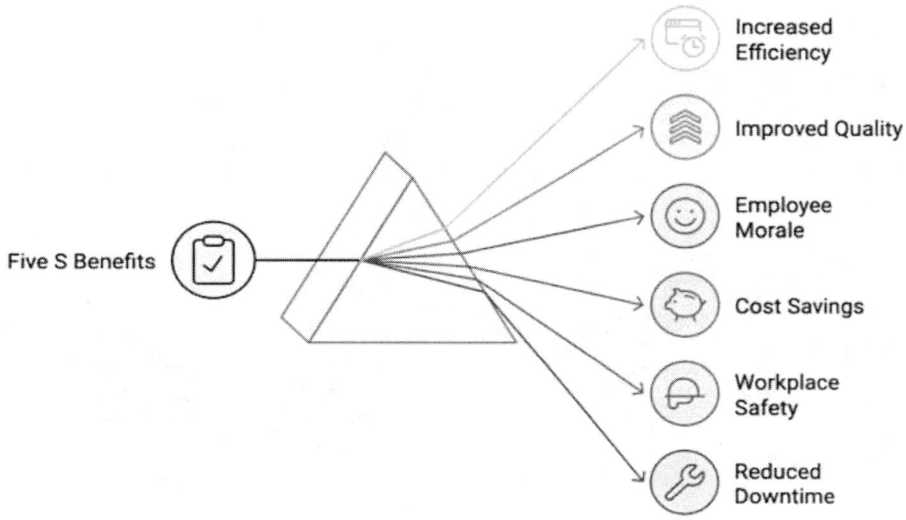


**Figure 11.1 Reasons for implementing 5S Concepts in Construction Site**

The next sections describe the five steps necessary for conducting Five S.

## 11.3 FIVE Steps of 'Five S'

Five S involve carrying out five specific steps. These are explained below with some relevant examples. We have also provided commonly used English vocabulary for the same Five S.

### 11.3.1 First 'S' of 'Five S': SORTING (Seiri)

The first step in the Five S methodology involves sorting through items in the workplace to distinguish between necessary and unnecessary items. The intention is to remove the 'unnecessary' items from the daily used space before they become problems. The items which are not required frequently (say on a daily basis, or hourly basis) items are removed from there, creating a clutter-free and organized workspace. This process helps optimize space utilization and reduces the time spent searching for tools or materials. The effort is to do things in the proper order. The removed items are kept either at a less useful space, if required infrequently, or are removed permanently, if those may not be required in the future as well. Figures 11.2 and 11.3 shows two examples of Sorting, which is self-explanatory.

**Figure 11.2 Example of Sorting (Seiri)**
**(Courtesy: Lakshmikashyap Mallavarapu, Savills India)**

**Figure 11.3 Another examples of Sorting (Seiri)**
**(Courtesy: Lakshmikashyap Mallavarapu, Savills India)**

### 11.3.2 Second 'S' of 'Five S': Set in Order (Seiton)

After removing the unnecessary items and things in the first step, the focus of the second step is to bring orderliness, neatness and proper arrangement of the necessary things identified. This means having the 'right things in the right place' so that they can be easily located. Tools, equipment, materials, documents etc. are stored in designated locations, ensuring easy accessibility and eliminating waste caused by searching and people's unnecessary motion. Visual enablers like labels and colour coding may be utilized to enhance clarity and facilitate quick identification. The emphasis here is on 'elimination of search activities'. Once everything has its allocated place, the workplace becomes neat and orderly, and anything can be located in very short time. Figure 11.3 provides one example of this.

**Figure 11.4 Example of Straightening (Seiton)**

### 11.3.3 Third S of 'Five S': Shine by Cleaning (Seiso)

The third step emphasizes on cleanliness and having spic-and-span working environment. This means getting rid of dirt, waste, dust, stains, filth and unwanted matter from every available thing in the working area. Cleaning is also a form of inspection. If the object is neat and clean, then many defects like cracks or missing parts become visible. Thus, the focus here is also on cleaning as 'inspection' and on creating a flawless workplace. Therefore, the recommended procedure for cleaning for items should be specified and regularly followed-up to ensure the adherence. Regular cleaning activities are performed to ensure a tidy and visually appealing workspace. Cleaning also helps identify and address potential safety hazards and equipment malfunctions, promoting a safe working environment. Figure 11.4 below shows the out outcome of this step in an office

### 11.3.4 Fourth S of 'Five S': Standardizing (SeiKetsu)

This step means continuous maintenance and retention of high standards and improvements achieved as a logical extension of first three steps of seiri, seiton and seiso. This can be obtained defining and using the standardized procedures, visual guidelines, and checklists

Figure 11.5 Example of Systematic Cleaning (Seiso)

ensure consistency and promote uniformity across the organization. The emphasis here is on 'visual management' (like, colour-coding management) and standardisation to attain and maintain defined conditions in order to act quickly when the need arises. Therefore, equipment and workplace should be kept in the best possible conditions, using all the best practices defined. Figure 11.4 provide some examples of this.

### 11.3.5 Fifth S of 'Five S': Sustaining (Shitsuke)

The fifth and last step involves sustaining through self-discipline, work-discipline; and periodic training. This also focuses on establishing a culture of continuous improvement. It involves

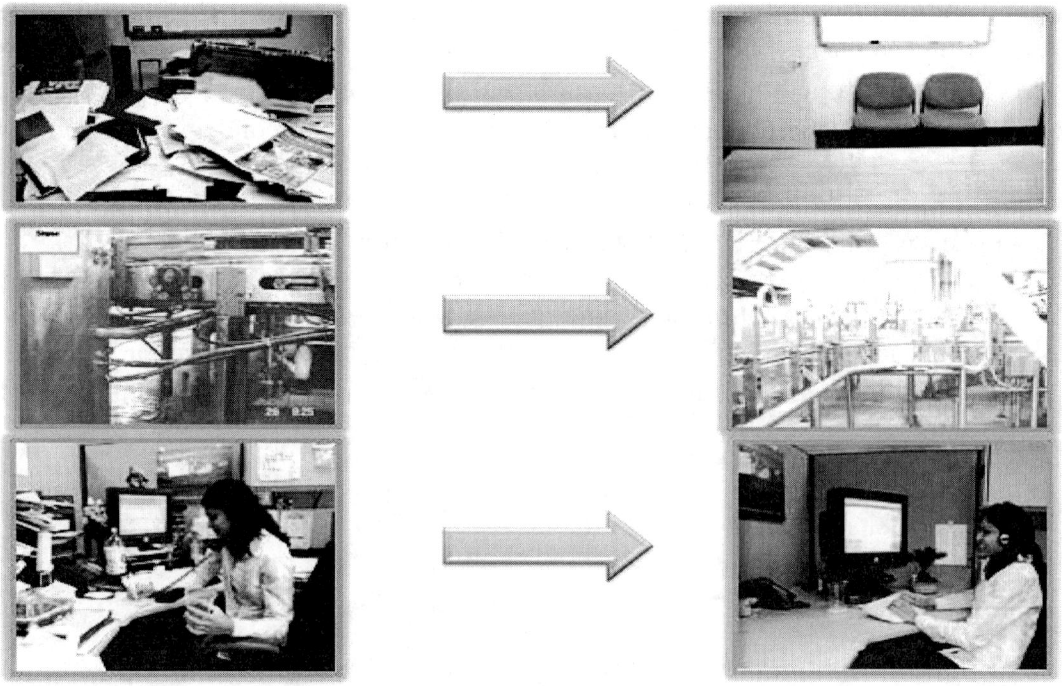

Figure 11.6 Example of Standardizing (SeiKetsu)

Table 11.1 below is a quick summary of Five S, with Japanese names, English equivalents and their main focus of action.

**Table 11.1 Summary table for Steps in Five S**

| STEP # | JAPANESE NAME | POPULAR ENGLISH NAME | MAIN FOCUS |
|---|---|---|---|
| 1 | SEIRI | SORT OUT | Get rid of Unnecessary / Untidiness |
| 2 | SEITON | SET IN ORDER | Organize everything in the workplace |
| 3 | SEISO | SHINE BY CLEANING | Keep everything pick-and-shining clean |
| 4 | SEIKETSU | STANDARDIZE | Standarize all locations & methods |
| 5 | SHITSUKE | SUSTAIN | Maintain diecipline to sustain |

## 11.4 Limitations and Implementation of 'Five S'

The following sections outline key steps for implementing the Five S methodology and its limitations.

### 11.4.1 Steps for Five S Implementation

- **Create awareness and secure management support**: Educate both employees and management on Five S benefits and gain their commitment to the process.
- **Form a cross-functional team**: Gather representatives from various departments to ensure a thorough implementation.
- **Assess the current state**: Evaluate the Construction Site and Office to pinpoint areas needing improvement and define the scope of implementation.
- **Sort and declutter**: Sort items, categorize, and remove unnecessary ones from the workspace.
- **Define storage locations**: Assign storage spots for tools, equipment, and materials based on usage frequency, accessibility, and safety.
- **Establish cleaning procedures**: Set cleaning schedules and protocols to maintain regular cleanliness.
- **Use visual management**: Apply labels, signage, and floor markings to improve organization and communication.
- **Standardize processes**: Create standard procedures, guidelines, and checklists for consistency across the organization.
- **Train and engage employees**: Educate employees on Five S principles and involve them actively in the process.
- **Monitor and sustain**: Regularly review Five S effectiveness through audits, feedback, and address any issues while fostering a continuous improvement culture.

These steps are summarised in the diagram on next page. Start with realistic expectations, as initial results may be modest, and early setbacks are common. Unlike VSM and Defect Prevention, the impact may not be immediate. Recognizing and celebrating employee contributions publicly will further motivate participation.

### 11.4.2 Limitations of Five S

The main challenges in implementing Five S are as follows and should be addressed accordingly:

- **Initial resistance**: Employees may resist changes in routines, as Five S assumes voluntary and positive involvement, which requires effective communication and training to overcome.
- **Cultural challenges**: In organizations with established work cultures that prioritize other aspects over organization and cleanliness, Five S implementation may be difficult. Many managers may struggle to see a direct link between Five S and business outcomes, requiring patience, perseverance, and confidence.
- **Maintenance requirements:** While easy to understand, Five S is difficult to sustain. Employee engagement through bulletins, posters, slogans, etc., supported by ongoing leadership encouragement, is often needed

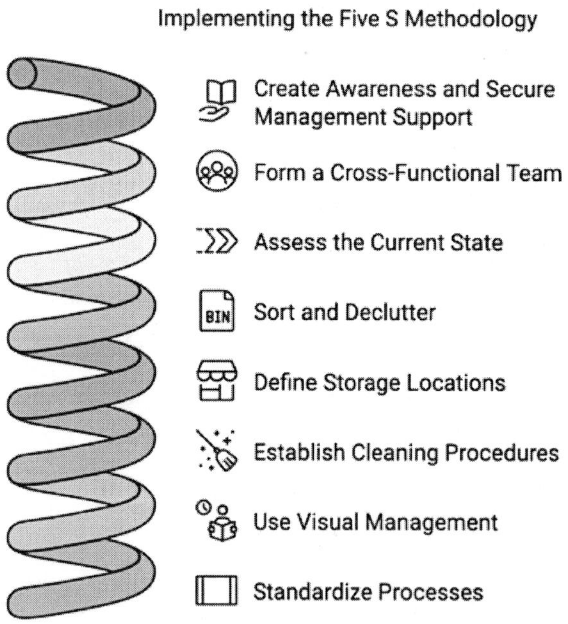


**Figure 11. 7: Steps involved in implementing 5 S concepts.**

- **Ongoing process**: Five S requires consistent monitoring, audits, and employee commitment to maintain improvements.

- **Incremental results**: While Five S shows results, they may not be dramatic or innovative.

In summary, Five S, as a core of Lean Construction philosophy, offers a structured approach to workplace organization and efficiency. By incorporating the principles of Sort, Set in Order, Shine, Standardize, and Sustain, businesses can improve productivity, safety, and engagement. While beneficial, organizations must recognize Five S limitations and invest in maintenance and employee involvement for long-term success.

## 11.5 ANNEXURE I : Case Study on Five S – A Modern Residential Complex

**The Project:**

This project was a residential complex with modern facilities, consisting of three towers of 22 floors each, a common Club House of four floors and lots of external open areas for Children parts, Jogging & Cycling Tracks and other sports activities, including Cricket. The total built-up area was about 20,00,000 SFT, having 264 Apartments. This was being made by a reputed Real-Estate developer of Maharashtra State Region, with the name Kumar Corp, Pune. All the three towers were being built simultaneously with common walls, and connected to the four-floors of the common Club House. These were being constructed by three different contractors. The total requirements of storing and transporting the bulk materials was very high. This included Steel reinforcement (25 MT of steel per slab, per Tower), 250 CuM (Cubic Meters) of Concreting, and 2200 SFT of Aluminium Shuttering materials among other things. It was really very difficult to store the required materials in terms of space requirements for each of the Contractors separately and also avoiding potential Safety hazards. The material movement was also high, as the slab-to-slab cycle time in each of the three towers was scheduled to be 10 days, and slab included concealed electrical wiring.

**The Objectives of Five S in this Project:**

Because of the problems stated above, about space shortage, multiple transactions of materials movements and potential accidents, the key objectives of these projects of embracing Five S concepts, were agreed as following:

- Workplace safety: Minimize the safety hazards and potential accidents at the project-sites to achieve a goal of Zero Hours lost in accident for the whole Complex.
- Workers and Yard's Productivity: Use the space Optimally and ensure easy and quick identification of the required material to transport from Yard to the work-site efficiently. Minimize the searching for tools and materials, for improved workflow.

- Improved quality: Enabling identification and then resolving the defects issues, including missing parts, cracks etc. promptly, leading to improved Building quality. This also provided better Customer visits to sites, during construction
- Improved Employee morale: A well-organized and visually appealing workspace fosters employee satisfaction, morale, and engagement.
- **Cost savings:** Reducing unnecessary searching time and eliminating unnecessary inventory to optimize space utilization
- **Work Discipline:** Encouraging better discipline at the work place.

**Outline of the Methodology Followed:**

- Discuss with the Senior Leadership Team of the project and what, why (benefits) and how of Five S methodology and get their concurrence and support,
- Identify the Core Champion Team for Five S implementation. The team would be a cross-functional Team representing Site Engineering Team, Contractors Team, and Procurement Teams, besides Project and Quality Management Teams.
- Have a brief introduction to the team on the Five S principles with good examples. Have a high-level assessment of the present state and the problems they face due to that.
- Have one brief session of Step 1 for Sorting. After that went to Yard to jointly identify the shuttering materials, steel for reinforcement etc. which are unwanted and to be removed. Also identified the required material which needs to be stored near the right Tower of construction.

**Figure 11.8: Identification and removal of unwanted items (Red Tags)**

- After 1-2 days, have another brief session of Step 2 for Set in Order (Seiton). Soon after session went to Yard to jointly identify the location & quantified the shuttering materials, steel for reinforcement etc. This was also keeping in mind the unloading of right quantities of materials for replenishment at right place and designed & suitable checks were provided to ensure compliance. This took half day of more exercise. A centralized Stores was designed & appropriate racks, bins etc. were provided. Also created the Zone maps based on the logistics plan to ensure receiving, unloading & storage of materials at the right place. Below are some sample pictures.

**Before improvement**

**Figure 11.9: Setting the Steel Bars in Order (Seiton)**

**Figure 11.10: Rearrangement of the Materials, before and after 5S implementation**

- After one week or so, have another brief session on Step 3 for Shine by Cleaning (Seiso). Then jointly prepare Daily and weekly cleaning schedule along with the Construction activities. Some sample pictures for this are given below:

**Figure 11.11: Defining Clearing activities and its frequency**

After some time, have another brief session on Step 4 and 5, that is for Standardizing (SeiKetsu) and Sustaining (Shitsuke). We focused on creating Visual Process Charts, different type and colours of Labels, Signage and Floor markings wherever possible, for all 5S standards, processes & outcomes already achieved were ensured to guide the site team, as illustrations for future training to more people. All this work was through general team consensus which became the reference points for replicating the system across the site.

Later, some audits were also carried out on twice a month basis, to ensure that the system agreed is being followed.

**Benefits of the application of Five S:**

**Following were the benefits achieved:**

- Workplace safety: Safe-working Man-hours (reported weekly) improved by 65 %
- Workers and Yard's Productivity: Material Search Time reduced from average 12 minutes to just 15-20 seconds. Waiting for material as the main waste reduced significantly, and ensured much better workmen productivity. Slab-to-slab Cycle time of 10-Days was achieved 80% of the time, compared to earlier only 50% of time.
- Reduced overstocking and Stockouts and space became better utilized and also lesser problem of material transportation manoeuvring.
- Improved quality: Customers' visiting the site to see sample flats appreciated neatness & cleanliness at site
- Improved Employee morale: Work place became clean, in good order, and a comfortable place to work.
- Cost savings: Cost saving was not quantified, but the Client was quite happy with the noticeable changes achieved, and overall faster progress of the Project.

- Work Discipline: Workers and others behaved in more disciplined way

**Some problems of implementing Five S in this Project:**

There was an initial resistance, even with the management. But after 3-4 rounds of discussions with them, and explaining the main benefit of better adherence to the planned 10-Days Slab-to-Slab cycle time convinced them. They then supported in a big way, which reduced the workers and others resistance. Every work was carried out during the working time only.

- We could not come out with complete ROI type of justification for the whole initiative. However, we did increase the adherence to 10 days Slab-to-Slab Cycle time from 50% to 80%. This meant roughly 2-3 Lakhs of INR saving in every slab casting.
- It took lots of efforts persuasion and sometimes small reward (like some tea & snacks) to sustain the initiative

## 11.6 Acknowledgement:

The authors would like to profoundly thank Mr. Piyush Pandey, Founder & Principal Consultant of Kakoli Consulting, Pune, India, for providing the details for the effective Case Study Also we would thank the Builders for this Project, Kumar Builders, Pune who allowed to use the data and the pictures of the project.

## 11.7 References:

1. ▯ *Seth, D., & Rastogi, S. (2009). Global management solutions demystified (2nd ed.). Cengage Learning.*

2. *Osada, T. (1991). The 5S's: Five keys to a total quality environment. Asian Productivity Organisation.*

3. *Talawadekar, S. (2002). Visual management through Five-S: A Japanese tool of Kaizen. Quality Management Systems.*

*5S in Lean Construction is more than tidiness—it's discipline in action. By sorting, setting in order, shining, standardizing, and sustaining, workplaces become safer, efficient, and inspiring, turning everyday order into lasting productivity and pride.*

# CHAPTER 12

# RECOMMENDATIONS FOR EFFECTIVE & RAPID IMPLEMENTATION

# 12 Recommendations for Rapid Implementation

Based on our total combined experiences of about 100 years in construction projects, we have observed several good and poor practices. We are sharing some of these with our readers, for their benefit. We are confident that you will find them useful in your projects.

## 12.1 Key Lessons Learned from Past Successes & Failures

**For CPS (Collaborative Planning System):**

- The signed-off project schedule made for tender or contract should be replanned to reduce the overall duration by 10%. This will help transfer all the time buffers towards the end of the project, ensuring consistent work pressure from the start and preventing the last-minute rush often seen with the 'Student's Behaviour Hockey Stick Curve.'
- Ensure that the DESIGN and PROCUREMENT teams participate in the PHASE PULL (at Rock Milestones level) and do Collaborative REVERSE PLANNING. They need to discuss any constraints related to their supplies, like Shop-Floor Drawings, materials, manpower requirements for the previous operations, Equipment etc.
- Shop-floor drawings should be signed off at least 7-8 days in advance of when are actually required. Teams must ensure their presence in all weekly 'MAKE-READY' meetings at least via video conferencing, if they can't come physically. The 'same' representative should attend every meeting for better continuity.
- Plan the supply of Drawings and materials according to the 'CAN-DO' (Pebble level activities plan) for the 3rd to sixth week Look Ahead Plans. Identify and resolve all other resource constraints, like money, manpower, Equipment, work space etc. .
- Clearly define and stick to 'Hand-over commitments' for MEP work with civil finish work, allowing enough time to complete finishing tasks.
- Each team member (from design, procurement, contractors) who fails to meet commitments must conduct a root-cause analysis periodically to understand why and how to improve next time.

**For Vendor Selection:**

- Vendor selection should consider historical data from the region, not just the lowest bid (L1) but also vendor performance elsewhere. Look at their past project records, exposure to Lean concepts, and their willingness to adopt changes for the client's benefit.

- Closely monitor vendor performance in terms of quality and delivery, and escalate issues to management whenever necessary.
- Conduct pre-bid meetings with vendors and the design/execution team to clarify major assumptions made for the project. If possible, invite all bidding vendors together to ensure everyone understands the same.
- Encourage vendors to use more off-site fabrication and mainly assemble at the project site, like popular Pre-Fab technology.
- Urge designers and architects to reduce variety in designs for similar elements, including column cross-sections.
- Stress the importance of minimizing frequent changes in workforce among vendors.
- Encourage vendors to provide on-the-job training for critical and slow processes, which will help reduce rework and boost subcontractor profits.

**For Paperwork Reduction:**

- Use digital apps like Quality-as-a-Service to improve quality and significantly reduce time-consuming paperwork while enhancing safety, health, and environmental sustainability.
- Try using apps like ACONEX for quick document access to eliminate paperwork, allowing for faster and more accurate information sharing while maintaining environmental sustainability.
- Prepare the team to implement concepts like Earned Value Management (EVM) to track project progress effectively.

## 12.2 Additional Clauses required for the EPC Contractors & Sub-Contractors

Often, Lean Experts join projects in the middle, making it hard to get Contractors, Consultants, and Suppliers to adopt a different project management style. They may see this as an extra burden. So, it's better to include certain clauses:

- Contractors will agree to follow Lean Design & Construction Practices or other effective management methods for better project execution.
- Contractors should use software like BIM (Building Information Modelling) available with their engineering & design team. This can help evaluate design alternatives and identify potential errors early, ensuring better quality for the client.

- Contractors need to submit a 1-page report every two weeks on:
  - Total man-hours spent on creating, developing, and evaluating improvements for bottleneck Boulder & Rock Milestones, as well as Pebble and Dust Activities. This ensures continuous problem-solving and recovery of lost time.
  - Key recommendations made for design or construction improvements to address reasons for not meeting commitments from the previous period.
  - Value created by the contractor in the current period and cumulative value from proposed changes that benefit the client.
  - Status of past improvement solutions implemented through design or execution method changes.
- At the project start, create a 'future reserve fund' by conducting Value Engineering, Value Stream Mapping, and waste elimination exercises. These should enhance design, procurement, or construction processes without compromising necessary functions and characteristics, such as service life, reliability, operational efficiency, maintenance ease, aesthetics, fire protection, and safety features. This proactive fund can later be used to accommodate small changes suggested by the team, without asking for more budget and more delivery time. This will make the customer happy and may give a preferred vendor status

## 12.3 Role of HR (both Client's & Main Contractor's) for Success of Lean Construction 4.0 Success

Implementing Lean Construction 4.0 is a "Transformation" initiative for the organization. It changes the 'way of working' and requires a cultural shift with better soft skills for project success. This is not just another Quality Management Process. Therefore, Human Resources Development (HR) must play a crucial role in building this new culture over time. HR teams from the Client's side, EPC (Engineering, Procurement, and Construction) Main Contractor's side, and key specialty Contractors' side need to be actively involved for the initiative to succeed. Here are some recommendations towards this:

- The Senior Leadership Team should define the long-term Lean Construction 4.0 Vision, Strategy, and implementation Blueprint for the Client. HR must understand this to act as an internal strategic partner.
- For any new project, the Senior Leadership Team should hold a special workshop on Lean Construction 4.0 concepts. This 90-100 minute workshop should involve the Core Lean Construction Team representing the Client, EPC Main Contractor, Lean Expert(s), and 2-3 major subcontractors like Civil and HVAC. Senior

Management should emphasize the Lean Vision and the roles and responsibilities of Core Team Members.

- The HR Department should arrange regular monthly Open House meetings, include updates in their Newsletter, or share Video Messages from senior management to promote a common goal of project success.
- The Senior Management Team (Steering committee members) should conduct a monthly 45-minute review with the Core Lean Project Team. This review should assess the effectiveness and issues of Lean implementation and provide direction and support. The team should also spend 15 minutes each month with Lean Expert(s) to discuss effectiveness and support needed, which can be part of the monthly Steering Committee meeting.
- HR should monitor the development of an 'Internal Customer' culture and offer support as needed.
- Finally, ensure there are Incentives/Rewards and penalties for adopting or failing to practice these concepts. These can be monetary or innovative non-financial incentives that HR can develop.

## 12.4 Type of Incentives & Penalty for practicing Lean

Some Reward & Recognition along with some Penalty System will be mutually agreed upon, for Main and Design Contractors for bringing greater value to Client's Project, practicing Lean philosophy. Some of our recommendations are given below.

Implementing a Reward & Recognition system, along with a Penalty system, can effectively encourage Main and Design Contractors to adopt Lean principles in the construction sector. Here are some ideas for both:

### 12.4.1 Reward & Recognition Systems

**1. Performance Bonuses:**

- Offer some meaningful monetary bonuses for teams or Contractors that meet or exceed project milestones, demonstrate exceptional efficiency, or significantly reduce waste. This mutually agreed at early stage and should be mentioned in the Contract. The reward amount should be attractive.

**2. Recognition Programs:**

- Create awards for teams or individuals who exemplify Lean principles, such as "Lean Champion of the Month" or "Innovative Solution Award." Also, one award at the end of the project for the Contractor, if they have done a good job. Some of them may also be recognized as the "preferred vendors".

**3. Public Acknowledgment:**

- Recognize outstanding contributions during team meetings, newsletters, or company-wide events, boosting morale and encouraging others.

**4. Career Development Opportunities:**

- Provide training or workshops on Lean practices, or sponsor attendance at industry conferences for teams that show commitment to Lean methods.

**5. Access to Resources:**

- Offer resources like software tools, consulting services, or expert training to teams that successfully implement Lean strategies.

**6. Profit-Sharing Agreements:**

- Implement profit-sharing arrangements based on the project's overall success, rewarding all contractors for achieving financial efficiency.

**7. Improved Work Environment:**

- Acknowledge teams by improving their work environment or providing better equipment and tools for exceptional performance.

**8. Client Testimonials:**

- Offer contractors the opportunity to receive formal testimonials from clients, which can help in future bids.

### 12.4.2 Penalty Systems

**1. Financial Penalties:**

- Impose fines for failing to meet Lean performance metrics, especially the project delays. These penalties are very significant in Middle East Construction Project, and almost meaningless in Indian Government Project. Unless these impact the Contractors' profit margins, these will not be effective.

**2. Reduced Payment:**

- Withhold a portion of payment until specific Lean objectives are met, such as timely delivery of materials or adherence to design specifications.

**3. Contractual Obligations:**

- Include penalties in contracts for failure to comply with agreed Lean practices, like excessive delays or failure to provide necessary documentation.

**4. Project Delays:**

- Implement consequences for repeated failures to meet deadlines, such as extending project timelines without additional payments.

**5. Negative Impact on Future Bids:**

- Consider past performance in Lean implementation when evaluating future bids, leading to a lower score for those who fail to comply.

**6. Mandatory Training:**

- Require teams that do not meet Lean objectives to undergo additional training sessions to improve their understanding of Lean practices.

**7. Performance Reviews:**

- Conduct regular performance reviews that include Lean practice adherence, and take corrective actions if objectives are consistently unmet.

**8. Increased Scrutiny:**

- Increase oversight and reporting requirements for teams that fail to meet Lean standards, leading to more frequent reviews of their processes.

### 12.4.3 Implementation Considerations

- **Clear Communication:** Ensure all parties understand the criteria for rewards and penalties.
- **Measurable Metrics:** Establish clear, measurable metrics for Lean performance to assess success objectively.
- **Collaborative Approach:** Involve contractors in designing the reward and penalty systems to foster buy-in and commitment to Lean principles.
- **Flexibility:** Be open to adjustments in the system based on feedback and changing project dynamics.

By implementing these systems, you can create a culture that encourages Lean practices and drives greater value for clients in construction projects.

## 12.5 Long-Term recommendations for Government, Professional Bodies, Educational Institutions

### 12.5.1 Codes & Standards for DESIGN & TESTING Guidelines:

Design is key in determining the lifecycle cost and quality of construction, making it essential to prioritize. While the government has developed many codes and standards, we suggest they re-evaluate these in light of rapidly changing design and construction technology. Implementing following changes could have a significant national impact.

a. **Increase in R&D Funding:** Invest heavily in innovative designs for all construction types and focus on developing local, eco-friendly materials. This could also include reusing industrial waste (e.g., from debris to all rejected materials). Budget should be 3-4 times higher than the current and must include a centralized database for local materials and workforce, updated by local authorities.

b. **Improved Testing Codes:** Regularly update testing codes to adopt faster, digital methods, including non-destructive testing. Portable testing equipment could move with project progress, enabling remote or virtual testing, crucial for infrastructure projects. Focus on developing such portable testing equipment.

c. **Collaborative Research:** Allocate R&D funds for joint research with technical and professional institutes (like IITs, IIMs, NITs) and government bodies. Designated institutions can maintain a centralized database for their regions.

d. **Management School Involvement:** Include management schools to introduce engineers to concepts like "Lean in Design", "Design Thinking." and "Value Management". Both management and engineering expertise are needed, especially for large projects.

e. **Institute-Industry Collaboration:** Foster closer partnerships between academia and industry. This creation and exchange of relevant knowledge will be very helpful.

f. **Paid Research:** Fund research on new codes and standards to ensure accountability.

g. **Frequent Updates:** Update codes and standards every 3-4 years to keep up with new technology and materials changes.

h. **User Feedback Mechanism:** Collect feedback at annual professional conferences, covering strengths and improvement areas for codes and standards.

The recommendations can then be passed to relevant committees, who can issue annual addendums based on feedback to maximize the impact of these codes and standards.

### 12.5.2 Codes & Standards for WORK METHODS STATEMENTS Guidelines

After design, the next major factor in determining construction quality and lifecycle cost is the construction method, documented in Work Method Statements (WMS). WMS are often crafted by Project Management, Quality Management, or Contractors, sometimes adjusted by client-contractor negotiations based on available equipment and skills. However, the specifics of WMS are often left to the contractor's discretion.

In many projects, especially smaller ones, methods are substandard and left to unskilled supervisors, resulting in poor quality. Even large firms may use non-optimal methods limited to internal expertise. Therefore, we recommend that the government also issue codes & standards for WMS, including testing, documentation, supervision, and checkpoints. Contractors could select from 2-4 WMS options per product based on project type and cost. Specific recommendations include:

a. **R&D Investment:** Fund research into innovative, efficient newer and better
b. construction methods (e.g., high automation, semi-automation), applicable to all project types, for better productivity and quality.
c. **Testing Codes:** Continuously update WMS testing methods for newer and better ways, as construction and digital technologies evolve. This might be done every 3-5 years .
d. **Collaborative Research:** Allocate R&D funds for joint studies by technical institutes (IITs, IISc), professional bodies, business schools, equipment manufacturers, and government bodies (NSDC, SDCs), with management school involvement as key for effective codes.
e. **Paid Research:** Fund research for accountability in developing new WMS codes, updating every 3-4 years to keep pace with technology.
f. **User Feedback:** Collect feedback through annual professional conferences with structured sessions. Committee summaries can inform annual addendums to maximize WMS benefits.

### 12.5.3 Focus on Skills development & Contractors' certification

Third very important area, where government must seriously consider is the development of the skills of the people who are involved in the construction sector. The skills have to be imparted on a mass-scale, but effectively.

Government's current initiative of National Skills Development Corporation's Program is not adequate for this, as it primarily covers only the workmen. Besides this, the programs there, primarily cover the domain and technology skills only, and do not focus on other equally important skills like Project Management Skills, People Skills and emerging technological skills. Besides NSDC, there are many national institutions and professions bodies, which conducts training programs for other higher-level employees. Some of these are listed below:

- **NICMAR:** National Institute for Construction Management & Research
- **NITIE:** National Institute of Industrial Engineering
- **ILCE**: Institute of Lean Construction Excellence
- **IIBE:** Indian Institute of Bridge Engineers
- **BFI:** Bridge Federation of India
- **IRC:** Indian Road Congress
- **IBC**: Indian Building Congress
- **ICI**: Indian Concrete Institute
- **IIMM:** Indian Institute of Materials management
- **IRT**: Institute of Rail Transport
- **IABSE**: Indian Group of International Association of Bridge & Structure Engineering
- **IAHE**: Indian Academy of Highway Engineers
- The Institution of Engineers
- Indian Institute of Architects
- **ICE**: Institute of Civil Engineering

Besides these there are IITs, and IIMs also. Of course, some of big corporate houses, engaged in Construction Projects, have their own skills development centres. For instant, L&T Construction, Tata Projects, Tata Housing, Shapoorji Pallonji, Godrej Construction etc. have invested reasonably in their internal skills development programs at different levels of employees.

There are three problems, we see with current system of skills development, which is far from being effective. We would like to mention following three main weaknesses in the system.

**a.** Indian culture basically is a hierarchical in nature. It is changing slowly however with the millennials, but still unless things are imposed on people, those don't work. Unless some relevant certification & some skills are made mandatory to the contractors, employees etc., knowledge will not reach a larger section.

**b.** Number of people covered in all these training programs is quite insignificant compared to the people who need to be trained under that. This is true for even for the large organizations.

c. The quality of the contents of the programs in skills development is questionable. It needs tremendous improvements and more frequent updates. This is also because every organization creates its own training material, for the same topics. There is a great variation of quality standards in those.

Therefore, Government should involve the relevant Engineering and Management Schools, for the appropriate 'Contents creation'. For greater effectiveness of learning and delivery, investments should also be done for preparing short interesting story-telling videos on several technical, domain, project management skills. The common contents for all the technical, project management, domain and people skills should be available to all at very affordable prices. Everyone should be allowed to use that for themselves, with any customization required. This would avoid multiple duplications of efforts, ensure that this attains a certain minimum level of quality of contents, as well as make it useful by localization of contents through customization.

It must be mandatory for Contractors to be certified in Lean Construction 4.0. Government must also prescribe a "Minimum % of Certified project managers, quality managers, workmen" in every Government Project. This will work very effectively, as Indian IT sector was asked a minimum % of PMP certified professional in each overseas project during 2000-2007, the big growth period for IT exports.

### 12.5.4 Focus on Support for Supplier's Eco-System Development

Construction SCM is underdeveloped compared to manufacturing, with much of the sector still unorganized. Even large companies like L&T haven't sufficiently invested in building SCM capacities with sub-contractors. As infrastructure becomes crucial for economic growth, it's time for the government to elevate SCM at an ecosystem level.

To compete globally and offer an efficient alternative to China-based supply chains, here are our recommendations:

1. Develop Construction SCM Parks with leased office and warehouse spaces, covering multiple supply chain levels, including suppliers, third-party logistics, software providers, and consultants in design, testing, and specialized fields.
2. Establish skill development and certification centres for subcontractors, suppliers, and testing consultants.
3. Authorize and certify testing consultants.
4. Create a network infrastructure allowing these stakeholders to reach customers across the country interactively.

5. Increase automation in construction practices and improve site work conditions to attract skilled talent. Emphasize training in automation and Lean Construction 4.0 for long-term productivity.
6. Provide specialized consultants to aggregate equipment from various suppliers, offering shared access like an "Ola for equipment."

These centres should be established in major cities with high construction activity, benefiting other areas as well. Collaboration with investors, professional bodies, and contractors will be essential for success.

### 12.5.5 Focus on Internet availability, reach and speed (Digital India)

Although, through Digital India program, our country has reached to a great height in internet infrastructure and usage, a lot more is required in terms or reach, speed, memory for storage etc. This is specially so considering that the construction work and especially Linear Infra projects. This is because these projects take place even in very remote areas and hilly sites also. There is need there for communications, and using various Apps for collaboration and information exchanges, virtual meetings etc. Therefore, there is need for internet to reach remotest parts in the country. A robust, secure, universally available network is needed for such collaborative platforms and tools to function optimally. Technologies such as 5G and infrastructure convergence of IT, Telecom and broadcasting are required to leverage the full transformative potential of such innovative management tools.

**To attain above objectives, we recommend the following:**

a. Strengthen and optimise Satellite Communications network throughout the country.
b. Create robust infrastructure directly or through public-private partnerships, to ensure secure Universal Broadband Access, especially covering uncovered areas, ensuring availability of broadband on demand.

For all the above initiatives, although the Government should take the leader's role, a big support would be required from almost all the stakeholders like Telecom Service Providers, Internet Service Providers, Infrastructure Providers, handset and equipment manufacturers, the academic community, investors and bankers, the innovators and start-ups. They all need to come together to forge a coalition to deliver this national policy and its missions. This is a huge task, but now is the time to perform as India also wants to be growing faster and be 'AtmaNirbhar'.

## 12.6 Long-Term Recommendations to Professional Bodies & Universities

Professional Bodies and Universities have key roles in bringing government recommendations to life, but in some areas, they should take the lead. Based on earlier discussions, we propose the following focus areas:

### 12.6.1 Improving Service Quality

- Define 'service level' standards for services offered by various groups (Consultants, Contractors, Financiers, etc.). Currently, these standards are often unclear, only discussed between Clients and stakeholders. Professional bodies could help define industry norms, similar to government service KPIs like Responsiveness, Reliability, and Customer Understanding.
- Educate stakeholders on the importance and national benefits of service quality.
- Support government in outlining a complete supply chain for various project types, recommending policy changes were needed.
- Advise government on forming Construction SCM parks or clusters, including location, group composition, and size.
- Assist in determining the types, sizes, and facilities needed for warehousing and logistics in these clusters.

### 12.6.2 Improving Process Quality

- Sponsor research to create user-friendly, accurate testing and validation methods that require fewer personnel, supporting mobile labs for simpler tests on-site.
- Promote research on non-destructive testing methods.

### 12.6.3 Encourage Skills Development and Certification:

Focusing on the advantages rather than just compliance, encourage skills development and certification which is essential for national growth.

### 12.6.4 Skills Development & Contractors' Certification

- Universities, professional institutions, and bodies should develop relevant, high-quality content for current and future skill needs across all employee levels.
- These organizations should create high-standard, industry-wide content that addresses both engineering and management skills, following recommendations in Chapter 9.
- Liberal use of Digital Technologies, including AI and ML for bringing personalisation and greater effectiveness in learning. These assets should also be sharable to all, preferably free of cost, so that no one hesitates in using those for national benefits.
- Leverage global best practices to produce digital content for easy, affordable access across the industry, promoting wide adoption.

## 12.7 Long-Term Recommendations to Investors & Banks

Investors and the Banks apparently may have very limited role for the various suggestions made for changes at the government. Still, they can play a strong supporting role for ensuring the implementation of the changes in most of the cases, starting from

Codes & Standards to Skills development of people and certification for the Contractors. Since they are the lenders and money providers and their money is at stake, they must ensure that the relevant stakeholders are playing their roles correctly. This will ensure that investors' ROI is ensured. Don't forget that the borrowers do listen to the Lenders, at the time of getting the approval for the finances.

## 12.8 Long-Terms Recommendations to Developers / Client Organizations

Client organizations and project developers play a key role in supporting government recommendations, but there are areas where they need to take the lead. Based on previous discussions, here are some recommendations for client organizations:

- Adopt design, testing, and service level standards sincerely to gain real benefits, not just for compliance certificates.
- Select certified Consultants, Contractors, and Suppliers with skilled manpower at all levels, avoiding lowest-bid selection, which often leads to cost revisions or quality compromises.
- Provide constructive feedback to the government on codes and standards, sharing positive and negative experiences.
- Introduce relevant automation in business and construction processes.
- Train workmen and employees to use effective tools and fixtures to enhance quality at a low cost.
- Create a modern, flat organizational structure suited to millennial culture, promoting diversity and equal opportunity across ethnicity, race, caste, gender, and age, with a focus on meritocracy.

Client organizations should also foster an entrepreneurial mindset, encouraging calculated risks for innovative improvements.

## 12.9 Recommendations to Main Contractors, Sub-Contractors & Suppliers

Client organizations and project developers play a key role in supporting government recommendations, but there are areas where they need to take the lead. Based on previous discussions, here are some recommendations for client organizations:

- Adopt design, testing, and service level standards sincerely to gain real benefits, not just for compliance certificates.
- Select certified Consultants, Contractors, and Suppliers with skilled manpower at all levels, avoiding lowest-bid selection, which often leads to cost revisions or quality compromises.
- Provide constructive feedback to the government on codes and standards, sharing positive and negative experiences.
- Introduce relevant automation in business and construction processes.

- Train key workmen and employees to use effective tools and fixtures to enhance quality at a low cost.
- Create a modern, flat organizational structure suited to millennial culture, promoting diversity and equal opportunity across ethnicity, race, caste, gender, and age, with a focus on meritocracy.
- Client organizations should also foster an entrepreneurial mindset, encouraging calculated risks for innovative improvements.

## Table 12.1 New Revised Role for Lean Project Team Leader in Quality Management

| QUALITY PARAMETER | KEY FOCUS AREA FOR CHANGE | PRIMARY/RESPONSIBILITY | SUPPORTING ROLE |
|---|---|---|---|
| **PRODUCT Quality** | Buildings Codes & Standards (Design & Testing) | Government's investments in R&D for relevant policy design & testing codes & standards formulations,updation &implemenrarion | Premium Technoly & Management institutions like IITs,IIMs,IISc,NITs etc;Other relevant R&D Organizations |
| | Linear Projects Codes & Standards (Design & Testing) | Government's investments in R&D for relevant policy design & testing codes & standards formulations,updation &implemenrarion | Premium Technoly & Management institutions like IITs,IIMs,IISc,NITs etc;Other relevant R&D Organizations |
| | (New) Builing Project Codes & Standars (Work Methods Statements) | Government's new investments in R&D for relevant work methods statements For KEY work packages.its formulations,upsation & implementation | Premium Technoly & Management institutions ;Key Global construction equipment manufactures/suppliers;Top class PMC consultants ;Relevant R&D organization;investors &banking industry |
| | (New) Linear Project codes & Standards (Work Methods Statement) | Government's new investments in R&D for relevant work methods statements For KEY work packages.its formulations,upsation & implementation | Premium Technoly & Management institutions ;Key Global construction equipment manufactures/suppliers;Top class PMC consultants ;Relevant R&D organization;investors &banking industry |

| | | | |
|---|---|---|---|
| | Skills Certification for Contractors/Sub-Contractors | Government's promotion &selection criterion for the vendors | Constaruction professional bodies of consultants(including PMC,QMC etc),Buildres,Developers,Contractors,Suppliers |
| | Skills Certification for Architects & Designers | Government's promotion &selection criterion for the vendors | Construction profssional Bodies of Architects& Designers |
| | Skills Certification for Facilities Maintenance experts | Government's promotion &selection criterion for the vendors | Construction profssional Bodies of Architects, Consultants(including PMC,QMC etc),Builders,Developers, Contractors and Facility Maintenance Experts,Investors |
| **SERVICE Quality** | Defining Service Level Measures & agreements for various types of Services | Relevant Construction professional Bodies,R&D Organizations | Universities,Suppliers,Contractors,Investor |
| | Supply Chain designs and definitions | Relevant Construction professional Bodies,R&D Organizations | Universities,Suppliers,Contractors,Investor |
| | Suppliers Eco-System development | Relevant Construction professional Bodies,R&D Organizations | Universities,Suppliers,Contractors,Investor |
| | Leveraging business process and warehousing &logistics automations or speed and Quality | Relevant Construction professional Bodies,R&D Organizations | Universities,Suppliers,Contractors,Investor |
| | Internet Connectivity even at remote areas (may be Satellite Network) | Governmnt | Professional Bodies |
| | Skills development at mega pace | Relevant Construction professional Bodies,R&D Organizations | Universities,Suppliers,Contractors,Investor |
| | Defining Service Level Measures&agreements for various types of Services | Client organization ,projects'Architects &designers,consultantors | Professional Bodies,Premium technology & management institution like IITs,IIMs,IISc,NITs etc.Other relevant R&D Organization |

| | | | |
|---|---|---|---|
| **PROCESS Quality** | Leveraging business process and warehousing &logistics automations or speed and Quality | Client organization ,projects'Architects &designers,consultant ors | Professional Bodies,Premium technology & management institution like IITs,IIMs,IISc,NITs etc.Other relevant R&D Organization |
| | Focus on Non-Distructive and speedy digitallized Testing and Validation Facilities | Client organization ,projects'Architects &designers,consultant ors | Professional Bodies,Premium technology & management institution like IITs,IIMs,IISc,NITs etc.Other relevant R&D Organization |
| | Compliance to Buildings Codes Standards (Design &Testing )& suggest areas for improvements | Client organization ,projects'Architects &designers,consultant ors | Government,Premium technology & management institutions like IITs,IIMs,IISc,NITs etc;Other relevant R&D Organizations |
| | Linear Projects Codes & Standards (Design & Testing) & suggest areas for improvements | Client organization ,projects'Architects &designers,consultant ors | Government,Premium technology & management institutions like IITs,IIMs,IISc,NITs etc;Other relevant R&D Organizations |
| | (New) Buildings Projects Codes & Standards (Work Methods Statements)&suggest areas for improvements | Client organization ,projects'Architects &designers,consultant ors | Government,Premium technology & management institutions like IITs,IIMs,IISc,NITs etc;Other relevant R&D Organizations |
| | (New)Linear Projects Codes & Stadards (Work Methods Statements)&Suggest areas for improvements | Client organization ,projects'Architects &designers,consultant ors | Government,Premium technology & management institutions like IITs,IIMs,IISc,NITs etc;Other relevant R&D Organizations |
| | **Better Usage of Tools and Techniques** | Client organization ,projects'Architects &designers,consultant ors | Professional Bodies,Premium technology & management institutions,Other relevant R&D Organizations. |
| **PEOPLE Quality** | **Talent/Skill Development** | Government and Professional Bodies' investments in national level Skills development Policies and investments for creating contents for Skills development in key areas for different levels | Professional Bodies,Premium Technology & management institution like IITs,IIMs,IISc,NITs etc.Other relevant R&D Organization |
| | **Organization Structure** | Client's and Main Contractors' organization | Professional Bodies |

| | | | |
|---|---|---|---|
| | **Organization Culture** | Clients' and Main Contractors' organisation | Professional Bodies |

## 12.10 Frequently Asked Questions (FAQ) about Lean Construction:

Following questions are often asked related to Lean Construction. We have suggested the answers to be given for that.

1. Why should CPS be considered a Lean method?
2. How is PPC measured? What is the right target for PPC?
3. How much should capacity be under-loaded?
4. How many weeks should we 'look ahead' when doing constraints analysis?
5. How to select the correct plan failures to analyse for countermeasures?
6. Should early finishes be counted as completions
7. Is CPS a scheduling system? Does it replace project controls?
8. How many more meetings and employees will we need if we do CPS?
9. Should we have crews do more work if they complete committed tasks sooner than anticipated?
10. Does implementation of CPS transfer power over project progress to subcontractors?

**1. Why should CPS be considered a Lean method?**

**Answer:** Lean is a management philosophy focused on increasing value for customers and stakeholders while reducing waste. Value grows when projects meet customer needs within their time, cost, location, and standard constraints, and when the result enhances customer goals. CPS is a method to achieve these objectives and guide projects toward them. In Toyota Production, three waste types are highlighted: Muri (overloading), Mura (unevenness), and Muda (unnecessary actions). CPS addresses all three by designing tasks to fit resource capabilities, releasing work predictably, and arranging tasks to reduce rework and increase resource use. Thus, CPS aligns closely with Lean principles.

**2. How is PPC measured? What is the right target for PPC?**

**Answer:** At the end of each commitment plan period (like a shift, day, or week), the team checks if commitments were met. A commitment is "met" if it started and/or finished as planned, simply answered with a "Yes" or "No." There's no partial credit. Either it is completed or not completed. It can't be expressed as 85% completed. 85% completed is also treated as 0% completed. This is to develop a culture of not leaving any job partially completed. PPC (Percent Plan Complete) measures how reliably a team plans and executes—not productivity or amount of work done. It's possible to have 100% PPC with low productivity if too much capacity is left unused. The ideal PPC target is 100%, aiming for reliable work release. This doesn't mean overloading resources; it's better to underload for steady workflow. As failures are proactively addressed, actual capacity grows. When PPC is close to 100%, teams can increase load and plan in smaller time slots, like half-day periods.

### 3. How much should capacity be under-loaded?

**Answer:** To ensure reliable workflow, we should generally underload so there's a 99% chance the assigned capacity will be enough to finish tasks on time. Achieving this with precision requires knowing the standard deviation of relevant operations. For example, 2 standard deviations mean a 95% confidence level, while 3 standard deviations mean a 99% confidence level, ensuring that capacity will be sufficient 99 out of 100 times. Reducing the standard deviation helps a lot! If the standard deviation is unknown, we adjust based on experience.

Low workflow reliability typically wastes around 30%-50% of labour capacity, which can act as a built-in buffer for underloading. While underloading may slightly reduce labour capacity, this reduction is usually less than historical losses, as underloading supports more reliable workflows.

### 4. How many weeks should we 'look-ahead' when doing constraints analysis?

**Answer:** The number of weeks needed to remove the constraint depends on the longest lead time. For instance, if a construction task enters the look-ahead window and design information is delayed, a 6-week look-ahead allows 6 weeks to expedite its production and delivery. If design resources aren't dedicated or have limited capacity, more time may be needed. Constraints like design information and materials are aligned with the construction schedule as their lead times often exceed 6 weeks. For items with production and delivery times longer than the look-ahead window, they should be included in higher-level schedules, like Reverse Phase Pull Planning.

### 5. How to select the correct plan failures to analyse for countermeasures

**Answer:** Analyse as many as your capacity allows. If capacity is limited, prioritize those with the greatest impact on project performance. Focus on areas with significant constraints, such as availability of drawings, complex or specialized materials, or rare skillsets required from personnel.

### 6. Should early finishes be counted as completions?

**Answer:** Yes, if tasks are finished within the committed time, they should be counted as completions. To improve the chances of meeting deadlines, it's recommended to under-load by assigning a bit more capacity (labour hours) than necessary, allowing for process variation. Early completion is preferred, but we should avoid tasks being completed "too early," which can lead to an overload of future tasks. Planning managers and supervisors should ensure balanced task loads. To prevent wasted capacity, commitment plans should include both priority tasks and additional fallback tasks. Extra capacity can also be used for problem-solving, like conducting a "5 Whys" analysis on plan issues.

**7. Is CPS a scheduling system? Does it replace project controls?**

**Answer:** No, CPS is a 'production' (read 'construction' if you wish) control system that includes planning, scheduling, and execution. Planning involves deciding "what" tasks to complete, in "what" order, "how" to do them, and organizing resources. Scheduling adds "when" to start and finish. LPS covers planning, scheduling, execution, and learning from any differences in planning and execution. LPS doesn't replace project controls, which set targets for cost, time, and other performance metrics. CPS helps guide project performance toward these targets set by project controls.

**8. How many more meetings and employees will we need if we do CPS?**

**Answer:** None, aside from those related to innovative meetings aimed at improving processes, such as value stream mapping. In fact, as workflow reliability increases, you may be able to reduce indirect costs by minimizing the need for firefighting.

**9. Should we have crews do more work if they complete committed tasks sooner than anticipated?**

**Answer:** Yes, but only if that work does not create more problems downstream than the benefits gained from utilizing otherwise lost capacity. It's important to specify this in the commitment plans. Each workgroup should have Plan B tasks available in case they complete their committed tasks early or are unable to perform them.

**10. Does implementation of CPS transfer power over project progress to subcontractors?**

**Answer:** In traditional contracting, main contractors aim to complete projects on or ahead of schedule, while subcontractors focus on productive crew use. With CPS in this structure, both parties work together to meet these goals. General contractors ensure tasks are ready in the right sequence and release them to a workable backlog, managing and controlling flows of design info, materials, and equipment, as planned by CPS. Subcontractors boost productivity by participating in look-ahead planning, which helps with labour planning, operation design, and only committing to well-defined tasks. When the project schedule, look-ahead, and commitment planning align, both progress and productivity improve.

# GLOSSARY

**3C (Communication, Coordination, Cooperation)** – The core principles enabling reliable availability of prerequisites and smooth workflow among stakeholders, encompassing the three collaborative behaviours required between clusters and sub-clusters to ensure readiness and avoid delays.

**3D Printing (Additive Manufacturing)** – Layer-by-layer construction of building components or structures directly from digital models, enabling complex shapes, reduced waste, and faster delivery.

**5 Whys** – A root cause analysis method that involves asking "Why?" repeatedly until the fundamental cause of a problem is identified.

**5W1H** – A questioning technique (Why, What, Where, Who, When, How) used to examine processes and uncover improvement opportunities.

**A3 Methodology** – A one-page structured problem-solving and improvement tool used to document issues, analysis, countermeasures, and action plans.

**Alternate Technologies** – Modern tools, materials, or methods (e.g., BIM, VR, advanced formwork) that improve construction quality and reduce defects.

**Augmented Reality (AR)** – Technology that overlays digital information on real-world views, enhancing project understanding, sales presentations, and site coordination.

**Bad practices** – Behaviours that undermine Big Room effectiveness, including multitasking, poor preparation, and disengagement.

**Bar Bending Schedule (BBS)** – A document detailing the type, quantity, and dimensions of steel bars required for construction; when integrated with scheduling, it also includes delivery timelines.

**Behaviour-Based Safety (BBS)** – An approach that focuses on observing and positively influencing worker behaviour to prevent unsafe acts and conditions.

**Big Room** – A collaborative physical or virtual space that brings together key project stakeholders to enhance communication, coordination, and decision-making, while supporting collaborative planning, displaying schedules and progress, and hosting multi-trade coordination meetings.

**Big Room mindset** – The cultural and behavioural shift towards openness, transparency, and teamwork required for effective Big Room functioning.

**Bottleneck process** – The step or activity in a value stream that limits the overall output or determines the pace of project delivery.

**Boulder Milestones/Boulder-Rock Processes** – The highest-level, visible physical units of progress in a Lean work breakdown structure, representing completion of all trades' work for that unit, and the high-impact processes in CPS that require targeted acceleration to meet project schedules.

**Breakout areas** – Spaces within the Big Room for smaller group discussions and focused problem-solving.

**Building Information Modelling (BIM)** – An intelligent 3D modelling process that supports planning, design, construction, and operations through shared, coordinated project data, enabling design analysis, clash detection, and cost/time integration.

**Cluster** – Groups of stakeholders or related tasks and trades, such as HVAC, civil, or electrical, often with sub-clusters for specific specialisations, coordinated to work together for a construction activity and enable effective collaboration within the CPS framework.

**Collaborative Planning System (CPS)** – The core of Lean Construction 4.0, this Lean-based, multi-layered production control system integrates planning, scheduling, execution, monitoring, and control across all stakeholders and project phases to eliminate silos, align milestones, foster coordination, create realistic and proactive workflows, improve reliability, and reduce variability and waste through continuous improvement.

**Collaborative Safety Planning System (CSPS)** – A structured, team-based approach to integrate safety into planning, mirroring the Collaborative Planning System (CPS) but with specific focus on hazard prevention and safe work readiness.

**Communication & Collaboration Technologies (CT)** – Digital tools like cloud platforms, mobile apps, virtual reality, augmented reality, and social media that enable real-time stakeholder communication using unstructured data such as text, images, and video.

**Configurator Software** – Digital tools that allow customers to personalise building and interior design choices through guided, real-time visualisation and cost estimation.

**Constraint List** – A list of missing prerequisites that prevents a construction activity from starting or finishing, such as absent materials, incomplete prior work, or unavailable resources.

**Constraint List** – A dynamic record of obstacles—such as approvals, materials, manpower—that must be resolved to keep work on schedule.

**Construction 1.0** – Labour-intensive early stage of construction with similar designs and limited specialisation.

**Construction 2.0** – Mid-20th century stage introducing specialisations, partial mechanisation, and early project/business software, but still low productivity.

**Construction 3.0** – Late 20th century stage with technology-driven design (e.g., AutoCAD, BIM), Lean philosophy adoption, and specialised equipment improving performance.

**Construction 4.0** – The current phase in the construction sector that integrates Lean philosophy with advanced digital, automation, and data-driven technologies to enhance productivity, cost efficiency, quality, speed, safety, and sustainability.

**Construction waste** – Any activity, material, or process in construction that consumes resources without adding value to the customer or project stakeholders.

**Continuous Improvement** – An ongoing effort to enhance processes, systems, and work environments for better performance and results.

**Corrective and Preventive Actions (CAPA)** – Structured measures to correct existing defects and prevent their recurrence.

**Cross-functional team** – A group of individuals from different disciplines or clusters collaborating to analyse and improve a value stream.

**Current state map** – A visual representation of the existing process showing all steps, delays, and wastes.

**Customer Approval Based Build (CABB)** – A quality management approach where customer approval is sought for sample components or finishes before full-scale execution.

**Cycle time** – The total duration required to complete all steps in a value stream, including both value-adding and non-value-adding time.

**Day-to-Day Issues (in Traditional PMS)** – Common operational challenges such as cash flow shortages, delays, budget overruns, manpower shortages, equipment unavailability, and poor quality control.

**Defect Density** – The number of snags identified per unit of measure, such as area (sq ft) or quantity of work items; used to forecast defects.

**Defect Prevention** – Actions taken to avoid the occurrence of defects, especially those critical, costly, or frequently recurring.

**Defects Lifecycle Management** – A structured approach for managing defects from identification through prediction, prevention, containment, and rework planning.

**Digital Twin** – A virtual replica of a physical asset or process that uses real-time data to simulate, predict, and optimise performance throughout its lifecycle.

**Drones (Unmanned Aerial Vehicles)** – Remote-controlled aircraft used for surveying, inspection, mapping, and monitoring construction sites efficiently and safely.

**Dust Tasks** – Weekly tasks detailed from Pebble Activities, ready for execution within 5–6 days.

**Earned Value Management (EVM)** – A project management technique for measuring project performance and progress by integrating scope, schedule, and cost data.

**Eight Prerequisites** – The essential resource categories and conditions that must be in place for an activity to start and finish, including drawings, materials, people, equipment, prior work, space, safety, and external environment readiness.

**Eight Wastes** – The Lean framework identifying eight non-value-adding activity categories in construction: waiting/delay, rework/defects, inventory, people motion, transportation, overproduction, overprocessing, and unutilised capability or talent.

**EPC Contractor** – A firm responsible for Engineering, Procurement, and Construction, delivering a project from design through to completion.

**External customer** – The end user or client who pays for and benefits from the completed product or service.

**Facilitation Workshop** – A structured meeting that brings together contractors, suppliers, and engineers to resolve supply, design, and execution challenges collaboratively.

**First-Piece Inspection** – A process where one completed unit of work is inspected and approved before full-scale execution, to set quality benchmarks.

**Five Keywords of Lean** – Core focus areas in Lean Construction: Value, Value Streams, Flow, Pull, and Perfection.

**Five Layers of CPS** – Progressive planning detail levels: Must-Do (Boulder Milestones), Should-Do (Rock), Can-Do (Pebble), Will-Do Weekly (Dust), and Will-Do 24 Hours (Task).

**Five Principles of Lean** – Fundamental Lean philosophy elements: Challenge the status quo, Go and See, Continuous Improvement, Respect the People, and Teamwork.

**Five S** – A Japanese workplace organization methodology consisting of five steps: Sort, Set in Order, Shine, Standardize, and Sustain, aimed at improving productivity, safety, and efficiency.

**Flow** – The smooth and uninterrupted progression of tasks through value streams, achieved by removing constraints, roadblocks, and causes of rework.

**Focus on the Whole** – The approach of prioritising project-wide value over individual contract or departmental goals.

**Foreign Direct Investment (FDI)** – Investment from foreign entities into domestic sectors such as real estate and infrastructure, often encouraged through specific government policies.

**Future state map** – The improved process design developed from the current state map to reduce waste and improve performance.

**Gemba Walk** – A structured on-site visit by leaders or managers to directly observe work in progress, engage with teams, and identify waste, hazards, defects, and improvement opportunities, fostering proactive safety, quality, and performance enhancement.

**GPS (Global Positioning System)** – A satellite-based navigation system used for accurate location, surveying, and positioning in construction projects.

**Green Construction** – Environmentally responsible building practices reducing environmental impact and preserving natural resources.

**Group Technology (GT)** – A collaborative work approach where interdependent trades coordinate to achieve common quality and schedule goals, reducing defects.

**Hazard Density** – A measure of safety risk based on the number of identified hazards per work package, work item, or area, used for predicting and preventing incidents.

**Hazard Identification and Risk Assessment (HIRA)** – A systematic process to identify potential hazards, assess their likelihood and severity, and define control measures to mitigate risk.

**Hybrid Big Room** – A combination of physical co-location and virtual participation, enabling geographically dispersed stakeholders to collaborate effectively.

**Internal customer** – The next trade, team, or function in the workflow that receives a completed work-front from a previous one to continue value-adding activities, e.g. another employee.

**Internet of Things (IoT)** – A network of connected devices, sensors, and systems that collect, share, and act on real-time data to improve building performance, safety, and maintenance.

**IPECC** – The five project management process groups: *Initiate, Plan, Execute, Monitor & Control, and Closing*, as per PMBOK standards; often compared to the PDCA cycle.

**Job Safety Analysis (JSA)** – A method of breaking down a job into sequential steps, identifying potential hazards for each, and defining specific preventive actions.

**Kaizen** – A continuous improvement approach involving small, incremental changes led by those directly involved in the work.

**Key measures of success** – Indicators used to assess Big Room effectiveness, such as participation quality, innovation, and trust levels.

**L1 Contractor** – The bidder quoting the lowest price in a tender; often selected in public procurement, though not always the most suitable for complex projects.

**Last Planner System® (LPS®)** – A collaborative planning approach developed by Glen Ballard and Greg Howell, focusing on foremen and supervisors ("last planners") to create reliable, realistic schedules based on field conditions.

**Lead time** – The elapsed time from the initiation to the completion of a process or value stream.

**Lean Champion** – An individual recognised for exceptional commitment to implementing and promoting Lean principles within a project or organisation.

**Lean Construction 4.0** – A systematic approach that applies Lean principles integrated with Construction 4.0 technologies and modern process innovations to define value, eliminate waste, and achieve faster, safer, more predictable, efficient, high-quality, and value-driven project outcomes.

**Lean Philosophy** – A management approach aimed at maximising value and eliminating waste in processes.

**Lean Problem-Solving Model** – A structured method for understanding safety-related problems, implementing solutions through the PDCA cycle, and embedding a continuous improvement culture.

**Look-Ahead Planning** – A short-to-medium-term planning window (often 3–6 weeks) where constraints are identified and resolved before work is scheduled.

**Machine Learning (ML)** – A subset of AI that enables systems to learn from data patterns and improve performance without explicit programming.

**Make-Ready Planning** – A proactive CPS pre-planning approach that removes constraints and ensures all eight prerequisites—drawings, materials, people, equipment, prior work, space, safety, and external environment—along with sequencing and safety checks, are in place before work begins for safe, efficient, and timely task execution.

**Mass Customisation** – Delivering personalised building features at scale, particularly in urban real estate projects.

**Mass Production** – High-volume, standardised construction delivery, often used for affordable housing in rural areas.

**Material Search Time** – The duration spent locating tools, equipment, or materials; reduced significantly through proper Five S implementation.

**Mini-Mock-up** – A scaled or limited construction trial of complex or defect-prone work to test and refine methods before full implementation.

**Mixed Reality (MR)** – A hybrid of VR and AR where virtual objects are anchored in real-world environments and can be interacted with in real time.

**Muda** – Non-value-adding activities in construction that result in wasted time, effort, or materials; further categorised into eight common types.

**Mura** – Unevenness or variability in workloads, methods, or outputs that disrupts smooth flow and lowers efficiency.

**Muri** – Overburden placed on workers, equipment, or processes that leads to fatigue, errors, and reduced performance.

**National Infrastructure Pipeline (NIP)** – India's five-year plan (2020–2025) to invest approximately INR 102 lakh crore in infrastructure across identified sectors.

**National Skills Development Corporation (NSDC)** – An Indian government initiative to promote skill development through training and certification, particularly in the unorganised sector.

**Non-Conformance Report (NCR)** – A formal record identifying deviations from safety or quality standards, requiring corrective action before work can proceed.

**Off-Site Fabrication** – Manufacturing components in a controlled environment before transporting them to the site for assembly, improving efficiency and quality.

**Operational Technologies (OT)** – IoT-based systems, prefabrication, 3D printing, robotics, and sensors that manage structured and semi-structured operational data to improve construction processes and outcomes.

**Overprocessing** – Carrying out more work, checks, or approvals than the customer values or requires.

**Overproduction** – Producing more than is required or earlier than needed, resulting in unnecessary use of resources.

**Parametric Estimation Model** – A data-driven approach to estimating project costs, quantities, and timelines based on historical project data and statistical relationships.

**Pareto analysis** – A prioritisation technique identifying the few causes that account for the majority of problems or wastes.

**PDCA Cycle** – A continuous improvement model of Plan, Do, Check, Act, used here as a reference for aligning PMS phases with iterative performance improvement.

**Pebble Activities** – Sub-activities of Rock Processes for a 4–6 week look-ahead period, usually within one trade or sub-cluster.

**Percent Plan Complete (PPC)** – A key performance metric in the Last Planner System® (LPS) and CPS, measuring planning and execution reliability by calculating the percentage of planned tasks fully completed as committed.

**Phase Pull Planning** – A reverse planning method where project teams define completion milestones and work backwards to identify all necessary tasks and constraints.

**Political Risks** – Risks linked to political influence, interference, or demands that affect project execution, procurement, or approvals.

**Positive Leadership** – A leadership style that inspires trust, openness, and proactive problem-solving in a Big Room setting.

**Preventive Maintenance** – Scheduled inspection and servicing of tools, equipment, and systems to prevent failures and avoid safety risks.

**Proactive Defects Prediction** – The process of forecasting the type and number of defects in advance, based on historical data or sampling, to enable preventive actions.

**Professional Bodies** – Industry associations that set standards, provide training, and influence policy in the construction sector.

**Project Lifecycle Phases** – Distinct stages of a construction project: Concept, Design, Procurement, Construction & Handover, and Facility Maintenance & Decommissioning (operations stage not covered in this book).

**Project Management System (PMS)** – A structured framework for managing projects to meet scope, time, budget, quality, and other client requirements, typically following the IPECC process groups.

**Public Private Partnership (PPP)** – A project delivery and funding model in which government and private entities jointly invest in and execute projects.

**Pull Planning** – A scheduling method where tasks are triggered based on downstream demand, allowing late starts, customer-driven changes, and timely completions.

**Reciprocal Tasks** – Construction activities requiring back-and-forth coordination between different trades, where the quality of one impacts the other.

**Red Tag Method** – A technique used during the Sort phase to label and remove items that are unnecessary or out of place in the workplace.

**Resource Efficiency** – Using materials, labour, and energy in the most effective way possible, reducing waste and conserving resources.

**Resource Levelling** – Adjusting and allocating resources to maintain a balanced workload across multiple projects or tasks without overburdening teams.

**Respect for People** – A key Lean principle emphasising the value of individual knowledge, experience, and contributions.

**Reverse Planning** – A collaborative scheduling approach that works backward from the end goal or milestones to sequence tasks, allocate resources, and resolve conflicts before finalising forward schedules.

**Robotics in Construction** – The use of automated machines for repetitive, hazardous, or precision tasks such as bricklaying, plastering, painting, and material handling.

**Rock Processes** – Trade-specific high-level activities needed to achieve Boulder Milestones, generally involving one main trade or cluster.

**Root Cause Analysis (RCA)** – A structured problem-solving method for identifying the fundamental or systemic reasons behind defects, plan failures, or poor performance—such as unrealistic corporate planning, poor collaboration, reactive culture, delayed contractor onboarding, and limited technology adoption—to prevent recurrence and improve outcomes.

**Safety and 5S Audit** – A combined assessment of workplace safety and 5S housekeeping standards to maintain an organised, hazard-free environment.

**Safety Checklist Method** – A structured list of safety requirements and inspection points to be verified before and during work execution.

**Seiketsu (Standardize)** – The fourth step of Five S, developing and applying consistent processes, visual controls, and best practices to maintain order and cleanliness.

**Seiri (Sort)** – The first step of Five S, involving the removal of unnecessary items from the workplace to reduce clutter and improve workflow.

**Seiso (Shine)** – The third step of Five S, maintaining cleanliness in the workplace to enhance safety, detect defects, and create a visually appealing environment.

**Seiton (Set in Order)** – The second step of Five S, organizing necessary items so they are easy to find, access, and return, minimizing search time.

**Set-up time** – Time spent preparing tools, materials, or equipment before actual value-adding work begins.

**Shitsuke (Sustain)** – The fifth step of Five S, fostering discipline and continuous improvement to maintain the established standards over time.

**Siloed approach** – A mode of working where individuals or groups operate in isolation, limiting collaboration and information flow.

**Smart Checklist** – A dynamic, updated quality checklist accessible digitally, containing essential specifications and inspection points for preventing defects.

**Snag** – A defect or deficiency in construction work that requires rectification before handover. Snag Density is same as Defect Density

**Social Risks** – Project risks arising from community or local stakeholder opposition, land disputes, or socially-driven demands impacting project timelines and costs.

**Stakeholder Engagement** – The process of actively involving all relevant parties—political, social, governmental, contractors, and suppliers including customers, government, investors, business partners, and employees.—in project planning and execution.

**Standard Operating Procedures (SOPs)** – Documented instructions detailing standardized methods to carry out specific tasks consistently and safely.

**To-Do Tasks** – Daily tasks derived from Dust Tasks, planned for the next 24 hours to ensure reliability and readiness.

**Unutilised capability** – Failing to use the skills, ideas, or decision-making abilities of available personnel to improve outcomes.

**Value (in Construction)** – Benefits perceived by stakeholders : agencies, clients, consultants; including quality, cost efficiency, sustainability, and user satisfaction, .

**Value Engineering (VE)** – A systematic method to improve the value of a project by optimising functions while reducing cost without compromising quality or performance.

**Value enhancement** – Turning waste reduction into tangible and intangible gains for customers, contractors, and other stakeholders.

**Value Stream** – The complete sequence of linked construction and business activities, including both value-adding and non-value-adding steps, required to deliver a project or service from concept to handover across all stakeholders.

**Value Stream Mapping (VSM)** – A Lean tool for visually mapping and analysing the macro-level flow of materials, information, and activities in a value stream to identify wastes, reduce cycle time, and improve overall process flow.

**Virtual Reality (VR)** – A fully immersive digital environment that allows stakeholders to visualise and experience construction designs before execution, improving decision-making and reducing rework.

**Visual Management System** – A coordinated use of physical or digital tools such as charts, boards, labels, signs, colour coding, floor markings, and process diagrams to make plans, progress, issues, workflows, standards, and metrics visible to all participants, guiding decisions, reinforcing workplace organisation, and ensuring transparency.

**Waiting time** – Idle time when work cannot proceed due to missing resources, approvals, or access; a major form of waste.

**War Room** – An alternative term for the Big Room, emphasising its role as an operational and problem-solving hub.

**Waste Elimination** – The proactive and systematic process of identifying, reducing, or removing visible and hidden inefficiencies in construction workflows to improve productivity, quality, safety, sustainability, and profitability, including minimising material waste and promoting recycling or reuse to support environmental goals.

**Work Breakdown Structure (WBS)** – A hierarchical decomposition of the project into manageable components, used to refine major milestones and support Lean planning.

**Work Method Statement (WMS)** – A documented, standardised best-practice procedure for executing specific construction activities that ensures consistency, safety, quality, efficiency, and defect prevention.

**Workflow Automation Technologies (WT)** – Software systems such as ERP, CRM, e-commerce platforms, project management tools, AutoCAD, and BIM that manage structured data and streamline business and project workflows.

**Workplace Organization** – Arranging tools, equipment, and materials in an orderly manner to improve efficiency, safety, and quality.

**Workplace Safety** – The practice of minimizing hazards and risks in the workplace, often improved through organized layouts and cleanliness from Five S.

**Zone Mapping** – A visual representation of designated storage or work areas to optimize space utilization and workflow.